Space of Service to Humanity

Preserving Earth and Improving Life

SPACE STUDIES

Volume 1

Managing Editor

WILEY J. LARSON

International Space University
Excellence in space education for a changing world

The International Space University (ISU) is dedicated to the development of outer space for peaceful purposes through international and interdisciplinary education and research. ISU works in association with a number of Affiliates (universities, research institutes, consortia..) around the world and in partnership with space agencies and industry.

For young professionals and postgraduate students, ISU offers an annual ten-week Summer Session in different countries and a one-year Master of Space Studies (MSS) program based at its Central Campus in Strasbourg, France. ISU also offers a Professional Development Program to provide training in the form of short courses and workshops to professionals working in space-related industry, government and academic organizations.

Independent of national and commercial interests, ISU is an ideal forum for discussion of issues relating to space and its applications. The network of alumni, faculty, guest lecturers, Affiliate representatives and professional contacts which characterizes the ISU Community makes it possible to bring together leading international specialists in an academic environment conducive to the exchange of views and to the creation of innovative ideas. ISU aims to promote productive dialogue between space-users and providers. In addition to the Annual Symposium, ISU supports smaller forum activities, such as workshops and roundtables, for constructive discussions which may help to chart the way forward to the rational international utilization of space.

Printing and distribution of these proceedings was made possible through the generous support of **Aérospatiale Espace & Défense**.

Space of Service to Humanity

Preserving Earth and Improving Life

Symposium Proceedings

International Symposium

5 - 7 February 1996, Strasbourg, France

Edited by G. HASKELL and M. RYCROFT

International Space University

KLUWER ACADEMIC PUBLISHERS
DORDRECHT / BOSTON / LONDON

Library of Congress Cataloging-in-Publication Data

The C.I.P. catalogue record for this book is available from the Library of Congress.

ISBN 0-7923-4344-1

Published by Kluwer Academic Publishers,
P.O. Box 17, 3300 AA Dordrecht, The Netherlands.

Kluwer Academic Publishers incorporates
the publishing programmes of
D. Reidel, Martinus Nijhoff, Dr. W. Junk and MTP Press.

Sold and distributed in the U.S.A. and Canada
by Kluwer Academic Publishers,
101 Philip Drive, Norwell, MA 02061, U.S.A.

In all other countries, sold and distributed
by Kluwer Academic Publishers Group,
P.O. Box 322, 3300 AH Dordrecht, The Netherlands.

Table of Contents

Table of Contents (continued)

Acknowledgements

ISU acknowledges with thanks the advice and support given by the following people as members of the Program Committee:

F. Becker, Member, ISU College of Teachers & Director, Ecole Nationale Supérieure de Physique de Strasbourg, France
M. Bied-Charreton, Chief, Environment Information Management Service, Food &Agriculture Organisation, Rome, Italy
M. Bousquet, Head, School of Engineering, Systems & Technologies, ISU & Director, Aerospace Electronics & Communications Programmes, Sup'Aero, Toulouse, France
H. Chasia, Deputy Secretary General, International Telecommunication Union, Geneva, Switzerland
G. Haskell, Vice President for Academic and Research Affairs, ISU
O. Hebin, Senior Liaison Officer, Telecommunications & Electronic Services, United Nations Environmental Programme, Geneva, Switzerland
R. Jakhu, Head, School of Management and Social Sciences, ISU
N. Jasentuliyana, Director, Office for Outer Space Affairs, United Nations Office at Vienna, Austria
S. Khorram, Dean , ISU & Director, Computer Graphics Centre, North Carolina State University, USA
T. Kuroda, Chief Engineer, NEC Corporation, Yokohama, Japan
G. Lindberg, Chairman, Committee on Earth Observations Satellites, Quebec, Canada
S. Madry, Member, ISU College of Teachers & Associate Director, Center for Remote Sensing & Spatial Analysis, Rutgers University, USA
J.N. Pelton, Member, ISU College of Teachers & Director, Graduate Telecoms Program, University of Colorado at Boulder, USA
U.R. Rao, Invited Distinguished Professor, ISU & Former Chairman, Indian Space Research Organisation, Bangalore, India
M. Rycroft, Head, School of Sciences and Applications, ISU
L. Stojak, Director, Summer Session Programs, ISU
R.Thomas, Chairman, MSS Team Design Project 1995-96, ISU

Symposium Convenor and Program Committee Chair: **G. Haskell**, ISU

Symposium Organizer: **E. Moyen**, ISU

Proceedings Editors: **G. Haskell**, ISU and **M. Rycroft**, ISU

Editorial Assistant: **E. Moyen**, ISU

Co-sponsors:
Committee on Earth Observations Satellites
Institute of Electrical and Electronic Engineers: Communication Society
United Nations Office for Outer Space Affairs

Financial sponsorship of individual Symposium participants was provided by:
Matra Marconi Space
United Nations Office for Outer Space Affairs.

Foreword

G. Haskell, Symposium Convenor & Vice President for Academic Services and Outreach, International Space University

By taking "Space of Service to Humanity" as the theme for the inaugural event in its series of annual symposia, the International Space University (ISU) is asserting that this application of space technology requires special attention at this time. Future symposia will examine the issues of the day from different perspectives.

In keeping with the fundamental principles of ISU, the symposium took a global perspective, as distinct from national or regional perspectives, and treated both technical and non-technical topics. Oral and poster presentations were delivered in the scientific, engineering, economic, legal, political and philosophical domains. All oral papers were delivered in plenary session to encourage cross-fertilization between specialities, and posters were readily available for viewing throughout the three-day event.

As an international and interdisciplinary forum for the sharing of experience and for discussion and debate, the symposium proved to be a stimulating and worthwhile event for the 135 participants from 25 countries and 5 international organizations with backgrounds in industry, government and academia.

In my introductory remarks I said that we should all expect to have our minds stretched by the symposium. One of these mind-stretching moments came for many participants when a speaker asserted that, whatever one's own list of fundamental human needs (food, energy, education....), it must be concluded that future space applications will have an essential role in meeting every single one of them. If this is indeed true, what a heavy burden of responsibility rests on the shoulders of the space community!

Some of the issues raised in the symposium have been taken further by ISU in the context of the Team Design Project undertaken by the postgraduate students and young professionals enrolled in the ISU Master of Space Studies programme. Their report will be completed in July 1996.

Through these proceedings, ISU offers the symposium to a wider audience in the hope of stimulating further constructive discussion and debate.

Space of Service to Humanity Opening Address

R. Doré, President of the International Space University

Ladies and Gentlemen, and dear friends of ISU:

The space world is changing.

During the first decades of the space age, the main motivating factors behind space activities were national security and national prestige. Space infrastructures (launchers, satellites, space probes) were developed under a technological thrust. Space applications were seen as valuable platforms for demonstrating national capabilities related to advanced technologies. Space was a highly visible activity. That was the epoch of the space race.

The present decade has seen the advent of a new political equilibrium worldwide and, with it, a major, sustained, economic world crisis. Faced with these new conditions, it is the "usefulness" factor in space activities that is slowly becoming the sector's main driving force. We witness space activities bridging ideological divides and geographical frontiers. A vivid example of the internationalisation of space is the International Space Station. The 14 countries involved in this project are prepared to invest tremendous effort in the task of demonstrating the feasibility of working and living together in space.

With the approach of the new millenium, we are entering a new space age in which space is put to the service of humanity. We shall judge the value of space technologies by their ability to serve human needs such as food production, medical and health services, rural education, disaster warning and mitigation, environmental protection, navigation services, communication, understanding the universe and extending humanity beyond the Earth frontier. Fulfilling the user's needs will be of prime importance in the coming years. This will trigger a shift from government-led programmes to activities initiated and fully developed by the private sector. It will lead to the creation of a multitude of applications on the ground and, consequently, a multitude of small enterprises to generate these applications. Launchers, satellites, space platforms and ground segments will be considered as tools to achieve other goals, and not as an end in themselves. The role of national and regional space agencies will change dramatically; their role will be confined mainly to the realms of space science and human activities in space, and to supporting the development of long-lead technologies.

The dreams of Verne, Tsiolkovsky, Clarke and Asimov are becoming reality. As the students of the International Space University Summer Session '95 in Stockholm put it, they

> *". . . envision worldwide benefits through space activities and their application; benefits based on the ideas of the common good of humanity and of shared global consciousness."* [1]

These students also maintained that the future success of world space activity would be closely linked to its capacity to serve humanity now.

We are all here today to share past experiences, to stretch our minds and to generate and discuss new ideas and visions of the future in relation to the very broad subject of this Symposium, Space of Service to Humanity. We shall also provide the first class of the ISU Master of Space Studies students with innovative thoughts and concepts on Space of Service to Humanity and give them the chance to discuss and exchange ideas with the speakers. In return, they will offer us - and the space sector -their views on this subject. A report on their Team Design Project will be produced by the end of July 1996. We wish them all the best in their endeavours.

The ISU Community is happy to welcome you here today at its Central Campus, and wishes you a most successful three days of exchange of ideas.

---

[1] International Space University: *VISION 2020. An International View of the Future, Executive Summary.* International Space University, Strasbourg, France, 1995

A Video Address by the Chancellor of the International Space University

A.C. Clarke, Arthur C. Clarke Centre for Modern Technologies, Katubedda, Moratuwa, Sri Lanka

Greetings to you all in Strasbourg! I wish that I could be with you.

First of all, I would like to pay a tribute to Todd Hawley, one of the Founders, with Peter Diamandis and Bob Richards, of the International Space University. I am glad that he lived long enough to see it become the remarkable institution that it is now, because ISU is playing an extraordinary role in turning young people on to space and in training them for what is going to be one of the most important professions of the next century. Of course, one problem is convincing the public that it will be an important profession. How often have we all heard the remark: "Why spend money on space when there are so many problems to be solved here on Earth?" It is a good question and there are many answers to it. We have to make sure that people understand them.

I am pleased to see that the theme of this Symposium is *Preserving Earth and Improving Life* and, I hope, not only human life, of which there is really rather too much on this Earth. Recently, I was involved in using space technology to save the few hundred remaining mountain gorillas. Nobody has ever been able to map their habitats properly because, as you know from *Gorillas in the Mist,* their terrain is completely cloud-covered. But one of the shuttle flights did produce a beautiful radar map which has helped to protect the gorillas, and I am very happy to see that in your Symposium you have a discussion of this. I was able to show the lovely radar photographs, incidentally, at the Première of *Congo,* in London, by satellite and I believe that the technology involved is actually shown in Michael Crichton's movie.

I hope to go to the shuttle myself by satellite in a few weeks,when the second attempt is being made to lower a payload by tether. (The first attempt failed because the tether jammed a few metres out.) This is the first attempt, you might say, to build a space elevator, with a cable from LEO all the way down to the Earth. That could eventually, one day we hope, give us really cheap access to space. This is a challenging idea and I used it as the theme of a novel, *The Fountains of Paradise,* which in fact the astronauts took up on the first shuttle mission. They all autographed it. At the moment, that autographed copy is in London at the Science Museum, where there is an exhibition of my work which I hope you will have a chance of visiting. You can visit it on the Web, I believe, but I hope you will have a chance of going there in person.

I would now like to tell you of an extraordinary coincidence about the space elevator. When it was conceived, there was no material which could make it possible. Obviously you need something of enormous tensile strength to lay a cable 36000 km long straight down through the Earth's gravitational field. Well, that material has now been discovered. It is the special form of carbon sixty, better known as Buckminsterfullerene. "Bucky" tubes would, at least in principle, make it possible to construct the space elevator. Some years ago I recorded sections from *The Fountains of Paradise* and the sleevenotes with the picture of the space elevator were done by Buckminster Fuller himself. I think that is the most extraordinary coincidence since Bucky didn't even live to see the discovery of the first Fullerenes. Well, the space elevator is an idea that I am very keen on. I don't know whether it will ever happen. There are various problems, but it is theoretically possible.

I would like to throw a few other ideas to you. I strongly recommend that you see a remarkable CD-Rom, *Visions of Mars*, which is going to go on the Russian Lander, we hope, later this year. It carries images and messages to future Mars colonists. I would like to show you a few minutes....

"My name is Arthur Clarke and I am speaking to you from the island of Sri Lanka, once known as Ceylon, in the Indian Ocean on Planet Earth. It is spring in the year 1993, but I am sending this message to the future. I am addressing men and women, perhaps some of you already born, who will hear these words when living on the Planet Mars."

An address to men and women who are living on the Planet Mars! Perhaps some of you are in this audience now. Do get hold of this CD-Rom and enjoy it. There is some beautiful artwork on it as well as other messages and extracts from many books including, I am happy to say, my own short story *Transit of Earth* and *Sands of Mars*, my early book about the first flight to Mars.

Finally, I would like to stretch your minds a bit by throwing some ideas at you. I think that half of what I am telling you now is going to be nonsense, but I do not know which half!

First of all, I think that we are probably on the verge of an energy revolution: some very strange things have been happening recently which remind me of 1895, when X-rays were discovered and the bottom fell out of physics. A lot of people are producing machines and I have some good evidence that some of them really are working. Apparently they produce more energy than goes into them. There is a great deal going on now, much of it confidential. There are various theories to account for this microcavitation-produced fusion, but it may not be anything to do with nuclear fusion as we know it. The latest theory is that these devices may possibly be tapping zero-point energy, the quantum fluctuations of space. This is an awesome thought because we know that the energy of space is there. It has been detected in that famous experiment called the Casimir Experiment, and the energy contained in the volume of a teacup is enough to boil all the oceans of the Earth. If we begin to tap that, we had better look out! I have often wondered if Supernovae may be industrial accidents!

Here is another thing that has come up. It looks as though it may be possible for the fictional space drive to be real. There are some studies of the properties of space and time which suggest that the rocket may not be the only way to go. Warp Five, anyone? It could be true.

Recently, astonishingly in the Quarterly Journal of the Royal Astronomical Society, of all places, there was a paper about "faster than light" travel. That raises a question. If you can travel faster than light - and it does seem that this is possible without violating Einstein's laws, strangely enough - the question is: "Where is everybody?"

Of course there are many answers to that question. A lot of people think that they are here already, which of course is nonsense! Recently, I had the privilege of addressing Mr. Ted Turner and his "merry men" in Atlanta and I ended my address considering this point. I am appalled at the quality of many of our TV programmes, the crime, violence, wars, real and fictional, which another civilisation, other intelligent beings, might have great difficulty in distinguishing one from the other. So there is a record of our crimes and bad behaviour occupying a volume of space maybe 100 light years across by now. I conclude from this that there is no, I repeat no, intelligent life within 50 light years of the Earth because, if there were, the cops would be here now, their sirens screaming right across the radio spectrum!

Thank you, and goodbye from Sri Lanka.

Session 1
Space of Service: Global Issues

Session Chairs:
P. Wood, Chairman-Elect of the International Space University Board of Trustees and Space Consultant, USA
J-B. Mennicken, Director General, DARA GmbH, Cologne, Germany

Space and the Fate of Humanity

P. Creola, Adviser For European Space Cooperation, Federal Department of Foreign Affairs, Eigerplatz 1, CH-3003 Bern, Switzerland

1. Space Benefits

I hate attending symposia! Instead of sitting in a meeting room and talking - however wisely - but just talking, we should be at one of the World's great launch ranges, living the inimitable suspense of the last few seconds of the countdown. And watch in awe as the mighty rocket rises from its pad, thundering starwards....

I know that you all share my own excitement about space and my deep conviction that exploring - and exploiting - space is not of less but, on the contrary, of more importance than ever.

Many people outside this meeting room, however, see the space endeavour as a thing of the past - a by-product of the Cold War, a piece of machinery left over from an earlier worldwide ideological confrontation that will soon come to a total standstill. That conclusion is demonstrably incorrect. Anyone who watches weather forecasts on television, telephones around the globe, or navigates with a GPS receiver is using space technology that has already become commonplace. And so are the people who, from a range of one hundred television programmes, delight in choosing the most stupid of all!

Indeed, forty years after Sputnik 1, satellites for meteorology, fixed and mobile telecommunications, navigation, positioning and observation are a part of everyday life. Whole branches of the economy live with - and through - space services. Whether they are operated by private or public entities, their value to the economy as a whole already far outweighs the cost of their development and operation. Yet without the conviction of the space pioneers and without the initial, and politically motivated, investments, these services would never have been conceived, let alone developed and realised.

The space sector as a provider of services to humanity, therefore, quite definitely has a bright future. It is inseparably and irrevocably part of our civilisation on Planet Earth - connecting instantaneously people, events and information. This is absolutely essential if we are to learn - at long last - and never to forget that everything that happens, happens just around the corner, and that everything we do has repercussions around the globe.

No category of spacecraft brings this home more clearly than remote sensing satellites. The eye of the satellite sees night and day, clouds or no clouds. It alerts us to icebergs, floods and locust swarms, and will soon be able to warn us of volcanic eruptions and earthquakes. It monitors the burning of the rain forests, detects oil slicks from leaking tankers and measures the shrinkage of the ozone layer. The European Space Agency's Long Term Space Policy Committee, which I have the pleasure to chair, has given considerable thought to these space missions as part of the overall concept of "Planetary Management". In our report *Rendezvous with the New Millennium*, we give some further examples of essential monitoring services from Earth orbit: surveillance of ocean topography and temperature leading to crucial insights into the mechanisms of the World's climate, or land use and erosion monitoring as one of the essential parameters determining food production. And, in general, more and more integrated, real-time data on the atmosphere, as well as on the liquid and solid surfaces of our home Planet will determine the input to environmental policies, economic activities, and their political and social consequences.

Planetary Management

- **Resource management**
  - Land use
  - Harvest forecast
  - Forestry
  - Fisheries
  - Mineral resources
- **Surveillance**
  - Weather
  - Ozone
  - Climate
  - Pollution
  - Peace
  - Disarmament
- **Warning**
  - Tropical storms
  - Earthquakes
  - Volcanic eruptions
  - Aggressions
  - Collisions

Figure 1. Planetary management

Figure 1 shows a menu of planetary management functions, ranging from harvest forecasting to peacekeeping. Planetary Management! I have, nevertheless, mixed feelings about this catchy expression which I helped to coin myself. It gives the impression of an integrated, well structured and efficient tool to assure a prosperous and happy future to an ever-growing number of humans inhabiting Planet Earth. The orderly use and development of mineral, fossil and living resources on the basis of objective space-based monitoring techniques seem to assure long-term peace and stability to Spaceship Earth on its course through the Universe.

2. Spaceship Earth

Spaceship Earth! - another catchword! Many of you have unfortunately not lived - in real time - the excitement of the first Apollo flights to the Moon. Humans not only set foot for the first time on an extraterrestrial body. Humanity also had, for the very first time, a look at its cradle: the blue Planet Earth shining in the infinity of the black, star-sprinkled Universe. Watching the living Earth rise in silent beauty over the Moon's barren horizon was to me - and many others - a moment of profound emotion, equalled by looking at the first human footprints on the dusty surface of our celestial neighbour.

Let us have a look at "Spaceship Earth" (Figure 2). Spaceship Earth is spherical in shape, constructed mainly out of stone and iron. It is covered by a very thin outer layer composed of water, soil and air. Its diameter is only 12,700 km. Its temperature varies with time and can change by 3 - 5 degrees Centigrade over a few decades. Its average speed is quite impressive: 100,000 km/h around the Sun and 900,000 km/h through our Galaxy.

From time to time, Spaceship Earth is on a collision course with other celestial objects. On the average, one body 10 m in diameter strikes it every ten years - there are about 150 million of these around in roughly our orbit around the Sun.

Earth is hit by 50 m objects every 200 to 300 years - the most recent, which struck in the Siberian taiga near Tunguska in 1908, flattened one thousand square kilometres of forest. Imagine the devastation that would be caused by one of those falling on one of the World's major metropolitan areas.

And, from time to time, 10 km objects strike Spaceship Earth, rocking it severely, as 65 mio years ago, almost extinguishing Earth's life. Some consider that it is because of the extinction of the dinosaurs that mammals and, therefore, we got a chance - until the next time around .... What that next time might look like, if we do not find ways to deflect celestial objects of that size from their course, we all saw on television in July 1994, when comet Shoemaker-Levy 9 struck our fellow planet Jupiter with the full force of 100 mio megatons of TNT.

Let us have a look at the crew of Spaceship Earth. The present crew took over the controls from the previous commanders, gradually, about 2000 years ago. They are now quite confident that they can handle the ship properly and steer it through the multiple dangers of its cosmic course. In fact, they are not! Although they are fiddling with the temperature control, frantically adjusting the air conditioning and spilling its precious onboard fuel into the atmosphere, they have not yet found the instruction manual explaining the principles - as well as the details - of the complex interrelation and interaction of the ship's systems.

Spaceship Earth

- **Spherical iron and stone construction**
- **Thin outer layer of water, soil and air**
- **Diametre**
  12,700 km
- **Temperature**
  Variable +/- 3-5°C/over a few decades
- **Speed**
  100,000 km/h around the Sun
  900,000 km/h through the Galaxy
- **Collisions**
  10 m object: every 10 years (150 million)
  50 m object: Tunguska 1908
  10 km object: Jupiter, July 1994
  Earth, 65 million years
  100 million megatons of TNT

- **Onboard consumables**
  Fertile soil: loss 10 million hectares/year
  Drinking water: rationed
  Natural gas, oil, uranium: 100 years
- **Crew**
  5.8 billion
  5 births/second
  2 deaths/second
  3 new crew members/second
  = 260,000/day, 100 million/year, 1 billion/10 years
  Within 500 years, 1 square metre will remain per person

Figure 2. Spaceship Earth

In fact, the status of Spaceship Earth is appalling! Instead of trying to improve their piloting skills, the crew members - particularly the males - indulge in endemic warfare and continuous skirmishes in all parts of the spacecraft. Their other hobby - this, together with their female colleagues - is reproduction beyond any reasonable rate. Two crew members die per second - from old age, disease, accident or war. But, each second, five new crew members are born. This amounts to a net increase of 270,000 per day, 100 mio per year and one billion every ten years. Every ten years, on tiny Spaceship Earth, the demographic equivalent of a new China must be fed, housed, educated and employed. If this trend continues, within only 500 years there will be just one square meter of ground per person on board our Spaceship.

A surprising number of people seem to believe that one of the more promising uses of space technology might be to export our surplus crew members to other worlds. Although we have had the privilege of witnessing, these very months, the detection of the first planets circling other Suns in the more immediate neighbourhood of our own solar system, this is, of course, utter nonsense. Even if we mastered the art of interstellar space travel - and I firmly believe that we will do this one day, if we do not eat or blow ourselves up first - 270 spaceships would have to take off per day, each one carrying one thousand passengers, just to stabilise our own World's population. This quite disillusioning calculation perhaps explains why those people and powers who still oppose birth control are not known to be among the more enthusiastic supporters of spaceflight. Otherwise, they would actually be the first to promote interstellar spaceflight and all the related marvellous technologies.

	Growth rate of one unit	
	2%	5%
After 50 years	3	11
After 100 years	7	130
After 200 years	50	17,000
After 300 years	400	2.2 mio
After 400 years	2,700	300 mio
After 500 years	20,000	40 mia

Figure 3. The dramatics of growth
Continuously growing productions and consumptions of any kind multiply dramatically in the long run. This is totally incompatible with a stable ecosystem on a single planet.

I went into some detail on the rather drastic example of Spaceship Earth's crew problems in order to illustrate what I call the dramatics of growth. Figure 3 shows the long-term consequences of two typical yearly growth rates: 2% and 5%. If you were plant managers or economic ministers, you would at least be respected for sustained growths rates of 2%. And you would be applauded for 5% more output, or profits or share value, whatever, per year. But look at what this leads to. One hundred years may seem far away. But some of your grandchildren will be around. Do you really imagine that the world will be nicer, gentler and more livable with everything (motor cars, commercials, Ariane launches....) having increased 130 times - with the exception perhaps of the population which, according to the present growth rate, will have increased from 6 billion today "only" to something between 20 and 40 billion people?

Think about 500 years from now. 20,000 times more Gross Domestic Product, if our economies expand at what is considered by currently prevailing economic doctrines to be a less-than-desirable growth rate. And by 40 billion times if the world's economy expands up to then by 5□%per annum - far below the growth rates which we presently admire in some boom regions of this World. But 500 years is really far off, some of you might say. No, it is not, I answer. Many of our more cherished institutions, like the church, marriage and symposia, are far older than that. But something will have been found before then to avoid my "pessimistic" scenario. Maybe I should ask: do you feel my scenario to be pessimistic? Why? How can it be pessimistic if it is only the extrapolation of today's prevailing optimisms? And my scenario is not even a scenario. It is simple evidence: continuously growing production and consumption of any kind multiply dramatically in the long run. This is totally incompatible with the limited resources and a stable ecosystem of one single planet, of Planet Earth.

The unavoidable conclusion seems to be then: zero growth! Remember the first report of the Club of Rome? It was all there. Even more refined world models clearly point to the imperative necessity to contain - and, in the end, stop - growth. But nobody knows how to do it. Nobody wants to do it. Even our beloved Greens like their dwellings at 20 degrees in winter and will not refuse modern medical care to save their own, and their children's, lives. And what about the world's poor, whose main worry is to have enough to eat, a roof over their heads and - if possible - a bed to sleep in? It is a distribution problem, one theory goes. If only the wealthy people shared with the poor! Are you kidding? Even if all of you in this room slashed your energy and overall consumption by half, you could support only twice as many of you at half your present lifestyle and not an ever-increasing number of them. And, by the way, the overwhelming majority of the poor people of this world want more than just food and shelter. Do we in the spoiled part of the world have the right to prevent them from buying television sets and cars, and from going on vacation? Of course not - and our companies are busy helping them to get there. The Cold War is over. Communism is defeated. Freedom and the market have won! New markets, new products, new jobs! Growth and prosperity everywhere!

Look at Figure 3! Look where the very growth which we are all dependent on, and addicted to, leads us! The human species seems to be condemned by its genes to grow and expand into every conceivable habitat. As it expands and grows as by far the most powerful species on Earth, with no enemy left but itself, it will, if it does not master the ill effects of this urge, continue ruthlessly to displace all other species except those on which it feeds. Then, it will literally eat up its own planet. Might this be the reason why so far the search for signals of extraterrestrial civilisations has not turned up a single coded 'beep' out of the vastness of the

Universe? Might it be that every species achieving our degree of evolution inevitably will self-destruct rather shortly after the respective Marconis have invented wireless communication? And - if it survives at all - will it fall back into the Stone Age after an unimaginably brutal struggle for the last fossil and living resources of its home planet? In fact, according to some of the more mysterious ancient artifacts that are dutifully ignored or explained away by official archeology, this might have happened on Earth once already.

I have already mentioned the Club of Rome. In the Long Term Space Policy Committee we had the privilege of having its president as a guest at one of our hearings. Do you know what he said? "*Man's destiny is not to look down to Earth - but up to the stars!*"

Indeed, since in the foreseeable future we shall need an ever-increasing quantity of resources in order to bring all the people on Earth as peacefully as possible to what we, the wealthy, consider to be a decent standard of living, and if there is no efficient means in sight to stop population growth on Earth before - let us say - the end of the next century, we shall then have no choice other "than to look up to the stars". Since, by definition, Planet Earth and its resources are finite then, obviously, space is infinite. We live on star matter anyway. Every chemical element in our bodies heavier than carbon was produced in those gigantic stellar explosions that are so dramatically shown in the breathtaking pictures of the NASA/ESA Hubble Space Telescope. We are indeed children of the stars - and we shall go back to the stars if we want to avoid the fate of extinction.

The difficulty is getting there. Every one of those spectacular space launches is a vivid illustration of the fundamental difficulty in getting out of Earth's gravity pit. Figure 4 shows the so-called gravity wells in the inner solar system. The height scale indicates to what distance from any of those bodies you have to throw an object in order to escape the gravity of that body and be in free orbit around the Sun. It is therefore a measure of the energy which you need and - in very direct correlation - of the costs involved to get there. Getting away from Earth is the most difficult. Getting to the Space Station orbit takes about half the energy of getting to geostationary orbit. But once you are in geostationary orbit, with very little additional energy, you can also get to the Moon. You all know that. What surprisingly few people realise is that, conversely, once you are on the Moon, it is very easy to get to the geostationary orbit. The launch cost barrier is high - but it is close to Earth.

Up to now, practically every space activity of immediate economic significance - the most prominent examples obviously being communications satellites - was and is confined to that region of near-Earth space that it is most difficult to get to. This is why we need one of those mighty launchers from Ariane to Proton every time that we want to put one of our commercial satellites into orbit. In fact, it is totally surprising that, despite the launch cost barrier, so many space services have already become commercially viable and competitive with terrestrial services.

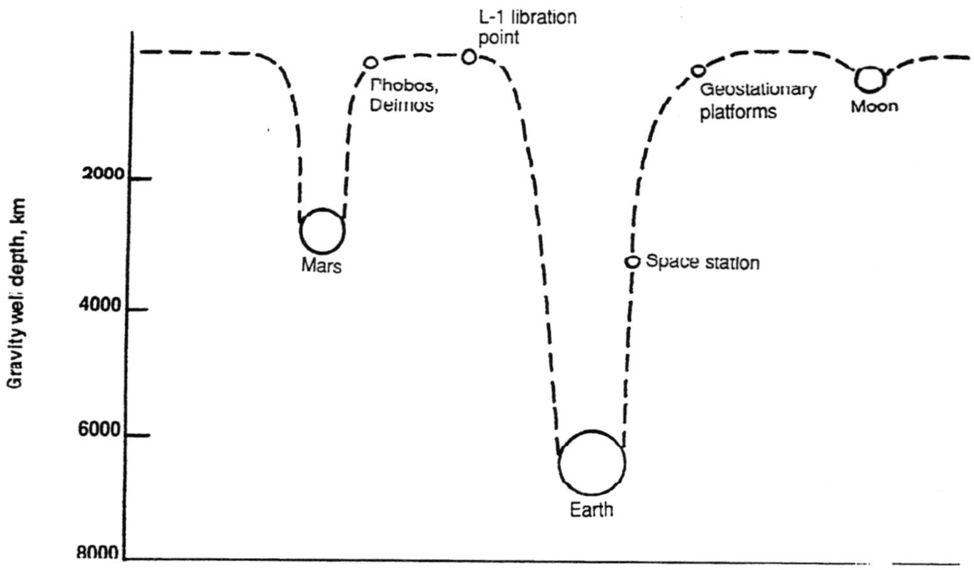

Figure 4. Inner solar system gravity wells

3. Three Imperatives

Thus the first imperative is certainly to overcome this launch cost barrier by going, as quickly as possible, to reusable launchers. I will not address this issue here. You all know that it is a fascinating subject, often at the limit of ideological battles. It is technically challenging and, therefore, costly to develop reusable launchers. And there is the chicken-and-egg question: will the market foster reusables when launch demand increases massively, or will the introduction of reusables foster a massive increase in launch demand? My humble view is that it will work both ways. Certainly I feel some resistance to the development of the reusable launcher from the launch industry itself. After all, for 40 years, the industry has been selling one complete launcher per mission. So why should it move too quickly - once the development phase is over - into the field of launcher maintenance solely?

The second imperative is to return to the Moon as soon as possible! Not the Apollo way, as an all-out warlike effort, but in a stepwise, internationally coordinated way, as was so constructively discussed at the first International Lunar Workshop held in Beatenberg, Switzerland, in 1994. The Moon is the key to the large-scale exploitation of near-Earth space. It is incredibly close: after the equivalent of only 10 orbits around the Earth, Space Station ISSA would be on the Moon if only it flew somewhat faster and in the right direction.... The Moon is the one and only natural and long-term Space Station. It is the key vantage point not only to reach out further into the solar system to the asteroid belt, Mars and beyond, but also to get back towards Earth, and particularly to geostationary orbit. And, above all, it has its own resources. You can fabricate things on the Moon itself, once you are there, instead of shooting them up from Earth at great expense! Remember that reusable launchers will not be exactly cheap, if you need numerous flights to put very large structures into orbit.

Let us turn, therefore, to the third imperative, getting at those resources which I am convinced we will need so desperately in the next century in order to avoid chaos and misery on Earth. The most needed resource will be energy, not only to equalise living standards, but also to produce more food for an ever-growing population and more drinking water in the face of dwindling reserves.

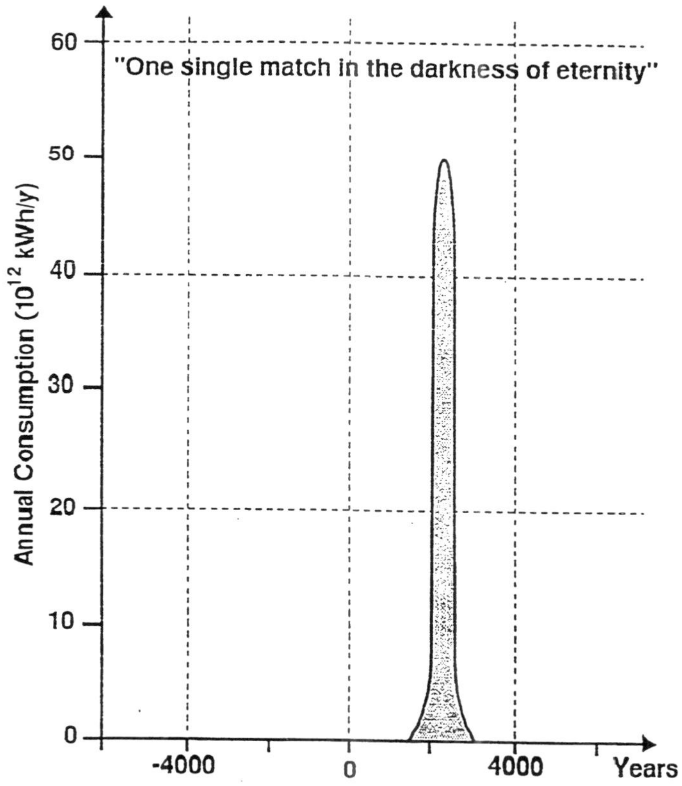

Figure 5. Fossil fuels annual world consumption: past, present, future (According to M.K. Hubbert, National Academy of Sciences - National Research Council Publication 1003)

Figure 5 shows the depletion of Planet Earth's fossil fuel reserves over a slightly longer time scale than is usually shown. Compared to the Planet's age, compared to the time that humans have existed, this happens practically in one single instant. We light them up - and they burn like "one single match in the darkness of eternity". People in the traditional energy business normally reassure us: oil and natural gas are still plentiful. You only have to look at current prices - new reserves are discovered at almost the same rate as old ones are depleted and new technology will allow us to exploit hitherto inaccessible reserves. So today's energy industry is fine: as long as fuel is plentiful, it earns a lot, because of rising consumption, and once it becomes scarce, it will obviously earn a lot as well. But, the present projections are normally based on present consumption per capita and do not take into account phenomena like the economic awakening of entire regions such as China. To come to the point: it does not matter whether fossil fuels - and uranium - are depleted in 100 years or 500 years, but one day we shall have to live without them. And we shall have to develop the technologies to tap other energy sources while there is still time. In the long run, we have no other choice than the massive exploitation of solar energy. We have to convert this Planet's industry, this Planet's society, this Planet's lifestyle to solar power.

For many years I opposed any concept of collecting that energy in space and beaming it back to Earth. After all, installing large photovoltaic farms on Earth would always be much cheaper than launching and assembling all those huge surfaces in orbit. I no longer believe this to be true. Let me give just three reasons. First: in the next century even desert land might become scarce, because we need it to house our exploding population or to produce food in vast hydroculture installations. Second: the total surface of the Earth collects far less than one billionth of the Sun's total radiated energy. So if we need lots of solar energy, we have to get out to catch it. Third: why should we launch solar power stations from Earth, if we can construct them right where they belong: on the Moon - or, as is easily accessible from the Moon - in geostationary orbit?

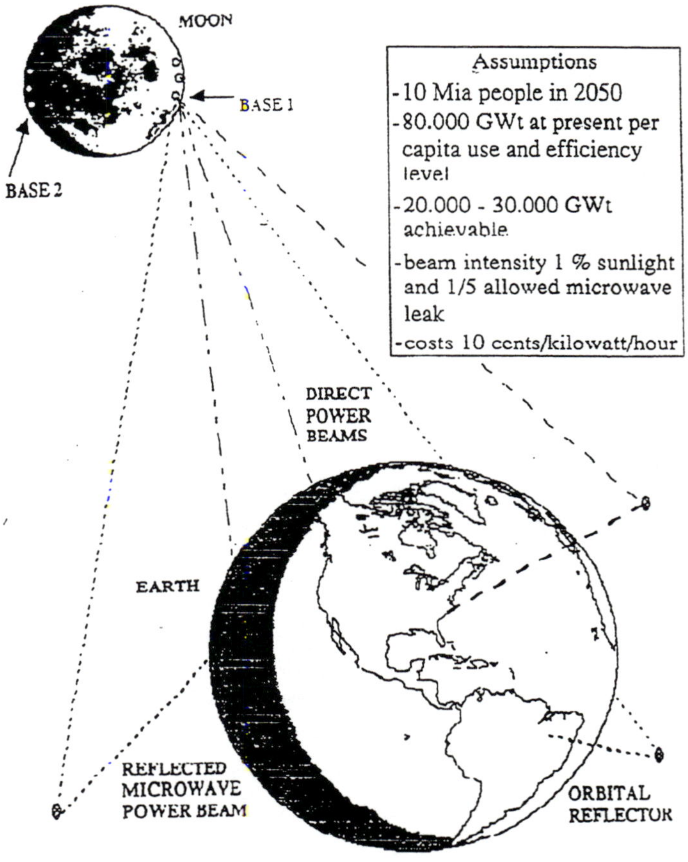

Figure 6. Space option. Example: lunar-solar power system

Figure 6 is David Criswell's famous Lunar-Solar Power System Concept of 1993, which is surely familiar to some of you. It is to be seen in the wider context of the so-called space option promoted by my compatriots Marco Bernasconi and

Arthur Woods. Actually building power stations on the Moon - easy once you have achieved permanent robotic and/or human presence there and you have mastered the technology of self-replicating machinery - certainly makes a lot of sense. This could be among the very few options for getting us out of the civilisation-threatening energy crunch.

4. Final Remarks

Well, time is running out. We are still on the Moon, constructing solar power stations. Will you help me to get back quickly to this meeting room so that the other speakers will also have a chance?

To do this, we will have to cut short further illustrations of large-scale integration of near-Earth space into the terrestrial economy. One is mining the asteroids to use their precious raw materials. Another is modifying the solar flux with a huge lens between our natural - fully operational and pollution-free - nuclear power plant and our Planet in order to mitigate natural or man-made climate change. Think about the tantalising scientific, technological, ethical, political and regulatory questions! We have just begun to decipher a few phrases of the chapter "life support system" in the operations manual of Spaceship Earth. We have a long way to go before we can model the regional and global effects of selectively heating or cooling certain areas with large space-based structures. But we know already - I mentioned it in the beginning - that, within the great climatic cycles of the past, there have been brutal variations of several degrees Centigrade up or down within two or three decades.

We know now, for instance, that the Gulf Stream could suddenly change its course, plunging northern Europe into a second Ice Age - ironically without changing the overall and probably man-induced global warming trend. Shall we then have to try to heat up that part of the globe? Once we have the means to influence climate in one region, other regions will inevitably ask for their share of a better climate, in order to win new lands to house their ever-increasing population and to grow more food for the starving masses. It is difficult to avoid the feeling that this might just be another of the potential future horror scenarios resulting from the combined long-term effect of the unmitigated overproduction of humans and relentless economic growth.

I feel uneasy about this kind of space technology. When I was still a student - though not at the International Space University - but writing my thesis on Space Law, I dreamed about exploration of the pristine wonders of the Universe and not massive exploitation to the sole benefit of one single form of Earthlife, calling itself intelligent, whilst being still too stupid to solve the problems which it was sufficently clever to create.

Let us hope for one thing: that economic expansion into near-Earth space will buy us time! One hundred, one thousand, or ten thousand years, time to find out why we are incapable of stopping to grow like cancer. We need time to find out why we always start destroying before trying to save, time to overcome the ill effects of our instincts without losing those that make us human beings: caring, loving and wondering where we came from and where we are going.

Only then can we confidently venture beyond near-Earth space to the stars, and discover why we are alone, or find out where the others are, and how they coped with their growth disease. But first, and this is the overwhelming priority, we have to earn our wings as the pilots of Spaceship Earth - otherwise, it will soon sail on without us.

The Space Myth: When Apollo Flirts with Gaia

J. Arnould, University Paris XI and in collaboration with the CNES, Direction des Programmes du CNES, 2 place Maurice Quentin, 75001 Paris, France

Abstract

Space missions offer new perspectives of the Earth and the planets. In this paper, these are discussed philosophically, from a historical viewpoint, from the era of Ancient Greece to the present day. Mythical gods have been replaced, in human society, by concerns for the environment.

1. Introduction

In this paper, the mythical god Apollo is identified with the programmes of planetary exploration and astronomy. These programmes bring to life the space myth as defined by J. Blamont:

> *"to harness the power, the nobility and beneficence of science for serving humankind in order to control Nature's laws"* (Reference 1).

Since the day that Earthlings equipped him with the means to leave the surface of the Earth, Apollo has unceasingly undertaken ever further and ever braver conquests. Millions of miles and months of audacious navigation no longer matter; he has gone on to discover one by one the planets that surround Earth, never failing to furnish his masters with ever sharper and more astonishing photographs - what could be more exciting than to penetrate the secret gardens of Venus! Then, one fine day, Apollo began seriously flirting with Gaia, the mother planet, Planet Earth. This new romance was called "Mission to Planet Earth". Gaia was not lacking in charm. From space, one admirer described her thus:

> *"This is a marvellous, harmonious planet, filled with peace, blue with white clouds, and gives one a deep impression of home, of being, of identity"* (E. Mitchell, quoted by P. Russel, Reference 2).

And Apollo, true to his usual self, sent the Earthlings lovely photographs of his new sweetheart. They were by no means the first: manned missions and meteorological satellites had already furnished some very beautiful ones. But Apollo's photographs, which arrived via LANDSAT or SPOT were even more astonishing, more intimate. Earthlings would undoubtedly appreciate these. Apollo was all the more convinced since, in addition to the interest that his fans back on Earth showed in faraway goddesses, there was now a growing concern for Earth's state of health. And yet Apollo still has not met with all the success that one had hoped for.... Why is this?

2. Between Greenpeace and Star Trek

Let us start with two explanations, or rather two remarks:

2.1 The Space Myth is in Crisis

Earth observation from space is only one aspect of the space myth, a myth which is now in crisis. No doubt the heavens still, after 40 years of space conquest and exploration, shelter enough mystery and secrets for humanity to project there more images than it has ever received therefrom. Contemporary mentality continues to see in the heavenly vault that Other Place, the Strange and the Transcendant without which humanity would no longer be quite the same. Would we still be humans the day we decided to stop "travelling the stars"? But space programmes no longer provide the fundamental response to these desires; they cost a lot for societies confronted with economic crisis and they are still often marked by the warrior spirit that motivated American and Soviet programmes in the 1960's and 1970's. In the words of one French weekly, we might well ask ourselves:

"Space - can the myth be born again?" (Reference 3).

Being no prophet, I will not risk answering that question. I have simply observed that, together with this dream, we need to take into consideration the fact that because space has become routine, and because of a demand for immediate utility, fascination with the space myth has been stunted. But these two factors do not explain the space crisis fully for, if they did, Earth observation would still be all the rage, its utility in furthering knowledge of our great planetary systems (meteorological, oceanographical, ecological) having been adequately demonstrated. But few people talk about Earth observation in this way.

2.2 The Space Myth is not Dead

One may be tempted to think that the new expectations and demands made on space programmes might benefit those that are marked less by a spirit of conquest than by an interest in environmental questions. In her study of *"L'Opinion Publique et l'Environnement"*, Michelle Dobré states:

"the attention given to questions concerning the environment is not simply a thing of fashion. It has to do with the state of public policies and with the state of social demands that come from them. Between these two poles there is a sound effect amplified by the presence of the environment theme in the media" (Reference 4).

Even space agencies, in their own way, confirm this interest. In studies conducted or sponsored by NASA, the heading "Satellites to Monitor

Environment" is high in American opinion polls favouring space programmes. In public requests for documentation from CNES, interest in environmental questions is real and growing.

Space launches, manned flights and interplanetary missions are talked about a lot. In a poll of members conducted by the US National Space Society, people replied that they would readily budget for a return to the Moon costing twice the sum allocated to Mission to Planet Earth. Such a difference signifies the continued dominance in this milieu of "space adventure" or what I call the "Star Trek" culture. In French-language science magazines aimed at the general public, pictures of the Ariane launcher are often used as a symbol of a successful political and economic conquest.

Between ecological fears and dreams of space adventure, between Greenpeace and Star Trek, the Earth observation satellites and their data are in a position which is almost as delicate to manage as it is to analyse. I shall put aside the technical, political or commercial aspects of the question to focus more on sociological, even more philosophical, issues. In the face of ecological concerns, what reality and what meaning do SPOT, LANDSAT, TOPEX-POSEIDON and their circumterrestrial kin offer us?

3. The World According to SPOT

There is more than one way to take in a reality such as that constituted by our planet. At the risk of over-simplification and caricature, I shall address two of these.

According to the philosophers of antiquity, the world is an organised and ordered whole, inside which each being finds and maintains its natural place. In this concept of the world, we have the ideas of order, of harmony (the world thus conceived displays an overtly aesthetic character), of rules and of natural law. But above all, this world is imposed *a priori* upon the human observer.

In a phenomenological approach (inaugurated by Immanuel Kant), the world is primarily a human event, the product, the construct of a collective conscience or subjectivity. The title of L. Ferry's book *"Le Nouvel Ordre Ecologique. L'Arbre, l'Animal et l'Homme"* (Grasset, Paris, 1992) is perfectly precise and to the point: ecologists' demands are founded on nothing but a "new" order of their own construction and certainly not on any ancient order (unless it be a mythical one).

Where in this scheme does one put the world according to SPOT?

The world, as the Spot Image Corporation tells us, for example, is a reality that one can feed into a computer.

> *"Spot puts down-to-Earth reality in your GIS. [...] SPOT is reality - not vectors or a map, or someone's abstract."* (Advertisement published in Earth Observation Magazine, January 1995)

The French brochure "Accéder à une Vision Vivante de la Terre" (Spot Image, 1993) has a more modest stance:

> *"SPOT images offer the dimension of reality, give all the richness of milieu and the density of the living world."*

These examples would lead one to believe that the literature distributed by space agencies, as well as by organisations whose mission it is to diffuse satellite images, use the word "world" in the sense known in antiquity. Certain phrases that echo today's ecological questions confirm this diagnosis. The following list of headings in the chapter "The Right to Know" of Spot Image's brochure *"L'Environnement Sans Frontières"* (in English, *"One Earth, One Environment"*) is a telling illustration of this:

- Global Change
- Desertification
- Our Forests are Dying
- Soil Degradation
- After Disaster Strikes.

Imbalance, degradation, pertubation - all of these are terms expressing phenomena that are quite real, largely caused by modern exploitation of raw materials and of energy sources using Western techniques. Do not all these terms, sometimes explicitly but more often implicitly, refer to a state of things that serves as a reference - an equilibrium or primary state that I might term, not without polemic, an "ecological Garden of Eden", in short, a "world" in the Greek sense of the word. No one can deny that some variations, even some states of imbalance, do exist at the core of the living world and the great cycles that are part of it, on whatever scale one wishes to look at it. But to what are these states of imbalance compared? To what order? That of the 1950's, of the beginning of the industrial age, of Cro-Magnon hominids?

The debates among experts or among politicians to which these questions give rise suffice to confirm the one truly correct answer: such a system of reference, a perfectly ordained world, has never existed except in human consciousness. We are condemned to observing variations around states in equilibrium. But they are ephemeral, especially on a cosmic time scale. The world of Antiquity has truly vanished; the world according to SPOT (one might say as much for LANDSAT, ERS, etc.) is one world amongst others, a more or less original construct of a reality born of the encounter between a certain technique and a certain human consciousness. The space image is not a map. A map is based on a collective consensus which gives it validity in the absence of the possibility to gauge resemblance with respect to an inaccessible reference point.

With the advent of aeronautical and space engines, cartography's imagined world has been replaced by "images", simply because humankind has reached the celestial spheres and begun to see with the eyes of the gods. But is our planet observed "in reality" for all that? Obviously not. When the camera lens is put away, SPOT's diodes, and even the astronauts' view of things, are just more places where a certain reality is constructed or elaborated, a reality which until now had never existed.

Given this perspective, one might venture to say that Earth was born once again in 1986 thanks to SPOT's instruments, and that the oceans were created anew in 1992 with the launching of TOPEX-POSEIDON.

All that I have just said applies not only to our planet but to all the others as well. Each time that a discovery mission reaches its destination and begins sending heretofore unsuspected information, we get the chance to build a new world which, added to earlier ones, increases our knowledge. There is, however, a very big difference when it comes to Earth. Gaia is not just another lover whose charms Apollo unveils with increasing savoir-faire. She is also a nursing mother. Put less poetically, that means that the world we build is not just one among many, it is our *environment*, even a part of our own selves.

The word "environment" is to be taken here in more than one sense. First, it designates that Nature in whose heart human societies have appeared. It is that Mother Nature to which the farmer, going over the ground with his plough, offered up sacrifices in atonement for his act of rape. It also designates a reality that is elaborated by humanity throughout its socio-economic, industrial and cultural history (see the works of René Dubos). In this case, observing the environment is as close to the "know thyself" of antiquity as it is to the order received by the Hebrews forbidding them to uncover the nudity of their parents.

One might object that all of this is a far cry from space technology and space policy. I quite agree, but I do not think that it is possible to develop the one or conduct the other while defying such powerful springs of (Western) human mentality. This is what I would like to show by looking now into the chances of Gaia being reconquered by Apollo.

4. Future Handicaps and Perspectives

The foregoing can allow us to identify certain issues of importance for the future of Earth observation from space.

4.1 Cartesian Ideology

Earth observation from space fits well with Cartesian and imperialist ideology: satellites share in the desire to control and take possession of Nature. Earth is, above all, One-World (Reference 5), an entity of which control must be gained; the word mission, as in "Mission to Planet Earth", can itself be immediately connected to the words "power, conquest, *imperium*," just as hero or male romanticism can (*"imperialist, masculine, and Christian dimensions of the American dream"*, see Reference 5). Therefore I am not sure that Earth observation respects that "Whole Earth" concept dear to environmentalists who defend a vitalist vision of Earth, an organic and spiritual unity of terrestrial life. Does it not lead instead to interrupting continuities, making incisions on the surface, emphasizing gradations and fractures? It defies walls and boundaries, travelling along beaches, rivers and highways. Henceforth the magic words in Earth observation are pixel and resolution: satellites break up in the space - and time continuum. On the contrary, the global approach would be one that goes around, limits, circumscribes, caresses the world: *"an erotic approach to the terrestrial body, the whole of it, one filled with sensuous feeling that follows every one of the body's contours without ever penetrating it"* (Reference 6). Yes, the SPOT programme might be presented as a compromise between extremely sharp vision and depth of focus made available on a single photograph; we get "synoptic and exhaustive data necessary for the study of these wide expanses..." when it comes to agricultural zones. Can we say the same thing when it comes to things on a smaller scale? There is no ideal scale of things in this domain, there is only a necessity: the ability to integrate the increasing quantities of information thus gathered. The person who takes a machine apart should know how to put it back together again in order to claim that he fully understands how it works.

4.2 Space Agencies

What do space agencies say on the subject of environment? This is a loaded question, no doubt, but one that ought to be addressed. The position of these organisations is scarcely a comfortable one. Accustomed as they are to the hard sciences and to engineering know-how, they are not always quick to adopt either an "environmentalist" style or its underlying ideology. I have in mind, for example, James Lovelock's hypothesis, the ramifications of which reach well beyond geophysiology alone, not to mention the "haziness" of environmental sciences, which are not limited solely to ecology but also take in elements from as far afield as evolutionary genetics and the human sciences. It is particularly telling that a review of press articles from 1992 led to SPOT becoming news when its use during the Gulf War became known.

4.3 Geographic Information Systems

We might think about a global approach by starting with Geographic Information Systems (GIS) which seek to consider multiple dimensions for constructing reality. Interest has been aroused in *"The Social Implications of Geographic Systems"* (References 7, 8). This book is interesting for at least two reasons: (a) for the study of the social consequences of using GIS, a study that branches out in several directions (geography, economics, law, ethics); (b) for the apparent undercurrent which claims that GIS have negative effects. Both subjects require future study and debate. What is at stake seems clear: to what global vision can Earth observation from space contribute?

4.4 The Unknown

What has happened to the *terrae incognitae* of ancient maps, given our modern means of observation and modern cartography? Have we not forgotten what role these mysterious regions, left blank or drawn with astonishing imagination, played: that of preserving and stimulating the imagination, the creative and constructive aptitudes of humanity? Should we not encourage

> *"primary emotions, the appeal of the exotic and the wonderful one finds in far-off horizons, the shock of natural or cultural differences; being awed, dreaming, being captivated by a geography that one discovers like a tale of adventure, like an exploration trip during which one regularly bumps into danger"* (Reference 9)?

"Living a space adventure" does of course pander to public demand, but it is also like putting aside for a moment the responsibilities that are the burden of our anthropocentrism. I think that it is neither possible nor even desirable to avoid all form of anthropocentrism: examining the ecological crisis may lead to recognition of what Dominique Bourg calls "practical anthropocentrism". However hard I try, I can never separate myself from my position at the moment I pronounce it. This simple observation should also cover humanity's position in the heart of earthly reality. One can always dream and try to set up a sort of "natural contract" calling for a greater respect of one's natural environment, but one cannot escape the simple fact that our species continues to occupy the current number one position (Reference 10).

I realise that it is not easy to detect *terrae incognitae* among the Earth observation data. Perhaps, then, we ought to seek help from those other architects of space: artists. Satellite photographs do not replace an artist's work, rather they offer it new perspectives. Thus, the painter Mary Edna Fraser discovers "the endless horizon of design" and draws tissues with an aerial motif, together with help and inspiration from Asiatic methods. What she comes back with is a call to detect new facets and figures of our planet. The convoluted North Edisto River, outside Charleston, becomes a flaming dragon in *"Curling River"* (1988).

And how could we forget the "architects of time" - archaeologists who benefit from aerial and space photographs. The market value of their exotic dreams is not to be sniffed at. Did not satellite pictures, in the early 1990's, pick out the ruins of an ancient city in Oman, the Atlantis of the sands, the mysterious Iram destroyed by a cataclysm, according to the *"Thousand and One Nights"* ? Add to that the hypothesis whereby this city is the departure point of the Three Kings (the surrounding region of Dhofar is the only one where incense trees grow), plus the fact that digging started just after Christmas 1991, and there you have, without the shadow of a doubt, the stuff of which dreams are made!

I hope that this paper has asked more questions than it answered. Let me close with a citation from R. Debray (Reference 11):

> *"One simple question for the next millenium: how shall we see well all around us and not concede that behind, under and above us, there are "invisible things"? Not necessarily angels or heavenly bodies. Ideal realities, myths or concepts, generalities or universals, immaterial things or symbols that could never be translated visually, not even virtually, in some cyberspace. How could there be a Here without an Elsewhere, a Today without a Yesterday and a Tomorrow, an Always without a Never....?"*

References

1. Blamont, J.: *La Crise (?) de l'Espace*. Paper presented at "L'Aventure Spatiale, Acquis et Perspectives", Conference held in Paris, France, 25 November 1995
2. Russel, P.: *La Terre S'Eveille. Les Sauts Evolutifs de Gaia*, pp. 19-20. Editions Le Souffle d'Or, Barret, 1989
3. Special report, *Le Figaro Magazine*, Paris, France, 28 October 1995
4. Dobre, M.: *L'Opinion Publique et l'Environnement*, p. 9. Institut Français de l'Environnement, Orléans, France, 1995
5. Cosgrove, D.: "Contested global visions: one-world, whole Earth, and the Apollo space photographs". *Annals of the Association of American Geographers*, Vol 84 No 2, pp. 270-294, 1994
6. Jacob, C.: *L'Empire des Cartes. Approche Théorique de la Cartographie à Travers l'Histoire*, p. 394. Albin Michel, Collection Histoire, Paris, France, 1992
7. Pickles, J. (editor): *Ground Truth. The Social Implications of Geographic Information Systems*. The Guildford Press, New York, London, 1995
8. Roberts, S.M. and Schein, H.: "Earth Shattering Global Imagery and GIS." In *Ground Truth. The Social Implications of Geographic Information Systems*, pp. 171-195. Edited by J. Pickels, The Guildford Press, New York, London, 1995
9. Jacob, C.: *L'Empire des Cartes. Approche Théorique de la Cartographie à Travers l'Histoire*, p. 235. Albin Michel, Collection Histoire, Paris, France, 1992
10. Bourg, D.: "Modernité et nature", in *Les Sentiments de la Nature*, p. 227 on. Edited by D. Bourg, La Découverte, Collection Essais, Paris, France, 1993
11. Debray, R.: *Vie et Mort de l'Image. Une Histoire du Regard en Occident*, p. 506. Gallimard, Paris, France, 1992

The European Space Agency's Work with Developing Countries

V. Hood, International Relations, European Space Agency, 8-10 rue Mario Nikis, 75738 Paris Cédex 15, France

Abstract

The European Space Agency has, since its beginning, had relations with developing countries. In the 1970's, they were very much focused on introducing the METEOSAT programme to African countries, as these were in the coverage zone of satellites. Training courses were organised in African countries in order to present meteorological satellite data. Later on, when ESA was preparing the launch of its ERS satellite, training events were organised in developing countries in order to introduce the new technology and to show how satellite remote sensing radar data could be tools in the development process of these countries. These training events have been made on an ad hoc basis, together with national or international entities. Similarly, ESA contributes to the United Nations Office of Outer Space Affairs training programme for developing countries.

Later, increased awareness among ESA's member states concerning satellite technology for development purposes led to the following actions. At a Ministerial Conference in November 1995, the Ministers representing ESA's member states requested the Director General of ESA to look into space applications with developing countries in order to see how these countries could benefit from the programmes developed by ESA. In response, the Ad-Hoc Working Group on Developing Countries was created by ESA's International Relations Committee and asked to draft a report. Following this report, the International Relations Committee recommended that ESA's Council ask the Director General to set up a Coordination Group on Developing Countries. The Coordination Group has included representatives from the different ESA programmes and has been asked to produce recommendations on how developing countries could benefit from ESA programmes. These recommendations will be presented to the member states for their endorsement.

1. Early Activities Between ESA and Developing Countries

ESA's involvement with developing countries started on a bilateral basis with agreements for cooperation with India, Indonesia and Iran and more informal contacts with Brazil in the mid-seventies.

With the launch of METEOSAT-1, contacts were extended to Africa and the near-East. These were reinforced with the advent of OTS (the first European telecommunications satellite), where ESA became involved in the planning of a telecommunications satellite for Africa. In 1982, as part of ESA's contribution to Unispace'82 held by the United Nations in Vienna, a conference to highlight the potential benefits of ESA's applications satellites for developing countries was held in Frascati, Italy.

One of the results of the Unispace '82 conference was that ESA's member states agreed to provide five annual fellowships for on-the-job training in an ESA establishment to be awarded to people from developing countries. These fellowships were in addition to the fellowships which were already awarded to developing countries with whom ESA had bilateral agreements.

Because of METEOSAT, ESA became actively involved with training courses offered by the United Nations, the World Meteorological Organisation (WMO) and the Food and Agriculture Organisation (FAO). First contacts were made with DG VIII of the European Commission and led to the DG VIII paying for the activities of ESA's Maspalomas ground station, acquiring Landsat and later SPOT data in support of the *Freedom from Hunger Campaign* in the Sahel.

ESA's member states were fully supportive of ESA's early efforts to promote the use of its satellites and services in developing countries. They even voted a budget, albeit a small one, for the promotion of METEOSAT usage. When the construction of ERS-1 was decided, a budget for promotion was also foreseen but rising industrial costs swallowed it before it could be used; ESA therefore used the training budget to promote ERS in developing countries. Regional training courses were begun, initially in Africa, in cooperation with the remote sensing centres and the African Remote Sensing Council, later extended to Latin America and still later to the Far East. As far as possible, and in order to eke out its scant resources, ESA organised these training courses in cooperation with the French National Space Agency (CNES) and the German Space Agency (DLR). ESA also continued to participate actively in the training courses offered by the United Nations and its specialised agencies. Support from ESA member states was lukewarm during the late 1980's and early 1990's, but demand from developing countries was increasing and so ESA turned to the development banks and DG-I of the European Commission for financial assistance. The development banks each financed a training course in their region. DG-I went further and, at the request of four countries of ASEAN (Malaysia, Thailand, Indonesia and the Philippines), financed a three-year project to promote the use of ERS-1 in those countries. Eight pilot projects, two per country, were carried out. The Thai ground receiving station was upgraded to receive ERS data for the region, Malaysia was given an AVHRR receiving station in support of the Joint Research Council's project TREES, and ESA provided work stations to the Philippines and Indonesia, plus technical assistance to the European Commission and the countries of ASEAN.

In addition, following a request from the Asian Institute of Technology, ESA sent a radar expert to Thailand to give assistance to those carrying out projects in the region. His presence has proved invaluable.

2. Recent Activities

In November 1992, at the Ministerial Meeting in Granada of ESA's member states, the Director General was requested to look into space applications in developing countries in order to see how these countries could benefit from the programmes developed by ESA (ESA/C-M/CIV/Res. 2 Final). In response, the Ad-Hoc Working Group on Developing Countries was created by ESA's International Relations Committee (IRC) in 1994 under the able chairmanship of Mr. Pifaretti from Switzerland. The group produced a report in December 1994 entitled *A World Wide Approach: ESA linking to Developing Countries*. On 22

February 1995, as a result of this report, ESA's Council adopted a "resolution on ESA's programmes and activities which could assist in fulfilling the needs of developing countries". Following this resolution, an Interdepartmental Coordinating Group was established in ESA with a mandate to make recommendations in the short, medium and long term to facilitate the use of ESA's activities and programmes in developing countries, in close association with the European Commission, the United Nations and its specialised agencies and national aid agencies in ESA's member states. An interim report listed some short-term actions which could be undertaken with existing resources, but also recommended that a concerted effort be made with national aid agencies and the development banks to find resources to increase ESA's activities in developing countries.

In parallel, ESA was represented on the steering group for the *European Commission DG XII Study on the Analysis of Constraints and Opportunities for Cost-Effective Implementations of Earth Observation Techniques in Developing Countries* and the steering groups also set up by DG XII to look into business opportunities for European industry in China and India. In preparation for the fifth Framework Programme, DG XII has recently established three thematic groups to look into possibilities for devoting some funds to encourage the use of remote sensing in developing countries. ESA and Eumetsat are both members of a thematic group.

Following the success of the ASEAN project, the project has been extended to mid-1996 to permit the transformation of the more successful pilot projects into operational projects. Discussions are underway with all the ASEAN countries and DG I to develop an ASEAN II project aimed at encouraging the operational use of ERS in the region.

Although remote sensing is, by virtue of its global character, the most important area of ESA's work with developing countries, other disciplines are not neglected. International Ultraviolet Explorer satellite archives have been supplied to Costa Rica, Brazil, India, China and to South Africa, for use on a regional basis by Austral Africa.

The archives permit scientists from all over the world to work on data obtained from the satellites independently. Scientific collaboration between scientists in the developed and developing countries is thereby encouraged. China will be participating in the data evaluation from ESA's CLUSTER programme, scheduled for launch on the first Ariane V later in 1996. Space scientists from India and China come to ESA as visiting fellows on a regular basis. ESA co-sponsors the United Nations annual workshops on Basic Space Science held regionally for developing countries.

An early successful milestone in the Indian Space Research Organisation (ISRO) - ESA cooperation was the Apple communications satellite built by ISRO with technical assistance from ESA and launched on Ariane L03 experimental launch.

Joint feasibility studies on navigation satellites have been conducted by ISRO and ESA, and discussions are currently being organised with China and India on global navigational satellite systems.

ESA, together with European industry, has developed very small antennae which have great potential for improving communications in developing countries. ESA was also involved in the development of DIANA and COPINE - a United Nations initiative to use communications satellites to network its offices in Africa.

ESA experts travel the world to provide technical and scientific assistance to those in need on a case-by-case basis. Unfortunately, ESA does not have an aid agency and receives more requests that it can reasonably meet with its manpower and limited resources. Modest success has been achieved in transferring training courses to other entities who then continue on a routine basis, and there is increasing willingness among the educational institutions in Europe to take fellows from developing countries for postgraduate training in space-related disciplines. ESA hopes to be able to expand this technology transfer in order to explore new avenues as new programmes are launched.

International Space Cooperation For Service To Humanity

G. Gibbs[1], Canadian Space Agency, 409 3rd Street SW, Suite 206, Washington, DC 20024, USA

Abstract

In December 1994 the American Institute of Aeronautics and Astronautics hosted the Second Workshop on International Space Cooperation. Its theme was: "Getting Serious About How". A carefully selected, diverse group of more than sixty space sector experts from fifteen countries contributed their knowledge and experience to develop findings and recommendations pertinent to how international space cooperation projects might be implemented. Five areas of potential cooperation were selected for discussion by individual working groups; these covered *Global Space Systems Services, International Cooperation for Peacekeeping, Cooperative Human and Robotic Exploration of Space, International Cooperation in Space Transportation* and *Solar Power to Earth.*

Since the Workshop, the participants under a Follow-on Action Committee have worked to ensure that the results of the Workshop are considered by the world's leaders, in the latters' quest to find ways to expand the use of space for the benefit of all the people of Earth.

The paper begins with an outline of the Workshop, the resulting key findings and recommendations, and what has been done since the Workshop. The reaction of the world community to these efforts is reported.

Progress in three areas of potential cooperation pursued during and after the Workshop are discussed in greater detail in this paper. These are: *Global Space Systems Services, International Cooperation for Peacekeeping,* and *Solar Power to Earth.* These areas were chosen for greater scrutiny at the ISU Symposium since they provide readily recognizable tangible services to humanity.

1. Introduction

The American Institute of Aeronautics and Astronautics has hosted two Workshops on International Space Cooperation. The theme for the first, held in December 1992, was: "Learning From The Past - Planning For the Future". The second Workshop on International Space Cooperation was held in December 1994. Its theme was: "Getting Serious About How". A third Workshop, this time co-hosted by the AIAA and the Confederation Of European Aerospace Societies will be held in May 1996, at the European Space Agency's ESRIN in Frascati, Italy.

These workshops have been by invitation only, and each had a diverse group of more than sixty space sector experts from fifteen countries representing industry, government and academia. They contributed their knowledge and experience to develop findings and recommendations pertinent to how international space cooperation projects might be implemented. For both Workshops, the participants were divided into five working groups.

[1] Graham Gibbs is Head of the CSA Washington Office. He was the Chairman of the Planning Committee for the AIAA Workshop on International Space Cooperation, held in December 1994 and since then has been Co-Chairman of the Workshop Follow-On Action Committee. This paper is based on the results of the Workshop and the follow-up that has taken place to help ensure that the results of the Workshop are considered by the world's leaders for the benefit of humanity.

The first Workshop focused on generic approaches to international cooperation in space activities, under five working groups that covered Space Science, Space Exploration, Space Applications, Supporting Infrastructure, and Policies and Approaches for International Cooperation.

For the second Workshop the AIAA International Activities Committee, through the Workshop Planning Sub-Committee, decided to focus on specific projects. After one year of debate and reviewing proposals, the five projects selected for discussion by individual working groups, were: *Global Space Systems Services, International Cooperation for Peacekeeping, Cooperative Human and Robotic Exploration of Space, International Cooperation in Space Transportation* and *Solar Power to Earth.*

Since the second Workshop, the participants, under a Follow-On Action Committee, have worked to ensure that the results of the Workshop are considered by the world's leaders, in the latter's quest to find ways to expand the use of space for the benefit of all the people of Earth. The paper summarises the resulting key findings and recommendations of the second Workshop and what has been done since the Workshop. The reaction of the world community to these efforts is reported.

Progress in three areas of potential cooperation pursued during and after the Workshop are discussed in greater detail in this paper. These are: *Global Space Systems Services, International Cooperation for Peacekeeping,* and *Solar Power to Earth.* These were chosen for greater scrutiny at this ISU Symposium since they provide readily recognizable and tangible services to humanity.

2. Findings and Recommendations

The report from the first Workshop held in December 1992 noted:

"We have all seen that over the past few years our world has been rapidly changing. These changes are characterized by not only the end of the cold war that once polarized nations and diverted critical resources, but also the shifting economic forces around the globe and the demand to apply resources to long neglected problems, such as budget deficits and the environment.

The nations of the 1990's face formidable challenges - political, economic and technical - but also are presented great opportunities through improved international relationships.

It is against this backdrop of change, that the space faring nations of the world can collectively look upon new opportunities for cooperation and the products of space science and technology to address problems ranging from the environment, global security and exploration that will benefit all people of the Earth."

(Throughout this paper, text in italics denotes findings or recommendations from the reports of the first or second AIAA Workshops on International Space Cooperation).

Space Science was singled out as being the real success story for international cooperation in space. The reason for this success was judged to be in part because international Space Science projects are being conducted with a common set of goals and objectives, are modular, incremental and continuous. It was recommended that other international projects use the experience of the space scientists. The Space Exploration working group at the second Workshop recommended (additions in [ ] are those of the author):

For the foreseeable future, large scale, long duration space exploration [and other] initiatives should be addressed as evolutionary in nature, consisting of a series of interlinked, but programmatically independent phases. Each phase must be justifiable on its own merits.

In the author's opinion, this recommendation should be applied to any international space cooperative venture. The recommendation recognises the realities of the political cycle, fluctuating budgets, and national needs.

Both Workshops highlighted the facts that national space agenda, with the appropriate coordination, could be accommodated within an international context, and that cooperative space projects should be given, and supported by, greater political visibility and emphasis.

Several of the key findings and recommendations resulting from the second Workshop further substantiate these overall observations. For example (again, additions in [ ] are those of the author):

The global, political, industrial and socio-economic climate makes wide ranging cooperation possible, but, what is ***new*** *is that international cooperation may now be a* ***necessary*** *strategy to achieve the goals set out for space-related projects.*

Definitive actions need to be taken to establish a mechanism by which the international community can foster and promote beneficial projects of international cooperation in space.

In today 's environment, other space-faring nations are no longer prepared to let the space superpowers take the lead in establishing the agenda for future exploration of space [or other space ventures]. International cooperation in space exploration [and other] will, therefore, be based on seeking to coordinate among various national agendas.

Coordination of national agendas does not require a single overarching rationale. Different countries will [can] follow their own rationales for engaging in exploration missions [or other international cooperative ventures].

As a first step, space agency heads should ensure that actions are taken to collectively review their respective exploration [and other space] agenda, with the view toward identifying opportunities for coordination.

Coordination should emphasize complementarity and avoid duplication of individual mission objectives in order to maximize overall return for resources collectively expended.

These recommendations coupled with the one, stated earlier in this paper, concerning the need to implement long range cooperative programs in an evolutionary manner with each phase being justified in its own rights, provide an overall framework by which international cooperation in space projects of any type could be implemented.

Also, we need to find ways for an international partner in a multinational project to participate in some phases but not others. This would help match national agendas, budgets and changing priorities to the overall international project. Where a project has stand-alone, but related, phases, such as exploration of the solar system, participation in one phase but not another is not difficult to achieve. Where the problem becomes most difficult is in a project like the International Space Station program with an extended operational phase. All partners are committed to sharing the common operations, even if they do not use the Station, and this means a long range commitment, one that may not be sustainable. If we can solve this problem we will have solved what may become a significant hurdle to future long term, multilateral, and large scale projects.

Both Workshops highlighted the most significant problem facing the exploration and exploitation of space in the future:

Availability of efficient, low-cost, reliable space transportation will play a pivotal role in the successful accomplishment of future space missions.

It was further concluded that truly low cost space transportation will most likely only result from market need - the market pull. It is unlikely that low cost access to space will become available from the private or public sectors in the expectation that a market will develop - the market push. As the working group on Space Transportation stated in its report, such an investment by industry or government would require a "leap of faith", and it is unlikely that this will be made.

The advent of Solar Power To Earth would provide the required market pull. To be economical, the construction of the space infrastructure for Solar Power To Earth will require access to space at a cost of between $100 and $500 per pound. It is only at the lower cost that Solar Power To Earth would be competitive with terrestrial energy sources - for as long as they last.

No longer considered so fanciful, space tourism may be another market pull for low cost space transportation.

3. Space Serving Humanity

Of the five space projects discussed during the second AIAA Workshop on International Space Cooperation, three represent the most readily recognizable and tangible services to humanity.

The three projects and their respective working group mandates are:

- **Global Space Systems Services.** Public application of emerging new, near term commercial products in remote sensing, navigation, and telecommunications might usefully be expanded through international cooperation.

- **International Cooperation for Peacekeeping.** Changing geopolitical relationships and the growing commonality in national security interests might set the framework for the collective use of current and future national security space systems to lessen global threats.

- **Solar Power To Earth.** An international effort could lead to the demonstration that space-based collection and distribution of solar power might be a viable option for meeting the increasing demands for sustainable sources of energy.

This is not to suggest that space transportation and exploration do not provide tangible benefits. Transportation is a means to another objective. And the benefits of space exploration often are not immediately evident to the general public.

4. Global Space Systems Services

Today the need for Global Space Systems Services in telecommunications, remote sensing, and navigation is growing. These services are becoming accessible to all nations and are increasingly commercial and international.

Commercial services, particularly communications satellites and more recently meterological and earth observation satellites, are providing a valuable service. These are recognized, or at least used, by the general public. One of the purposes of the International Space Cooperation Workshop was to promote discussion on "how" these and other services could be used for greater benefit to humankind.

Governments are becoming less involved in the establishment of these services. Today there are several examples of Global Space Systems being built only with private sector capital investment. The sale of the resulting service justifies the investment. Some of these services are being built through international consortia comprising teams of a satellite manufacturer, launch provider and service provider. Industry is showing the way. But are these services being fully utilized in a global context? The answer is "no".

In addition to the services provided for commercial reasons, these services have an enormous potential to serve a wider public. They could, in addition to their private sector service, provide a "public service" by supporting public safety, disaster warning and relief, peacekeeping, search and rescue, telemedicine and education services.

This was a major finding of this working group.

The questions that now need to be addressed by the international public sector community are:

- What are the public service needs around the world?

- How can the interoperability of these private sector services be enhanced to serve the public service needs better?

We must recognize that many of these services have been made available, through private investment, for commercial purposes. The owner/operators have no obligation to provide a public service. However, a trade-off is possible. A trade-off was proposed by the working group which included representatives from some of the major companies in this business. These systems are not operating to their full commercial potential because of less than optimum orbital slots and frequency spectra allocations. A condition for more optimum allocations (licensing criteria) could be a requirement to provide a "public service". The systems are, or could be, designed with minimal incremental cost, to provide public services in addition to their primary commercial service.

The working group concluded that the AIAA, working with other international professional organizations and the appropriate national and international public service entities, should address the questions raised above with regard to determining the Public Service needs.

Also, the National Telecommunications and Information Administration, with the cooperation of the US Department of State, International Telecommunications Union of the United Nations and others should be asked to review and consider improved orbit and frequency allocations in return for enhanced public service.

This latter recommendation of the working group may prove to be the most difficult to implement. The recent World Radiocommunication Conference (WRC-95) did little, as far as can be determined by the author, to make progress on more optimum frequency allocations. The Head of the US delegation, Ambassador Bryan Fontes, was quoted in Satellite News (December 11, 1995) as saying: *"From what we've heard from this conference, [countries] hold sacred their broadcast spectrum. The opportunity to use that spectrum for any allocation, regardless of sharing, is going to be a very difficult obstacle to overcome."* His remarks were in the context of the battle that took place with regard to the spectrum requirements of the broadcast services versus those of the evolving Little LEO systems. However, the same "sacred cow" mentality may apply even when the public interest is at stake. We must hope for some progress in this area before WRC-97.

Since the second AIAA Workshop on International Space Cooperation, some of the participants in the Global Space Systems Services working group have started to identify, in more detail, the global public service needs.

National governments and international public service organizations should take advantage of the emerging capabilities of global space systems services.

5. International Cooperation For Peacekeeping

Peacekeeping operations around the globe, ranging from crisis prevention, to crisis monitoring, to crisis resolution, are increasingly international. Precedent setting discussions conducted during the workshop indicate that opportunities may now exist for cooperatively using national space assets for multinational peacekeeping operations in ways that have never been done.

The subject of International Cooperation For Peacekeeping was the most politically complex of the five potential cooperative projects reviewed at the second AIAA Workshop on International Space Cooperation. The author is not qualified to improve on the discussion that took place at the Workshop - the working group's report speaks for itself. Also, when researching additional information for this paper the author found it difficult to improve on the comments made by the working group's co-chairs in their paper presented at the 46th International Astronautical Congress, held October 2-6, 1995 in Oslo, Norway. The following four paragraphs are extracted, verbatim, from this paper (Reference 5), and are also the author's own observations of the discussions that took place.

"The authors [Fuhrman and Wild] believe that the discussions that took place during this conference [second AIAA Workshop on International Space Cooperation] were precedent-setting. This conference was the first occasion for representatives from major space-faring nations to engage in an open and far-ranging dialogue on space cooperation for peacekeeping, covering the broad array

of mutual assets that might be involved in cooperation and the breadth of issues that will have to be addressed. There had been - and continue to be - government-to-government discussions involving cooperation on specific systems and specific missions, but there had never been a such general, multi-lateral review and discussion of both the technical and the political issues involved and, more importantly, on how to proceed to resolve them. Even though the views expressed by government representatives during this conference were unofficial, we were impressed by the degree of sincerity by all parties and extent to which there is conviction that the recommendations set forth in this report were both prudent and achievable.

Secondly, the authors were frankly surprised at the scope of issues on which the participants were able to agree. Cooperative use of the assets of the major space-faring nations for peacekeeping will involve declassifying and opening up some of the most highly sensitive national assets ever devised. These assets represent collective investments of hundreds of billions of dollars and are derived from cutting-edge technologies that individual nations will continue to leverage to their competitive advantage for years to come. It would be no small matter and, indeed, would represent a major reversal of policy for some of these nations to attenuate long-standing national security and economic concerns for the promised benefits of cooperation. Yet, not only did the status quo have less support than we might imagined - though probably for different reasons for each participant - common ground for change was less difficult to define than we had anticipated. In this area, perhaps, more than any other area of national security and defence, the geopolitical upheaval of the last five years has caused a serious rethinking of national policies and priorities, and the inevitability and necessity of change is more widely perceived than we had thought.

Finally, and most importantly, the fundamental recommendation made by the group was itself remarkable. We had anticipated that a group with broad technical expertise as well as government and industry experience might be able to reach consensus on a series of findings with relative dispatch. However, the central question AIAA had posed to the group - how do we proceed? - required a recommendation on process, often as contentious an issue in international frameworks as substance. Yet, the group reached a shared vision: to make progress in space cooperation for peacekeeping will require leadership; that leadership should come from the two nations with the vast majority of space assets - the United States and Russia; these two leaders, along with other nations, should devise a forum initially outside any pre-existing framework to begin discussions and exchange information.

Our belief is that this first step could be implemented with relatively little political difficulty and could lead to meaningful and highly beneficial space cooperation efforts for all nations. The difficult part of this initial process will be bringing a sometimes complex and often unfamiliar subject to the attention of the highest levels of leadership of all involved nations. This would have to involve, at the outset, the United States and Russia and, we would hope, the

European nations which could make a unique contribution given their experience in cooperative issues. Lacking such high-level attention and leadership, cooperation will probably proceed, but it will proceed piecemeal and in fits and starts, jeopardizing much of its potential in the next decade and perhaps even longer. Our hope is that the report of this AIAA working group will help in that education process."

The principal providers of military, civil, and commercial space capabilities should establish an international forum to investigate the feasibility of improved use of current and future space systems for peacekeeping operations.

The United States and Russia, as two of the key providers of existing space capabilities, should take the initiative to invite the other providers and other users to participate.

6. Solar Power To Earth

Provision of clean, low-cost energy is an environmental, economic, and international imperative.

An international body needs to be formed and tasked with responsibility of determining the feasibility of a space-based solution for the collection and distribution of solar power to earth.

To quote Erb (Reference 8),

"The possibility of supplying Earth with abundant clean energy imported from space has been appreciated for many years. [The modern concept was articulated in 1968]. Much solid work has taken place on studies and technology demonstrations [principally in France, Russia and Japan], the results of which have persuaded many that this concept is the enabler for a sustainable energy future. [In late 1995 the United States NASA initiated a broad study of Solar Power To Earth]. As with any new large-scale technology, however, valid questions can be raised and it is timely to determine whether power from space is needed for the Earth and if the approach suggested by its advocates is sound."

The second AIAA Workshop on International Space Cooperation gave the advocates of Solar Power To Earth another opportunity to further refine the concept. The author is a "reporter" and so writes without bias. As Chairman of the Workshop Planning Committee, Co-Chairman of the Follow-On-Action-Committee, and Co-Chairman of the Planning Committee for the third Workshop, the author has had the opportunity to observe at first hand the deliberations of the growing group of advocates for Solar Power To Earth. What was first considered as either a fanciful idea, or one too complex to implement, is now being treated as a serious possibility.

The modern concept of power from space involves a four step process:

- Capture solar energy in space and convert it to electricity
- Transform the electricity to radio frequency energy and transmit it to Earth
- Receive the radio frequency energy on Earth and convert it back to electricity
- Provide the electricity to the utility grid.

The questions most commonly asked by people outside the community of space power advocates are:

- Is power from space technically feasible?
- Is it needed to meet the increasing demand for energy?
- Is it needed to replace environmentally troublesome sources of energy?
- Is it safe for humans, their artifacts, for other life forms, and for the Earth's environment?
- Can it be cost-competitive, especially compared to Earth-based solar power?

Space power advocates will answer "yes" to all these questions, and there appears to be good reason for their optimism. These questions are addressed in detail in Erb's IAF'95 paper (Reference 8). The second AIAA Workshop provided a forum to address these questions in the context of international collaboration. Before, and since, several other fora have provided an opportunity for Solar Power To Earth to be debated, discussed and planned. In short, the space power advocates are active and they are gaining the attention of a broader audience.

Solar Power To Earth is becoming technically feasible, but there is a considerable spread in the projected global needs for energy. The most likely scenario indicates that an additional 6.7 TW of global energy will be needed by 2020. Projections of the Earth's future energy sources indicate that only coal, solar energy, wind energy and nuclear sources will serve past the middle of the next century.

Perhaps the most significant inhibitor for Solar Power To Earth is the cost of access to space. The working group at the AIAA Workshop recommended:

Recognition should be promoted that solar power from space for use on Earth is an application which can help meet a pressing human need and provide a market for a larger and more economical space transportation capability.

Solar Power to Earth has the potential for solving our energy requirements and providing the market pull needed to reduce the cost of access to space by orders of magnitude.

Solar Power To Earth advocates understand the complexities of space systems needed to implement this energy source. A key finding and related recommendation of the working group at the AIAA Workshop were:

A program of demonstrations and pilot operations is the best way to advance understanding and acceptance. This is particularly true in the area of the safety of the wireless power transmission. Although studies uniformly find no significant biological effects, studies alone are insufficient. Demonstrations and extended pilot operations will be necessary to convince the public that space solar power systems are both technically feasible and environmently safe.

Governments and industry in all interested countries should support the development of programs and fundamental technologies for space solar power to Earth.

7. Conclusions

International cooperation in space can have a broad range of benefits to humanity. Some are less tangible than others, but no less important. During the second AIAA Workshop on International Space Cooperation, five working groups each addressed selected international projects and how they may best be implemented. Three of the projects studied (discussed in this paper) had easily recognizable and tangible benefits to all humankind. The three projects also covered near, medium and long term requirements.

- Global Space Systems Services can be readily modified to provide no, or low, cost Public Services; the commercial owners of these systems are granted, in return, more optimum orbital slot and spectra allocations.

- With political barriers falling throughout the world, national space assets can be used, in combination, to provide effective support for peacekeeping operations.

- During the next century there will be a growing need for energy and to replenish non-renewable sources. Solar Power To Earth is becoming technically feasible, and has the potential to be an unlimited, economical, and clean source of energy. To be economical the cost of access to space will have to be reduced by orders of magnitude. However, Solar Power To Earth may be one of the few projects that could provide the necessary market pull to justify the investment in the development of low cost space transportation.

We have the tools, technology, and expertise to make space a true servant for the benefit of humanity. A significant by-product would be greater peace throughout the world - peace comes to those who work together. And most future space projects can only be carried out through international cooperation.

Do we have the will and the courage to move forward? Let us hope so.

Principal References and Suggested Further Reading

1. *International Space Cooperation - Learning From The Past, Planning For The Future,* Report on First AIAA Workshop, December 1992. AIAA, Paris, March 1993
2. *International Space Cooperation - Getting Serious About How,* Report on Second AIAA Workshop, December 1994. AIAA, Paris, April 1995

The following papers were presented at a session, dedicated to the second AIAA Workshop on International Space Cooperation, of the 46th Astronautical Congress, October 1995, Oslo, Norway.

3. Aldridge, E.C. (The Aerospace Corporation, USA): "International Space Cooperation - Getting Serious About How", reference IAA-95-IAA.3.3.01
4. Comet, E. (Vice President, Booz-Allen & Hamilton Inc., USA) and Vallerani, E. (Chairman, Alenia Spazio, Italy): "Global Space Systems Services", reference IAA-95IAA.3.3.03
5. Fuhrman, R. (Chairman of the Board and COO [retired], Lockheed Corporation, USA) and Wild, J. (Director, NATO Advisory Group for Aerospace Research and Development): "International Space Cooperation for Peacekeeping", reference IAA-95IAA.3.3.04
6. Egan, J. (Egan International, USA): "International Cooperation in Space Transportation", reference IAA-95-IAA.3.3.02
7. Clayton, J. (National Research Council, Washington DC, USA), Smith, M. (Library of Congress, Washington DC, USA), and Azcarraga, A. (SENER, Madrid, Spain): "Cooperative Human and Robotic Exploration of Space", reference IAA-95-IAA.3.3.05
8. Erb, B. (Canadian Space Agency): "Power From Space - The Tough Questions. The 1995 Peter E. Glasser Lecture", reference IAF-95-R.2.01

Space Activity and Comprehensive Security

G. Khozin, Center for Global Problems, Russian Diplomatic Academy, Bolshoi Kozlovsky Per 4, Moscow 107078, Russia

Abstract

The Post-Cold War political environment opens up new perspectives for cooperation, integration and large-scale endeavors by countries which in the recent past were irreconcilable military and ideological rivals. While developing the conceptual and institutional foundations for large-scale international programs and projects aimed at the resolution of pressing global problems and the facilitation of transition to sustainable development, the rich legacy of space activity should be carefully scrutinized. The paper puts forward the concept of comprehensive security, which comes to replace the "zero-sum", military-dominated security of the Cold War era. This concept combines at least five necessary elements, which may be considered to be the basic prerequisites of survival of the human race at the threshold of the XXIst century and a sustainable civilization of the future. Comprehensive security of individual nations and of humankind at large comprises political security, military security, economic security, humanitarian security and ecological security. In the future, new components of security may be included into the system for more adequate support of survival and sustainable development of the civilization.

The paper argues the relevance of space activity to all the elements of comprehensive security both at present and in the foreseeable future. The interdisciplinary essence of motives of space activity, the close interrelation of space programs with issues of national and international security, foreign policy, domestic and world economy, together with most advanced decision-making procedures and management methods used in national and international space programs, make the legacy of space activity an invaluable tool for resolving the large-scale problems facing humankind. The most urgent of these is comprehensive security, which could be more adequately ensured with the space assets available to individual nations and the international community. Examples of using space technology for conflict resolution, peacekeeping, environmental monitoring, disaster prediction and control, and national and international decision-making in crisis situations are analyzed in the context of the theory and practice of comprehensive security and sustainable development, as well as in the framework of new paradigms of an integrated, demilitarized, and eco-rational civilization of the future.

1. Introduction

The end of the Cold War along with the sweeping social and political change in individual countries and the world community at large produced strong incentives for reassessment of the well-established patterns of space activity both in the framework of national space programs and in the realm of international cooperation in exploration and practical use of outer space. In the prophetic words of the "father of the theory of space flight", Konstantin Tsiolkovsky, " *Humankind will not stay on the Earth forever, but in the pursuit of light and space it will first timidly emerge from the bounds of atmosphere and then will conquer for itself the entire circumsolar space*" (Reference 1), which so recently were frequently used as a rationale for military-political rivalry in pursuit of essential national interests of the "space powers", may now become the motto for sustainable cooperation in space for the benefit of the living and the future generations of all the nations comprising the interdependent world community on the verge of the Third Millenium.

This paper focuses major attention on the new socio-philosophical, political, environmental and ethical foundations which will determine space development by individual nations and - to a greater extent - the joint efforts in space of international organizations and private corporations throughout the world. Space activity is analyzed in the context of the pressing global issues, with special impact on the changing essence of national and international security.

The Second United Nations Conference on the Exploration and Peaceful Uses of Outer Space (UNISPACE '82) pointed out a unique feature of space activity as a motive force of shift of paradigms of human consciousness and humankind's vision of itself both on the planet Earth and in the Universe. Its final report stated: "*Space technology has made dramatic progress in our lifetime. It has changed our concept of distance, made each person the neighbour of all others and given us a new measure of ourselves in relation to the cosmos. Can we now rid ourselves of our obsolescent biases and concepts and move foreward to the more equitable, humane and co-operative society that the image from space conjures up?*" (Reference 2).

As the space era approaches the end of the fourth decade of its history the rich legacy of the world-wide space activity proves itself an ever more adequate means of understanding the basic phenomena in the world of nature, technology and society. The "overview effect" or "look at our planet from outer space"- a unique experience of humans observing our planet from space - provided the wide public and the decision-makers with a "collective consciousness" of the passengers of the "Spaceship Earth". The "overview effect" - one of the major changes in the consciousness brought about by the progress of spaceflight - supports social transformation and serves as the foundation for a series of new civilizations evolving on Earth and in space (Reference 3). Following the moral and ethical principles of this future-dominated consciousness will help to implement a quantum jump in positive transformation of world politics and international relations after the Cold War. Rejecting the negative experience of policy-making in the recent past, correcting the mistakes originating in military and ideological rivalry, the space-faring nations contribute to the transition of the world community to a new system of political, socio-economic and ethical values reflecting the urgent needs of the closely interrelated and integrated holistic humankind.

2. The Global View

The major positive shift in foreign policy concepts of many of the states comprising the world community may be characterised first of all by the nations' desire not to rely on military power as the primary instrument of implementation of their national interests. Despite outbursts of nationalism, ethnic conflicts and other crises in different regions of the globe, the world community is persistently pursuing a course aimed at large-scale international cooperation, at integration in many fields of political, economic, scientific-technological and cultural activity as well as at the transition to sustainable socio-economic development. Breakthroughs into space have radically changed the human perception of life on the planet Earth, given convincing evidence that humankind faces a complex system of large-scale, closely interrelated, and essentially multidisciplinary global issues which can be resolved most effectively only by the joint efforts of individual nations and international organizations.

The collective efforts of Russian, American and European thinkers, scientists and prominent political figures have put at the disposal of the world community an elaborate body of theoretical concepts, which explain the origin, the scale and the forms of manifestation of global problems. The results of space exploration, primarily the development of space science, in combination with remote sensing data from space and new methods of planning, management and implementaion of space projects have contributed to a better understanding by the decision-makers and the wide public all over the world of the solid fact of the growing interdependence of all the states and nations, which comprise the world community. The theory of global problems, or globalistics, explains the complicated relationship of natural, technical, political, social, economic, moral, ethical, religious and other factors of national and international scale, which were not previously given due attention by the governments, leaders of the international organizations, national and international community, and became the source of complicated crisis situations of global scale.

The concept of global problems became an integral part of the vision of the contemporary world and the immediate trends of its natural, technological and socio-political development (References 4, 5). Practically all the strategies of political and socio-economic development of individual nations, geographic regions and the world community at large, which were put forward by authoritative scientists, international commissions, groups of experts of the United Nations and other international organizations, take into account the state of global problems and propose different methods for easing the tension in the course of cooperative projects and programs of different scales and purposes. Such an approach incorporates the view of the planet Earth as a cradle of reason in the Universe, maintained by K. Tsiolkovsky and other "space philosophers" of different countries. The successful solution to global problems is one of the major prerequisites for humankind's expansion in the Universe.

As of the middle of the 1990's the system of global problems, which occupy top places on the agenda of the governments, highest legislative bodies, political parties, trade unions, public organizations and movements in the majority of the nations may be described as the following hierarchical closely interrelated structure:

- safeguarding peace and ensuring irreversible disarmament;
- protection of the environment;
- ensuring the rational supply and use of mineral resourses;
- development of balanced and efficient energy infrastructure;
- exploration and rational use of space and the World Oceans;

- stabilizing the world population;
- prevention of hunger and disease;
- bridging the development "gap" by means of transition to sustainable development.

3. The International Perspective

The experience accumulated in the course of implementation of national and international space programs, in combination with the space assets operated by military and civilian agencies from year to year, becomes an ever more valuable and efficient instrument for the solution of pressing problems on national, regional and planetary scales. Adequate understanding of the urgency and importance of global problems for the world community would hardly be possible without the incorporation of this view of the planet Earth from space into the human mentality. This is another proof of the importance of space activity for the progress of civilization.

The conferences, discussions of experts and public hearings held in recent years have made it possible to develop important recommendations and to put forward constructive ideas concerning the role of space activity as a means for gradually transforming humankind into a progressive harmonic civilization of the planet Earth. Space activity helps to bring the nations and peoples together in their desire to protect the ecological integrity of the planet using among other means spaceborne remote sensing assets. The view from space of the planet Earth as a holistic natural-technical-social system became a strong motive force, which helps to form a humanistic outlook for the present and future generations and contributes to establishment of the harmonious interaction of man and nature.

Successful transitions to peaceful, economically feasible and socially efficient space activity, serving the interests of all humankind, will depend among other factors on how soon and how radically the governments review the priorities of their national space programs and begin large-scale cooperation in the exploration and practical use of outer space for the benefit of the Earth's civilization. It is vitally important to shift the growing technical potential of space activity from serving primarily destructive military goals of individual nations to the maintenance and safeguarding of comprehensive security of the entire international community. Space assets can and must become a reliable means of strengthening confidence-building measures in relations between nations, an instrument of conflict-resolution and monitoring the practical measures in the field of arms control and disarmament. It is very difficult to overestimate the role which can be played by observation and remote-sensing satellite systems, including those controlled by the military, in warning about natural disasters and man-made catastrophes, in planning and implementing programs of socio-economic development for individual states and for entire regions, in controlling the efficient use and reproduction of natural resources.

4. Security Issues

In the same way as visions from space contributed to deeper understanding of the global problems, the experience of national and international space programs helps to rationalize the concepts of security, to shift them from the "zero-sum" principle, which prescribes violent suppression of the interests of "hostile" states by ever more destructive military means including weapons of mass destruction, to more adquate norms of comprehensive security. The first important move in this direction was the rejection of unlimited politico-military rivalry by the major "space powers" - the USA and the Russian Federation - and their turning to a closer coordination and integration of their national space programs. Implementation of this mutually beneficial practical course has brought the first positive results, reducing the degree of militarization of space activity of the two countries and putting an end to large-scale projects aimed at the development of space weapon systems.

The concept of comprehensive security, as is approved in general by scholars and political figures in Russia and many other countries of the world, including the author of this paper (Reference 6), postulates that essential security in the post-Cold War era cannot rest only on the military capabilities of any nation. In the conditions of radical political, socio-economic and cultural change all over the world and the growing urgency of global problems, security of the nations becomes to a greater extent the issue of survival and sustainable development than the mission of protection of "national interests" by military means.

In full conformity with the logic of "space consciousness", which ranks humankind as an integral part of the biosphere of the planet Earth, the concept of comprehensive security concentrates major attention of the world community and the individual nations not on increasing their capability to protect their interests in military operations of different scales and gravity, but on avoiding conflicts of any kind, on seeking harmony and mutually beneficial forms of interaction between themselves and with the biosphere of the planet Earth. From this perspective security becomes a combination of conditions and factors, which keep the nations in a peaceful, mutually acceptable mode of relations excluding the necessity to threaten the use of military force or to enter into military conflict for resolving the tensions between states or nations. In the new conditions on the world arena, the "military dimension" of security does not enjoy the "sine qua non" status in foreign policy doctrines, but becomes one of the instruments of protection of "national interests" in specific conditions. The role of military force as an instrument of foreign policy is not considered any longer to be of high priority and universally efficient.

Since the concept of comprehensive security is not limited, as it was in the Cold War era, to the protection of a nation's people and territory from physical attack by military means, including the most sophisticated weapons of mass destruction, it also addresses transnational issues of global interdependence, which may be qualified as "external threats" to the survival of humankind and to the stability of the international system. Comprehensive security is not a goal in itself, but a means of meeting such value-oriented human needs as personal freedom, survival, societal welfare, international order and human dignity. Our analysis is limited to general characteristics of the following major components of comprehensive security: political, economic, military, humanitarian and ecological (environmental).

Political security involves a system of carefully planned measures taken by national governments in cordination with the international community. These measures are aimed at sustaining the socio-economic systems of individual states in balance and at ensuring peaceful, non-violent social change on national and international levels in full conformity with democratic principles. An adequate level of political security gives the national leadership an ability efficiently to control the functioning of society in the state and to protect its "national interests", values and cultural traditions in relations with other nations, which should be based on mutually acceptable terms. Political security may be reliable only if the principles of national sovereignty will not be violated and at the same time the "national interests" will not be implemented at the expense of universal interests of survival of humankind.

Economic security deals with the fundamental conditions of the functioning of a national economy as an organism sustaining everyday life of the society in any given state and as an organic component of the world economic infrastructure. It involves not only the issues of industrial and agricultural development, the problems of ownership, finance, investments, money and banking system, but also the mechanisms of the stimulation, management and control for the utilization of natural resources both on the territory of the state and in the framework of international trade, commerce as well as different forms of economic, scientific and technological cooperation of individual countries and international organizations. The global scope of economic security makes the elimination of disparities in the economic development of the different countries of the world an especially urgent task for the ever more interdependent world community. The states where people live in poverty have very little economic security. The socio-political situation in such countries promotes instability and crises capable of undermining the foundations of the state and society, not only in the appropriate nations but also on the regional scale and even on a global scale. The most promising ways and means of improvement and safeguarding economic security will be the collective efforts made by governments and international organizations as well as by national and international corporations aimed at the positive transformation of the world economy on the principles of due account of the vital interests of individual states and of humankind as a whole, whose survival depends on the successful transition of states and regions of the planet to sustainable development, mutually beneficial forms of cooperation and integration between themselves and with the biosphere.

Military security is the degree of physical protection against external threats to national territory, population and its welfare, integrity of national economy, natural resources of the state as well as to the cultural heritage of the peoples, which the nation can afford in the specific conditions inside the state and on the world arena. Military security is ensured by the military arsenals available at the disposal of the state. The government may use them as a means of intimidation of hostile nations to follow the desired political course in foreign or/and domestic policy. But the essence of military security of a state is measured by its actual capabilities to use military force in offensive action or for defense of its territory, population, natural resources and cultural heritage as well as an element of international collective defense force against a hostile politico-military coalition.To obtain such a capability, the states created impressive military establishments and for several decades up to the beginning of the 1990s have been engaged in an unprecedented arms race, which among other dangerous implications resulted in the militarization and weaponization of outer space. The end of the Cold War and the disintegration of the Soviet Union has put an end to the unlimited military rivalry of the two "superpowers" and created favourable conditions for the orientation of the foreign policy doctrines of active participants of the arms race and ideological rivals of the recent past on more adequate principles of "sufficiency" and "build-down" in the development of their military potentials. Active bilateral and multilateral measures aimed at curbing the arms race and the limitation of the most dangerous types of weapons have contributed to an improvement of the general international political climate. One of the first positive results of this favourable trend became visible reduction in the scope and scale of space weapons projects in national space programs and the greater access for civilian authorities to information gathered by space systems controlled by the military establishments. Greater restraint on the part of the major military powers and the space-faring nations in setting the priorities for their military policies in the immediate future will be a reliable factor for safeguarding the military security of not only individual states but also the entire international community.

Humanitarian security deals primarily with ensuring favourable conditions for education, health, personal development of individual citizens, provision of food, clothes, shelter for the population of a state as well as with the protection and enrichment of the socio-cultural heritage of the nationalities and ethnic groups comprising the population of the state. In conditions of growing interdependence and vulnerability of humankind in the face of the global problems, humanitarian security obtained a new important dimension - the dissemination among all the nations of information about the state of the international community including such aspects as human rights, democratic change, peacekeeping, prevention of terrorism, family planning, exchange of cultural and ethical values, etc..

Ecological (environmental) security involves all the practical and potential actions required to protect, sustain and improve the quality of the biosphere. Environmental dimensions of comprehensive security appeared on the political agenda of both individual governments and the world community as a whole after the critical reassessment of the fundamental principle of the concept of military-dominated national security: the treat to a nation comes from the military might of other nations. As far back as 1977 Lester Brown, president of the authoritative Worldwatch Institute, stated: "*...the threats to security may now arise less from the relationship of nation to nation and more from the relationship of man and nature*" (Reference 7). Leaving aside the specific environmental issues, which stand on the national and international agenda, in the context of our analysis of comprehensive security it is necessary to point out that the environmental dimension of security becomes the central element of the concept which puts together all the elements of comprehensive security for the primary goal of sustaining the health of the biosphere and setting the standards to space activity which prescribe that the use of outer space did not adversely affect or alter the environment on Earth and outer space in the vicinity of our planet.

The concept of comprehensive security is a dynamic system of norms and values, which in combination should provide for non-violent social, political and economic change in individual states and in the world community as a whole alongside the protection, rational use and reproduction of the resources of the biosphere sustaining the phenomenon of life on the planet Earth. This circumstance speaks for the fact that the main elements of comprehensive security in any situation in a particular state or on the planet in general will have a different "specific gravity", which will determine the priority and the amount of resources required for keeping them in a stable condition. Moreover, changing socio-political, military, economic and natural environment on the planet may make it necessary to incorporate in the concept of comprehensive security new elements, which will demand the attention of the governments and international organizations, as well as the material resources of individual nations and the world community (for example, scientific and technological security, information security, socio-cultural security, "genetic" security, etc.).

5. Conclusions

Discussing the issues of space of service to humanity and education for a changing world it is especially important to point out the relevance of the rich and multi-faceted experience, accumulated by humankind during the space era, to practically all the pressing problems facing both individual states and the world community. Any grave and complicated problem of national or global scale which will be put on the agenda of practical action may be better understood and effectively resolved on the basis of a "space mentality", with the proper use of methods of planning and management available to space community, relying on the assets of space technology operated by individual states and international organizations. From this perspective it is very difficult to overestimate the importance of space activity for the present and the future of the civilization.

In conclusion, it is worthwhile to stress once more that the most serious task of our time, the solution of which should be put at the top of the priority list of governments, parliaments, leaders of national space programs as well as the general public of many countries, is the elaboration and putting into practice of the principles of equal and mutually beneficial cooperation in space, where the interests of a holistic civilization rank higher than the short-term pragmatic "national interests" of the space-faring nations. While accepting the principle of large-scale cooperation as an imperative for the post-Cold War space activity, it is necessary to pay due account to one of the major findings from the assessment by the experts of the Space Policy Institute, at the George Washington University, and the Association of Space Explorers. This states that when the interests of the participants of cooperation in space are similar and their technical capabilities are equal,"*reaching an agreement that distributes the benefits of the collaboration in a way satisfactory to all participants can be quite difficult*" (Reference 8).

The famous British philosopher Bertrand Russell once said: "*The good life is one inspired by love and guided by knowledge*". Konstantin Tsiolkovsky believed that outer space is dominated and will be dominated by the highest reason and intelligence as well as by the perfect forms of social organization. Space activity of humankind after the Cold War regains the best ideals and values from the thinkers of the past and at the same time incorporates all the rational concepts of the progressive, sustainable and harmonic civilization, which are elaborated on the principles of space consciousness and planetary ethics formulated with due account of the vision of the planet from space. The contemporary state of human knowledge is indeed a solid foundation for ensuring comprehensive security on the planet Earth and for the continuation of the space activity for the benefit of the living and future generations of all the peoples and nations.

References

1. Tsiolkovsky, K.: "Letter to B. Vorobiev, August 12, 1911", cited in *K.E. Tsiolkovski. Selected works,* p. 482. Nauka, Moscow, 1962
2. *Report of the Second UN Conference on the Exploration and Peaceful Uses of Outer Space, August 1982,* A/CONF.101/10, p. 7. Vienna, Austria, 1982
3. White, F.: *The Overview Effect,* pp. 4-5. Houghton Mifflin, Boston, 1987
4. Khozin, G.: *Talking about the Future.* Progress, Moscow, 1988
5. Gudozhnik, G.S.: *Civilization and Global Problems.* Progress, Moscow, 1990
6. Khozin, G.: "The problem of ecological security and international relations", in *In Search of Balance. Ecology in the System of Social and Political Priorities,* edited by B. Makliarski, pp. 131-204. Mezhdunarodnie Otnosheniya (International Relations) Publishers, Moscow, 1992
7. Brown, L.: "Redefining National Security", in *Worldwatch Paper 14,* p. 6. Worldwatch Institute, Washington, October 1977
8. Space Policy Institute and Association of Space Explorers: "International cooperation in space - new opportunities, new approaches."*Space Policy,* Vol. 8, No. 3, p. 200. Butterworth-Heinemann, United Kingdom, August 1992

Objectives and Roles of Space Development for Social Challenges in the Post Cold War Era

H. Yoshida, CSP Japan Inc., 2F YS Kaigan Bldg., 2-2-6 Kaigan, Minato-ku , Tokyo 105, Japan
M.E. Vaucher, CSP Associates Inc., 55 Cambridge Parkway, Riverfront 2, Cambridge, MA 02142, USA

Abstract

The space age was born at the height of the Cold War - a period of intense political, economic and military competition between the United States and the Soviet Union. In this charged atmosphere, government-supported civil space efforts were structured as one facet of this competition. Other objectives, such as the acquisition of scientific information or the development of practical applications of space technology were also important but, ultimately, secondary goals.

Conditions have changed dramatically within the past five years. The conclusion of the Cold War, worsening public sector budget deficits and closer scrutiny of the appropriate role of the public sector in national economic and social development are leading policymakers to reappraise the size, scope and objectives of space activities.

The future course of space programs can be expected to further emphasize the role of space activities in addressing societal challenges. As society enters the next millenium, we must find new ways to meet the food and energy requirements of our modern industrial age, while at the same time preserving the fragile environment which sustains us. Space technologies may help to address these challenges. However, a new set of objectives for civil space efforts(and public consensus) will be required.

This paper provides a forecast of the evolution of the political and societal underpinning of civil space efforts in the world as well as the influence of the global long-term social challenges to Japan in the post cold war era. The emphasis of the study is on the identification of appropropriate objectives for civil space efforts and the technological priorities for realizing them.

1. Introduction

The aftermath of the Cold War has forced national Space Agencies to shift their perspectives on the support of space activity from one of pursuing space as a vehicle of scientific discovery and demonstration of technological/ military prowess to one in which space activities address concrete problem on Earth. Space agencies must redefine their roles to match new domestic agenda focusing on more direct public benefits. Space activities that were once largely removed from day-to-day life will increasingly be linked to the challenges that we face as a global society. Already, the well-publicized benefits in life sciences and materials processing research, and Earth observation activities indicate a shift to more direct benefits-oriented space programs.

In this paper, a forecast of the evolution of the political and societal underpinnings of civil space efforts in the post-Cold War era will be provided. The emphasis of the study is on identifying the objectives for civil space programs linked to expected societal challenges. Section 2 summarizes a range of literature forecasting the major societal challenges of the next 30-50 years (the FY94 NASDA Consignment Study *"Long-term objectives of civil space programs after The Cold War"* contributed to this paper). Section 3 illustrates the Japanese perspective on those challenges. Section 4 provides a summary analysis of the identification of technology responses and space linkages.

2. Identification of Global Socio-Economic Challenges

2.1 Methodology

The assessment of long-term objectives for space exploration and development requires a comprehensive analysis of the leading socio-economic challenges confronting humankind over the next 30-50 years. To compile an as complete as possible list of future socio-economic challenges, the futurist literature addressing the future problems for humankind was surveyed.

In selecting these studies, the reports that were comprehensive in viewpoint (e.g. which covered the complete range of socio-economic and political factors) and long-range in projection were sought.

Finally, the following seven studies were selected for content analysis;

- The Brooking Institution, World Resources Institute, and the Santa Fe Institute: *Project 2050* (see Reference 1). The 2050 Project addresses the challenge of sustainable growth; it aims to identify the conditions necessary for a sustainable global economic system and to suggest potential policy actions to create such a system.

- Carlson, R. and Goldman, B.: *Fast Forward: Where Technology and History Will Take America and the World in the Next 30 Years* (see Reference 2). This book presents a comprehensive list of key political, social and economic issues confronting the US and the world over the next 30 years.

- Conway, M.: *A Glimpse of the Future: Technology Forecasts for Global Strategies* (see Reference 3). This study includes a set of technology forecasts for the upcoming 10, 20 and 30 year periods as well as a comprehensive list of technology and scientific developments expected to influence significantly the social, economic and political developments of the future.

- Meadows, D.H., Meadows, D.L. and Randers, J.: *Beyond the Limit: Confronting Global Collapse, Envisioning a Sustainable Future* (see Reference 4). This book is a sequel to the 1972 Club of Rome study, *The Limits to Growth*. The authors used a computer model to revise earlier projections of population and economic growth. In addition, they employed new policy frameworks and current statistical data. The study concludes that the global industrial system has surpassed some limits; the authors suggest that critical changes during the 1990s are required to meet sustainable growth in the future.

- Sheffield, C., Alonso, M. and Kaplan, M.A. (editors): *The World of 2044: Technological Development and the Future of Society* (see Reference 5). This book includes a collection of essays written by scientists, technologists and science fiction writers on what the world will be like in the next 50 years. The book addresses such issues as new technologies (e.g. robotics and biomedical technologies) as well as future colonies in space and the oceans.

- Singer, M.: *Passage to a Human World: The Dynamics of Creating Global Wealth* (see Reference 6). *Passage to a Human World* investigates the future development of world economic growth. The author asserts that many so-called poor nations such as China and India will make the transition to wealthy societies over the next 70-75 years.

- United Nations, Global Outlook 2000: *An Economic, Social, and Environmental Perspective* (see Reference 7). This study uses historical trends in agriculture and food supply, energy use, population growth and the distribution of wealth among nations to make projections about future global conditions and societal challenges.

In addition to reviewing and analyzing the contents of these reports, leading experts in the areas of sustainable development, international security and space science were interviewed. The experts interviewed included:

- Dr. Rob Coppock of the World Resources Institute in Washington, D.C. and Director for The 2050 Project.

- Dr. John Steinbrunner, Director of foreign policy studies at the Brookings Institution in Washington, D.C.

- Dr. Bruce Murray, the former Director of the Jet Propulsion Laboratory in Pasadena, California.

2.2 Definition of Major Future Challenges

We identified and grouped the socio-economic challenges into five categories (see Figure 1). These five categories (population and natural resource factors, environmental factors, technology factors, social, economic and political factors, and natural disasters) broadly set the parameters for the definition of future socio-economic challenges. They are defined as follows:

(a) **Population and natural resource factors** include issues related to sustaining the current and growing populace, and resource scarcity. Important concerns associated with population include: current and expected growth rates, migration and distribution of the populace. These demographic factors will be critical to utilization of natural resources. Natural resource factors include renewable and non-renewable inputs required to support the human population as well as its industrial and agricultural activities.

(b) **Environmental factors** include issues related to climate change and human overuse and pollution of the environment. The phenomenon of increased "global warming" due to increasing manmade emissions of "greenhouse gases" (methane and carbon dioxide among others), and the deterioration of the ozone layer of the Earth's atmosphere resulting from the use of chlorofluorocarbons (CFCs) have been noted as leading environmental concerns. Also, exploitation of the land and oceans for cash crops and human habitats have led to deforestation and desertification. Finally, direct pollution of the environment and its impact on habitat and species diversity is an ongoing concern.

(c) **Innovation and diffusion of new technologies** was noted as a key factor for the future, especially in terms of the risks to humans and the environment. The dangers that accompany new technology development, however, are not always known at the time of technology innovation and therefore present hidden risks to society. Recognized hazards, such as population exposure to dangerous micro-organisms or the moral/ethical risks of genetic engineering, could pose negative impacts to society. Similarly, the diffusion of current and new technologies to developing nations and globalization of the associated industrial infrastructure place ever increasing pressure on the environment in the form of growing resource utilization and pollution.

(d) **Social, economic and political factors** were often cited as leading concerns. Ethnicity issues, such as racial tensions and tribal/regional conflicts, represent a destabilizing force for society, the extreme of which is exemplified in the Bosnian-Serb war in the former Yugoslavia. Support for education is considered a critical ingredient to national economic productivity and international wealth distribution. As well, international security structures (e.g. the degree of cooperative engagement or isolationist practices) and the proliferation of "smart" weapon systems are also salient issues with which society must grapple over the next 50 years. Furthermore, the distribution of wealth associated with a global, interconnected economy is expected to have a continuing impact on the social, economic and political structures of the future. Similarly, national and regional political structures, namely the degree of centralized or decentralized organization of the government (i.e. spread of democratic practices), has been recognized as a central ingredient to empowering individuals to contribute to and participate in all aspects of society. Finally, health and social welfare will continue to have an important impact during the next 50 years. Specifically, such issues as a growing elderly population, crime and drug abuse are expected to drain societal resources and destabilize the social, political and economic climate.

(e) **Natural disasters** include forest fires, earthquakes, hurricanes, tidal waves, asteroid/comet impacts, volcanoes and floods and have been noted as major factors influencing future societal challenges and regional stability.

The process of defining the major factors that will significantly influence the future of society is complicated by the interconnections among them. These factors cannot be considered as separate, but rather serve as a useful organizational construct in the consideration of how to frame future challenges. In fact, all the factors are interdependent. Moreover, issues are often perceived differently; agriculture can be identified as an arable land issue (non-renewable resource) or as a natural renewable resource concern (level of yields based on plant technology). However, these categories serve to illustrate the major issues which were identified within the studies that shape the nature of future socio-economic challenges, and are important in understanding the nature of these challenges.

POPULATION/ NATURAL RESOURCE FACTORS	ENVIRONMENTAL FACTORS	TECHNOLOGY FACTORS	SOCIAL, ECONOMIC & POLITICAL FACTORS	NATURAL DISASTERS
• POPULATION - GROWTH - MIGRATION - DISTRIBUTION • NATURAL RESOURCES - AGRICULTURE/ FOOD - ENERGY - WATER USE - MINERALS - LAND	• CLIMATIC CHANGE - GLOBAL WARMING - OZONE LOSS • OVERUSE - DEFORESTATION - DESERTIFICATION - DESTRUCTION OF FISHERIES • POLLUTION - HAZARDOUS WASTE - WATER QUALITY - HABITAT/SPECIES LOSS - AIR QUALITY	• TECHNOLOGY INNOVATION AND DIFFUSION - RISKS TO HUMANS - ENVIRONMENTAL IMPACTS	• ETHNICITY - RACIAL/ REGIONAL CONFLICT • EDUCATION - LITERACY AND WORKFORCE SKILLS • INTERNATIONAL SECURITY - SECURITY STRUCTURES - PROLIFERATION • DISTRIBUTION OF WEALTH - GLOBALIZATION OF ECONOMY • POLITICAL STRUCTURES - CENTRALIZED VS. DECENTRALIZED • HEALTH & SOCIAL WELFARE - HEALTHCARE - CRIME - DRUGS - FAMILY - AGING	• FOREST FIRES • EARTHQUAKES • HURRICANES • TIDAL WAVES • ASTEROID/COMET IMPACT • VOLCANOES • FLOODS

Figure 1. Major factors defining future challenges

2.3 Results of the Content Analysis

Using the categories established in Section 2.2, information gathered in the individual studies to quantify the relative ranking of these factors as future societal challenges was assimilated. A content analysis for each of the studies was compiled and a comparative analysis of the seven studies is presented in Figure 2. This approach resulted in the categorization of these factors into three major groups.

In the first category, the content analysis revealed five primary issues (cited by all studies), which can be expected to "drive" the long-term social, political and economic challenges of the future. These are:

- Population growth and its impacts on natural resources and the environment.
- Energy use and access to new energy resources, and its impact on the economy.
- The impacts of climate change/global warming for humans.
- Continuing growth in hazardous and toxic waste products from economic activity.
- Democratization and the spread of citizen empowerment and its impact on political structures and the international economy and distribution of wealth.

In the second category, eight factors were mentioned by a majority of the studies, and therefore represent a secondary set of major challenges (although without the level of uniform concern of those in category 1). These include:

- Agricultural productivity/food distribution.
- Access to adequate water resources and water use.
- Habitat/species loss(decreasing biodiversity).
- The need to support education of a growing populace and its impact on the economy.
- Proliferation of military technologies and destabilization of international security.
- Globalization of the international economy and their impact on the distribution of wealth among nations.
- Growing health care requirements and its impact on the economy.
- Risks and impacts of new technologies and the diffusion of technology.

For the most part, the remaining factors which fell into the third category (not cited by many studies) are either subsets of one of the major factors identified in the first two categories, or were clearly of lesser significance on a relative basis. The single exception in this is the potential impact of natural disasters, which can cause significant regional (or, in the case of a large asteroid impact, global) destruction.

3. Identification of Japanese Socio-Economic Challenges

3.1 Currents of Change

Major transformations such as those noted below are occurring in Japan and abroad, challenging the country to come up with policies to deal with its various structural problems.

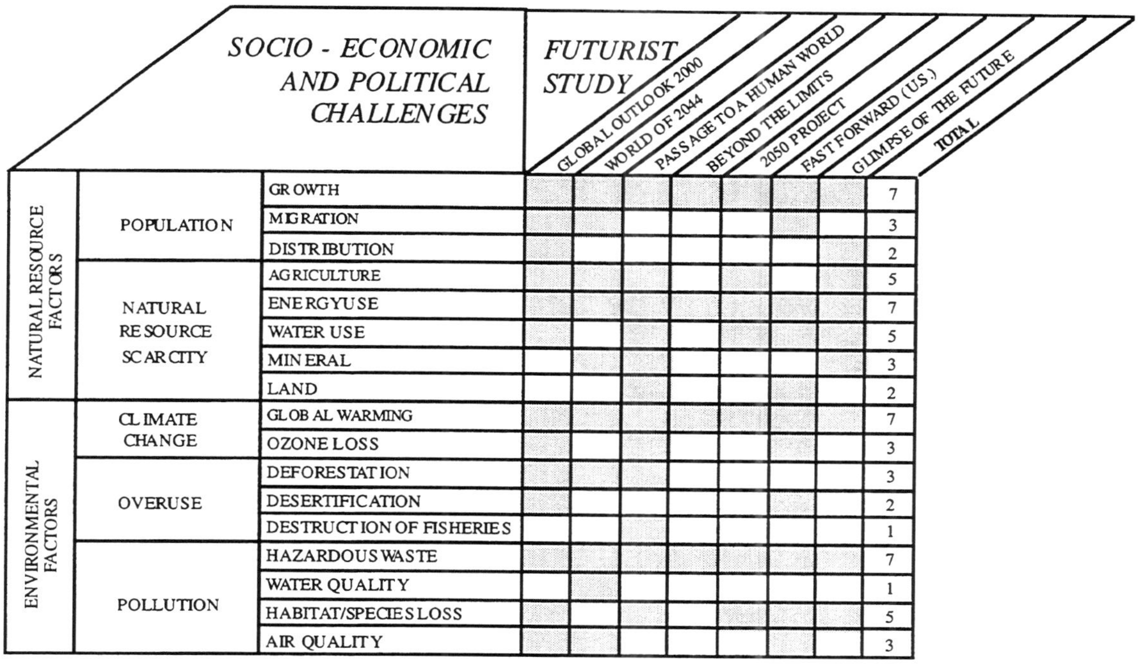

Figure 2a. Content analysis

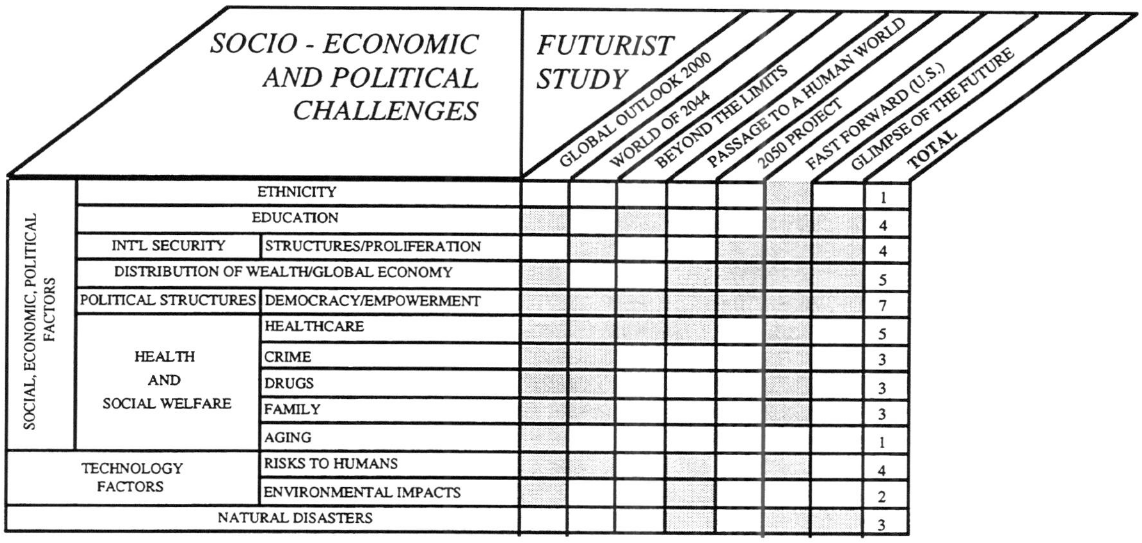

Figure 2b. Content analysis (continued)

- **Accelerating globalization.** Along with advances in information and telecommunications technology, improvements in modes of transportation, and expansion of the free-trade system, a global movement of people, material, capital and information has begun to occur. As the previous section dealing with global challenges has also made clear, borders are being erased from people's consciousness as well, as they recognize that not only economic problems but also population, environment, food and other problems must be dealt with on a global scale.

- **Transition to a highly mature economic society.** Japan's traditional pattern of development - to import technology from abroad and sell mass-produced products based on adaptations of that technology in domestic and overseas markets - is under intense competitive pressure from other countries, particularly those of East Asia. Consequently, the country is being challenged to break away from enterprise-centered and group-based patterns of activity.

- **Shift to an elderly society with few children.** At the beginning of the twenty-first century, Japanese society will enter a phase of declining population growth, leading to a shrinkage in the working population during the next century (those over age 65 are forecast to rise sharply from 17 percent of the total population in 2000 to 25.5 percent in 2020; by way of comparison, predicted percentages for other countries in 2020 are: US 16.1%, UK 18%, Germany 20.9%, and France 19.7%). Also, longer postponements of marriage by women, and their growing tendency to stay single, probably continues the trend of their having fewer children.

- **Advances in information and telecommunications technology.** Advances in information and telecommunications technology not only herald a drastic change in the flow of information and materials and an enhancement of industrial productivity. They also hint at the creation of new related industries and employment opportunities.

3.2 Structural Problems Requiring Action

The four transformations noted above interact mutually, and in the process are affecting Japan in a major way. The current socio-economic structure is failing to respond satisfactorily to this change. Specifically, problems such as the following are becoming increasingly evident.

- The lagging development of new industries and the reduction of current industries.
- Uneasiness about long-term employment opportunities.
- Uncertainty about life in a society of old people and few children.
- Dissatisfaction about the lack of tangible benefits from affluence.
- Increasing responsibilities and roles in global society.

Before structural adjustments dealing with these problems could make a sufficient impact, the collapse of the bubble economy, further yen appreciation and distrust of the financial system brought these problems into the limelight. At the same time, the Great Kobe Earthquake in January 1995, the sarin poison-gas attacks on the subways in March 1995 and other incidents aroused concerns about safety in general, exacerbating the uncertainties and anxiety about the prospects for Japan's economy and society.

Figure 3 summarizes the challenges facing Japan in a more concrete fashion, by area of concern.

Economy	International	Society/Daily life
•Reduction of industry •Domestic-foreign price differential •An aging society and fiscal burden (national health care) •The need for deregulation	•Regional integration and conflicts •Downward shifts in growth rates of industrialized nations •Rapid growth of East Asian economic sphere •Concepts for regional economic blocs becoming reality •Rising world population and increasing gap between rich and poor •Global scale environmental problem	•The ageing of society and its impact on the working population •Increasingly educated population and growing acceptance of lifelong education • Collapse of the myth of safety -Social abnormality -Increasing variety of crime • Bringing about a comfortable pace in daily life • Developing social infrastructure
Market	**Industry**	**Technology**
•The development of value-creating products •Internationalization of the market place •Creating businesses out of insecurity -Security -Crisis management •Creating new markets -New frontiers	•Greater international division of labor (Asia) •The creation of leading industries -Information-related industries -Infrastructure-related industries Environmental Protection Recycling Area development •High value-added industries -Technological pre-eminence -Systems-related products -High technology areas -Information technology areas	•Technology-related friction and increasing globalization •Virtual reality •Eco-materials

Figure 3. Issues confronting Japan by category

3.3 Fundamental Directions of Structural Reform

From the above, three basic actions concerning paradigms for economic society follow.

(a) **Creating a free and vigorous economic society.** To spawn vitality, it is essential that every effort be made to create employment opportunities - through fostering deregulation and encouraging the correction of commercial practices which obstruct competition, remedying the economy's high-cost structure, and encouraging the development of new industries - so as to allow the creative abilities of individuals and enterprises to function fully.

(b) **Creating an economic society that is affluent and secure.** For the Japanese, affluence is not a tangible reality. To dispel such dissatisfaction, the fruits of economic affluence must be clear in the daily lives of citizens.

(c) **Taking part in global society.** In order for Japan to fulfill its international responsibilities in an age of globalization, it must become an open society, both internally and externally. It must also participate actively in the international harmonization of systems and mechanisms, and in the creation of international rules. A motto encapsulating these themes and the issues confronting Japan is "creating human values for a global age".

3.4 For the Solution of Social Issues (Establishing a Foundation for Progress by the Next Generation)

The following must be urgently nourished as foundations supporting a new economic society.

- **The promotion of science and technology.** The fountainhead of the new economic society's frontier is science and technology. It will be vital to enrich the intellectual capital by which society promotes creative research and development as well as an efficient expansion of an economic frontier linked to social needs. Science and technology offer potential benefits in solving societal problems. However, they are not solutions by themselves. For example, education and the expansion of societal values such as trust and teamwork are equally important.

- **Advances in information and telecommunications technology.** Advances in information and telecommunications technology make possible the creation of new employment opportunities. Consequently, the rapid realization of a society well-versed in information and communications technology is vital.

In this way, the two points cited above represent important keys for the successful realization of a vital industrial structure that can also adapt flexibly to changes in people's values and needs.

3.5 The Significance of Japan's Space Program - and What it Needs to Accomplish

In a report issued in November 1995 entitled "*Our Country's Challenges and Basic Policy for Economic Management*" (see Reference 28), the Economic Planning Agency identified "the creation of science and technology" as an exceedingly important task - something vital to Japan's creation of new industries. Furthermore, it touched on the point that such science and technology had to be linked with society's needs. It concluded by saying that economic growth was already unfeasible under the traditional pattern of development, and that the country needed to aim to become "*a nation founded on sccience and technology*" at the frontier of a new economic society. In order for the space program to figure in this science and technology, and to be identified as a leading industry in a nation based on science and technology - and to play the role which it was capable of in terms of creating new industries - we believe that it is important that it catch up with social needs.

What are such social needs? They include science and technology that allows humankind to prosper by helping to protect the global environment, develop and use energy, recycle resources, and deal with the problem of food production. They also range over areas related to the enrichment of society and everyday life - the maintenance and promotion of health, the maintenance of the environment surrounding everyday life, the enhancement of social infrastructure, policies for disaster prevention and safety. In short, the technology that allows nature and society to exist synergetically is the technology that society requires. Indeed this must be carried out to return benefits to society through the development and use of outer space.

The significance of Japan's space program lies broadly in the following two areas.

- **Space as a strategic industry.** According to one publicly announced figure (research by Mitsubishi Research Institute), the space industry in Japan accounts for about 5 trillion yen in revenue, including not only companies involved in equipment production for outer space, such as rockets and satellites, but also industries that use outer space (for satellite communications, broadcasting, etc.) and those that manufacture related equipment for ground use (antennas for satellite broadcasts, mobile communication equipment, etc.). Satellite communications and broadcasting (including mobile communications systems), satellite-based surveying systems, remote sensing, etc., will be considered indispensable parts of the social infrastructure of life in the 21st century.

- **The space industry as a catalyst of advanced technology development.** The percentage of employees involved in research and development in Japan's space industry is high in comparison to the average for all industries. As a prototypical knowledge-intensive industry, the space industry contributes to elevating the level of industrial technology (in fiscal year 1993, employees involved in R&D in the space industry accounted for about 40 percent of all employees, compared to an overall industry average of about 10 percent).

  In this way, technology's roles in society and everyday life are expanding. In such an age, a widening of the sphere of "socio-techno impacts" through the space program will be broadly sought. And it is an age in which the expansion of these impacts will be possible.

4. Identification of Technology Responses and Space Linkages

Section 4 is focused on describing the role of technology in addressing thirteen primary and secondary challenges identified in the previous section. It considers (a) how technology can help solve future socio-economic problems, and (b) how space activities and technologies can support the identified technological solutions. It should be noted, however, that while technology is a critical tool for solving many societal challenges, it is neither the best nor the only solution for all issues. Often, social problems are better solved by using legal, cultural or educational instruments.

There are four general means by which technology (and space activities linked to those technologies) can be used to alleviate socio-economic challenges. These are:

- activities aimed at monitoring or characterizing the dimensions of a problem in order to better understand and guard against its effects,
- solutions aimed at improving the efficiency of production or consumption of resources in order to extend the life of a resource base,
- solutions aimed directly at mitigating a problem (e.g. negative impacts of human and industrial activity), and
- solutions aimed at providing a new resource (opening up an otherwise closed system).

These four categories of response are used in identifying and characterizing a practical set of technology solutions and possible space linkages for the listed challenges.

Several sources were examined when determining the technology solutions and space linkages. These included the seven studies mentioned in Section 2, technical reports and a wide range of futurist literature (see References 9 to 27).

Figure 4 depicts the number and kind of space linkages which address the defined societal challenges. The greatest number of direct benefits of space technology are seen in the categories of energy resources, agricultural productivity, pollution/hazardous waste, health care, and asteroid/comet impact. These appear to be areas that warrant additional study and may become focal points for a long-term space plan. Challenges which benefit least from space technology include population growth (none), democratization of political structures, education, globalization of the economy, and risks of technological innovation. These areas would seem to have less potential to be aided by space activities, although science/mathematics/engineering education can be indirectly influenced very strongly by space activity, which can inspire students to make careers in these fields. Also there are opportunities for tele-education using space.

Figure 4 also indicates where space technologies have the greatest potential benefit. Technologies that focus on monitoring/characterizing the problem, or increasing the efficiency of production or resource utilization, appear to have the widest impact on socio-economic problems. Monitoring technology is focused on space-based remote sensing systems, which are widely used for assessing problems on a regional or global basis. Communications systems, life sciences research, and closed-loop life support technologies support increases of the productive efficiency of current processes. Space technologies focusing on the mitigation of societal problems include hazardous, waste disposal systems, extraterrestrial mining and manufacturing technologies, and life support systems. There seem to be relatively few space linkages which can be expected to provide fundamentally new resources, although utilizing space-based energy resources (solar wind helium three) would seem to be an area of real potential.

SPACE TECHNOLOGY CATEGORY	NATURAL RESOURCE FACTORS				ENVIRONMENTAL FACTORS			SOCIAL, ECONOMIC & POLITICAL FACTORS					TECH-NOLOGY FACTORS	NATURAL DISASTERS				
	POPULATION	ENERGY RESOURCES	AGRICULTURE/PRODUCTIVITY	WATER RESOURCES	CLIMATE/GLOBAL CHANGE	POLLUTION/HAZARDOUS WASTE	HABITATS/SPECIES LOSS	DEMOCRATIZATION	EDUCATION	MILITARY TECHNOLOGY PROLIFERATION	GLOBALIZATION OF ECONOMY	HEALTH CARE	HUMAN/ENVIRO RISKS OF TECHNOLOGY	ASTEROID/COMET IMPACT	WEATHER	SEISMIC ACTIVITY	FOREST FIRES	TOTAL
MONITORING		X	X	X	X	X	X			X				X	X	X	X	11
INCREASED EFFICIENCY		X	X				X	X	X	X	X	X X		X	X	X	X	13
MITIGATION			X	X	X	X X X							X	X				8
NEW RESOURCES		X X										X						3
TOTAL		4	3	2	2	4	2	1	1	2	1	3	1	3	2	2	2	35

Figure 4. Mapping of space technology linkages to societal challenges

5. Conclusion - Toward Creating Human Values for a Global Age

In the post-Cold War present, international society is increasingly interdependent in the political, economic and cultural realms. Political borders are increasingly permeable to ideas and trade; nationalist policies are less efficacious in a world that requires global solutions. Japan, too, in the midst of a major transformation, prompted by the need to break away from traditional patterns of development. Such changes in Japan can by no means be viewed separately from changes in the international environment.

After developing since the 1980s as a production base, Asia has achieved dynamic economic growth, principally in East Asia. Today, with its attractiveness as a market also established, Asia is following the example of the industrialized countries. The world economy has suddenly entered an age in which a new international division of labor has become necessary, one which includes the developing nations. This means, in other words, that Japan and other industrial nations will have to pioneer new frontiers in order to sustain further growth.

The situation confronting Japan today reflects both the need for reforms stemming from these kinds of changes in the international environment and the need to deal with problems peculiar to Japan, as well as those shared in common with humankind - the global environment, energy, crisis management, etc.... Japan is being challenged to transform its entire socio-economic structure. As concerns shift from the pre-eminence of production to the quality of everyday life and the creation of self, Japan's success in this transformation will hinge on how well it grasps the fact that international agreements are no longer a function of power but of values that are universal - intellect, reason and virtue.

When considering the relationship between space development and use and social issues, the most important question is: "what kinds of contribution can space-related activities make to positive structural change". We consider that the greatest direct benefits from space technology are in the areas of energy resources, agricultural productivity, pollution/toxic waste, and medical care. These have the potential of becoming the focal points of long-term space planning. Under a contract of NASDA, CSP Japan and CSP Associates have been engaged in a detailed analysis of Japan's long-term social issues and a quantitative analysis of specific space missions that can contribute to solving them (see the FY95 NASDA Consignment Study *"Japan's Long-term Social Challenges and Objectives of Space Programs"*, in press). Although only a preliminary view at this time, we believe that the space program will demonstrate effectiveness in the area of creating new growth markets in pioneering fields and in the area of the global environment, including energy and resources.

References

1. The Brooking Institution, World Resources Institute, and the Santa Fe Institute: *Project 2050*. Study in progress
2. Carlson, R. and Goldman, B.: Fast Forward: *Where Technology, Demographics, and History Will Take America and the World in the Next 30 Years*. Harper, 1994
3. Conway, M.: *A Glimpse of the Future: Technology Forecasts for Global Strategies*. Conway Data Inc., Notcross, Georgia, 1992
4. Meadows, D.H., Meadows, D.L., and Randers, J.: *Beyond the Limit: Confronting Global Collapse, Envisioning a Sustainable Future*. Chelsea Green, Post Mills, Vermont, 1992
5. Sheffield, C., Alonso, M., and Kaplan, M.A. (editors): *The World of 2044: Technological Development and the Future of Society*. Paragon House, St. Paul, Maine, 1994
6. Singer, M.: *Passage to a Human World: The Dynamics of Creating Global Wealth*. The Hudson Institute, Indianpolis, Indiana, 1987
7. United Nations: *Global Outlook 2000: An Economic, Social, and Environmental Perspective*. United Nations Publications, New York, 1990
8. Angelo Jr., J.A. and Buden, D.: *Space Nuclear Power*. Orbit, Malabor, Florida, 1985
9. Boeing, Martin Marietta, General Dynamics, McDonnell Douglas, Lockheed, Rockwell: *CSTS Commercial Space Transportation Study*. May 1994
10. Bova, B.: *The High Road*. Houghton Mifflin, Boston, Massachusetts, 1981
11. Heppenheimer, T.A.: *Colonies in Space*. Warner, New York, 1977
12. Glenn, J.C. and Robinson, G.S.: *Space Trek: The Endless Migration*. Warner, New York, 1978
13. JSC Life Science Project Division: *Life Sciences Experiments in Space*. Lyndon B. Johnson Space Center, Houston, Texas, 1986
14. MacElroy, R.D. and Smemoff, D.T. (editors): *Controlled Ecological Life Support System: Regenerative Life Support System in Space. Topical Session of the XXVI COSPAR Meeting, Toulouse, France, July 1986*. NASA Ames Research Center, Moffett Field, California, 1987
15. Johnson Space Center: *Manned Space Flight Benefits*. Lyndon B. Johnson Space Center, Houston, Texas, 1987

16. NASA Science and Technical Information Branch: *Space Resources and Space Settlements.* US Government, Washington, DC, 1979
17. Mori, M.: *Evaluation of Influence through Introduction of Innovative Space Infrastructure, Executive Summary.* National Space Development Agency of Japan, Tokyo, October 1994
18. Rockwell International: *The Space Program, Down to Earth Benefits.* Rockwell, USA, 1988
19. Salvin, T., Liening, F., Oleson, M. and Olson, R.L.: *Controlled Ecological Life Support Systems (CELSS): Physiological Waste Management Systems Evaluation.* NASA Ames Research Center, Moffett Field, California, June 1986
20. Savage, M.T.: *The Millennial Project: Colonizing the Galaxy in Eight Easy Steps.* Little Brown, New York, 1992
21. JohnsonSpace Center: *Spacelab Life Sciences 1: First Space Laboratory Dedicated to Life Sciences Research.* Lyndon B. Johnson Space Center, Houston, Texas, 1989
22. NASA Lewis Research Center: *Space Power.* Proceedings of a workshop held at NASA Lewis Research Center, Cleveland, Ohio, April 10-12, 1984
23. NASA Space Systems and Technology Advisory Committee: *Advanced Technology for America's Future in Space.*
24. Stine, G.H.: *The Space Enterprise.* Ace, New York, 1980
25. Stine, G.H.: *Space Power.* Ace, New York, 1981
26. Stine, G.H.: *The Third Industrial Revolution.* Ace, New York, 1975
27. US Congress, Office of Technology Assessment: *Solar Power Satellites.* US Government, Washington, DC, August 1981
28. Economic Planning Agency of Japan, Committee on Economic Planning: *Our Country's Challenges and Basic Policy for Economic Management.* November 1995

Lessons Learned from the Viking Missions Which are Relevant to Mission to Planet Earth

G. Soffen, University Programs Office, NASA Goddard Space Flight Center, Code 160, Greenbelt, MD 20771, USA

Abstract

The Viking missions to Mars (1976) were pioneering soft lander/orbiters to search for evidence of life, map the planet, and combine several measurements and observations to characterize the surface. The missions lasted on the surface over five years and involved the dedicated work of hundreds of scientists for over a decade (1968-83). This massive scientific and engineering undertaking was comprehensive and educated many about thinking holistically about the planets. From these thoughts about Mars (our sister planet) came a new view of the Earth, and this gave rise to Earth System Science, a recognized interdisciplinary field of study. Earth System Science is an approach to understanding how the physical-chemical-biological processes interact and influence each other. This permits an understanding of how the whole system works (as contrasted by the disciplinary nature of each individual field of research). This systems approach produces a strategy between observation and modeling that allows an understanding of the "whole", in the same way as systems engineering reveals the understanding of the "black box" in counter distinction to a collection of individual engineering subsystems.

This line of thought eventually led to a program called Mission to Planet Earth (MTPE), and to the development of a series of spacecraft observation. MTPE will occupy many hundreds of scientists and thousands of engineers from around the globe for several decades. The demand for these broad overviews and planned comprehensive programs came mostly from scientists who recognized the need to cross disciplinary lines in order to understand their particular scientific problems.

1. Essence of the Viking Missions

The Viking missions were among NASA's most successful planetary exploration missions. Two identical unmanned spacecraft, each with an Orbiter and Lander, were launched in the summer of 1975 from Cape Kennedy, Florida, using Titan/Centaur rockets. The payloads were scientific instruments designed to map and characterize the surface of Mars and to search for evidence of life. The Project was initiated in 1968; spacecraft (launched in 1975) landed on Mars in 1976, and operated until 1981. Observations in Mars orbit were made for over two years, and measurements made on the surface lasted for five years. The missions proved to be a watershed of information about the red planet.

Upon reflection of the management of the Viking missions, the following lessons were learned. These may be helpful for other missions, particularly for Mission to Planet Earth (MTPE). The two Viking spacecraft required about eight years for design and construction, required about 20,000 man-years over the life of the project; every major subsystem on both spacecraft operated successfully within its design specification. Several engineering problems were particularly troublesome, the major one being the difficult remote landing on an unknown surface. A low-power budget, limited weight and limited telecommunications each required innovations.

2. The Lessons Learned

- **Money** is the fuel. The cost of the project over the course of its lifetime was a factor of three larger than the initial estimate. Some of this increase was due to a delay of two years stemming from when funds became available, some to unforeseen complexities of the scientific instruments, and some to the severe conditions imposed - such as the need to sterilize the Landers, so as not to contaminate the planet. For projects like Viking that have great visibility and where success is crucial as a matter of national policy, the funding must be adequate. There was always a struggle within the Viking program to keep to the schedule and to balance the available resources, but the central driver was always the success of the mission.

- **Politics** can stop the mission. Whenever large sums of money are needed, oversight of the project grows exponentially as the funds grow. Such oversight is not necessarily a bad thing, since public funds are being expended, but oversight that tends to micromanage the technical details can destroy a project. It is the responsibility of NASA to ensure that nothing is being hidden from public scrutiny while, at the same time, to develop a trust that the technical management is being carried out in a professional and efficient manner. This balance is not easy to attain and requires constant attention at the highest levels.

- **Technical management** is the life blood of the project. For a space program, often a once-in-a-lifetime opportunity, where a unique, one-of-a-kind spacecraft is being developed, including new technology, experienced and skilled technical management is the determining factor in success. A single-point failure is constantly the 'wolf at the door'. For planetary missions, the repair of hardware after launch is impossible; complete redundancy is impractical. An infinite amount of testing is not reasonable. All managers believe that quality must be built in to the spacecraft, by constant reviewing, yet without destroying the schedule.

- **The scientific community** must support the mission. The scientific community is made up of highly competitive individuals seeking recognition as well as the truth. Yet, all of the interested scientists cannot be Principal Investigators. For both practical and resource reasons, a representative group of science leaders must be perceived to have a strong hand in the decisions and the priorities of the mission. Trusted members of the US National Academy of Sciences and also Nobel Laureates among the selected Principal Investigators were enormously helpful during the Viking program.

- **Public affairs**: who 'owns' the mission? By its very nature, the US space program is dependent on an approving public. While the initiation of that space program was partly for political reasons, the heroism of the astronauts spilled over into the exploratory program, and so the US public became fascinated by the challenge of difficult robotic missions to explore the solar system. Conducting a complex mission, while having daily press conferences, required a great deal of awareness of matters of importance to the man-on-the-street. The attention to public affairs was developed over many years prior to the launch of the Viking spacecraft.

- **The time-line**, like the 'beat' in music, must be kept. Every project manager knows that the commodity in shortest supply is time. For a price, a little time can be bought with money - but not very much, and at great expense. Like the leader of an orchestra, the project manager must select the pace so that all elements arrive at the appointed place at the appropriate time. Schedule management is as difficult as hardware management or software management.

- **Teamwork** is essential. Successful projects are carried out by a team of people working towards a common goal. Every member's contribution must be seen as being a necessary and vital component. The upper-level managers must not be isolated from the technicians at the work bench. Good morale is built up over a long time by constant attention to this issue. The continuity of personnel, resulting from dedication to the project, is of enormous value in 'traceability' when the need arises.

- **Problems** must be identified early. On every project there are a few very difficult technical problems that must be solved. If these are put off, they can become a cancer. For Viking, the landing was clearly the hardest part of the mission. Both the surface and the atmosphere of Mars were unknown and the only experience of a successful remote landing was on the Moon. Unusual requirements, namely to maintain a pristine surface, and to ensure a virgin sample for testing, further compounded the problem. Other problems were found among the scientific instruments, many of which had never been developed prior to the mission.

- **Contingency planning** is essential. Regardless of any amount of care being taken in a space project, there are always some idiosyncrasies of nature. On every project there are areas where the margins of safety are exceeded, and systems that are thought to be fault-free; yet they can cause surprises. Most projects try to prepare for these instances by developing contingency plans. Frequently, the things that do go wrong are not those for which contingency plans were made, but the very act of planning prepares the team to be able to react to a sudden problem in a cool and thoughtful way. The formation of 'tiger teams' to deal with these problems is very helpful.

- **In a large project**, communications between many staff members become difficult. Usually, the size of a project is such that the people involved mostly know one another, as in a small town. As the project grows, communications become less efficient; the strain of information flow becomes more serious. The Viking missions took place well before the modern age of PC communications and did not benefit from e-mail, WWW or any of the contemporary conveniences. At the peak of the project, when all spacecraft hardware and software were coming together, upper-level managers felt limited in their ability to get the 'word out' across the project.

- **The legacy**: how does follow-on planning work? The results from the Viking missions and other planetary missions gave rise to the early planning of MTPE, but did not lead to another mission for a Mars Lander for over a decade. It might be said that no one planned for success.

3. Mission to Planet Earth

MTPE may be thought of as a descendant of the planetary missions. This program is designed to gain data by remote sensing which, when combined with ground-based data, will lead to a better understanding of how the various systems of the Earth interact on the planetary scale. The scientific basis of the MTPE is Earth System Science; this involves scientists from all disciplines and from all countries on the globe. The idea behind MTPE is to use spacecraft to make observations and measurements that will advance and test models of Earth subsystems. From these models, we should be able to make predictions and test the interactions between the various subsystems and their elements.

Compared with the Earth, in the global sense, Mars is a relatively simple planet. The atmosphere is more tenuous and simpler, both chemically and structurally. Currently, there are no bodies of water present, no biosphere and no rapid changes due to human activities. Only diurnal and seasonal changes occur, driven by the Sun. By contrast, the Earth is extremely dynamic, driven by plate tectonics, and with volcanic eruptions. There are circulating oceans and an atmosphere, severe storms, fires, and now the full force of contemporary technology.

Modeling the Earth on a global scale is a vast enterprise, but some of the methods for planning the missions, conducting the observations and synthesizing the data are similar to those used for the Viking missions. Two examples provide useful illustrations. One concerns data visualization, and the other is the 'science team' approach.

- Combining data is valuable. During Viking, we had an important need to combine data sets for a very practical purpose. The observations from the Orbiter were used to enable the selection of a safe and scientifically important landing site. The combined data from three instruments, one in the visible range, one in the ultraviolet and one in the infrared, were used together for determining the structure, character and possible soil moisture content of candidate landing sites. In combining the data sets graphic visualization techniques were used that are the forerunners of what is already planned in MTPE.

- Scientific teams are the way forward. Viking used teams of scientists rather than individual Principle Investigators to establish the scientific goals of each measurement, to set the criteria and to specify the instruments. Team leaders, together, formed a Science Steering Group that set the scientific priorities, allocated the resources, and interacted in order to maximize the scientific return from the mission as a whole. Currently, MTPE is using some of the same management techniques.

4. Summary

The Viking missions served as important pathfinders in NASA's learning of how to conduct large-scale scientific projects. These missions of twenty years ago had many hazards along the way, but now they provide milestones for all subsequent large missions. There is very little literature on this subject; yet passing these ideas on to the next generation is very important.

The Australian Five Year Space Plan - Public Good, Commercial Benefits or Internationalization?

M.E. Davis, Satellite Communications Research Centre, The University of South Australia, Adelaide, Australia & MSS participant, International Space University
M.J. Miller, Satellite Communications Research Centre, The University of South Australia, Adelaide, Australia

Abstract

In 1994, the Australian Government was presented with a Five-Year National Space Plan prepared by its newly created Australian Space Council. This paper reviews the key elements of the proposed National Space Plan with particular emphasis on its intended national benefits.

Scientific, economic and political factors all are shown to be key elements in the plan. The paper outlines ways in which a modest National Space Program could lead to public good benefits and regional opportunities for nations with limited economies like Australia.

1. Introduction

Australia made a significant contribution to the world's early efforts in space but its achievements were largely aided by the technological competition between the super powers during the Cold War. With the end of foreign involvement, Australia needed a new rationale for public expenditure on space programs which would take into account its relatively small population and limited economic resources. The establishment in 1994 of a National Space Program by the newly established Australian Space Council was a turning point for Australia in space. The public good and economic benefits to industry were the new rationales. Space is now used in Australia to underpin important services and scientific research in areas in which benefits for ordinary people are apparent. The Australian space industry provides modern telecommunications and broadcasting, contributes significantly to environmental, meteorological and oceanographic activities as well as natural resource assessment, mining exploration, mapping and navigation. It is also playing a role in the protection of life and property and in defence and civil security services. The Five Year Plan of the Australian Space Council is designed to focus and maximize the benefits of space for Australians, their neighbors and humankind in general. Proponents of the National Space Program are finding, however, that it is necessary constantly to remind political decision makers of the national importance of the program.

2. Australia's History in Space

Australia has a proud record of achievement in space activities. Its history in space can be traced back to 1946 when Australia and the United Kingdom decided to set up a joint experimental guided weapons testing range in the centre of the continent. Woomera, located 480 kilometres North West of Adelaide, was chosen as the location of the new launch facility. From there, rockets could be launched and tracked with little danger because the vast deserts they flew over contained very few inhabitants.

A large infrastructure of launch pads and ground support facilities was developed at Woomera to meet the needs of a number of countries, especially the UK and the members of the European Launcher Development Organisation. Australia became a member of ELDO after its formation in 1961 and Woomera was used as the launch site for the Europa-1 vehicle when 10 launches were made between 1964 and 1970. At the peak of its operation, Woomera was used for the development of major international launch vehicles, some of which placed satellites into orbit. With the launch of a scientific research satellite, WRESAT, in 1967, Australia became only the third country in the world to build and launch its own satellite into orbit from its own soil.

In 1987 the Cape York Spaceport proposal attracted considerable attention throughout the world. The plan was to create a commercial launch operation to compete with the US and European launch services using Russian made launchers to put payloads of between 1.9 and 2.4 tonnes into geosynchronous orbit. Unfortunately the proposal was never implemented because of the very large investment required for transportation and communications infrastructure in a very remote region of Australia.

3. Current Uses of Space

Today, Australia is a major provider of tracking and ground support facilities which have been used for NASA manned and planetary missions as well as for defence satellite surveillance and communications purposes. The Canberra Deep Space Communications Complex is a focal point for these activities.

Remote sensing is a particularly important space application for Australia. Examples of its applications include:

- weather forecasting,
- environment monitoring,
- regional planning,
- mineral exploration,
- mapping,
- environment degradation studies,
- rural sector land management,
- mapping grazing patterns and fodder availability,
- crop management and yield forecasting,
- soil salinity studies,
- crop disease monitoring,
- pest monitoring,
- weed surveillance,
- illegal cropping detection,

- irrigation planning,
- agricultural dispute settling.

Data are obtained from a variety of international sources including satellites operated by the United States, France, Japan and the European Space Agency.

Australia was a founding member of Intelsat and Inmarsat and makes extensive use of their international telecommunications services. It also has its own satellites for domestic telecommunications and broadcasting services.

In the area of launch services, Woomera had fallen into disuse as a launch site but, within the last 12 months, Woomera has been the subject of some important developments. Woomera has recently been used by NASA for a series of launches of sounding rockets carrying scientific experiments. Commencing in April 1996 it is to be a test site of ALFLEX (Japan's Automatic Landing Flight Experiment). In a series of tests a one-third size model of NASDA's reusable space vehicle, Hope, will be lifted into the air and dropped from a helicopter at a height of 1.5 km. GPS receivers will steer the craft towards Woomera airport.

Furthermore, it has recently been announced that an Australian company and a Russian launch services company are exploring the prospect for establishing an Australian space launch service at Woomera, using the START family of launch vehicles. It is anticipated that the first launch from Woomera could be within 18 months.

4. The National Space Program

Our national space agency, the Australian Space Office, has only been in existence since 1986. It is therefore one of the world's newer space agencies. In 1992 the Australian Government decided that Australia needed a plan to integrate the country's diverse space activities. In 1994 the Australian Space Council, comprising representatives from industry, research organizations and Government agencies, was established under legislation. The Australian Space Office was established as the executive arm of the Australian Space Council. Administratively, it is part of the Federal Department of Industry, Science and Technology.

The Australian Space Council's mission is to secure for Australia the net benefits from the development and use of space-related science, technology and industry. Its objectives include the securing of cost-effective space goods and services for the public good, developing space-related industries, fostering space-related research and a public educational role. It can therefore be seen that economic motives and the public good are strong driving forces.

One of the main tasks of the new Council was to develop an annually renewable Five Year Plan for a National Space Program. This plan was prepared for and received by the Government in May 1994. In preparing the plan the

Council saw its role as establishing priorities and work programs which would provide guidance for both public and private sectors involved in space endeavors within the constraints of a very limited budget by international standards. The underlying theme of the plan was to develop Australian industry capability and to develop a space project of national significance utilizing Australian space science, engineering and industry capabilities.

The elements of the plan were as follows.

4.1 Objective 1: Space Services for the Public Good

According to the plan, the objective of securing cost-effective space goods and services for the public good would be advanced by greater use of Earth observation and satellite positioning. Earth observation has four important applications in the public sector - operational meteorology, oceanography, climate monitoring and environment management. A set of strategies and activities related to maintaining and expanding Australia's capabilities in these areas was proposed.

Satellite positioning (GPS) was also seen as providing opportunities for Australian industry to participate in the development of value-added products and services for transportation, resource development and emergency services applications.

4.2 Objective 2: The Development of Space-Related Industries

This objective was to be achieved in the following four areas:

(a) Launch Services

The plan noted the emergence of a market for Low Earth Orbiting satellites and Earth observation activities. Certain competitive opportunities enjoyed by Australia including:

- its geography (which allows for both polar and equatorial launches),
- its political stability,
- its climate,
- its low population areas and its well developed human and physical resources capable of supporting launch operations.

The plan also argued that Government support would be an essential catalyst to the emergence of a commercial launch industry. In order to establish a launch service capability Australia would need to develop and promote its capabilities, secure the transfer of appropriate technology and develop strategic alliances with other countries.

(b) Small Satellite Industry Development

The plan noted Australia's advantages with respect to the sophistication and depth of its telecommunications and supporting industries and considered that Australia was well placed to be a gateway and supplier of telecommunications services to the Asia-Pacific region. The emerging market for telecommunications services using LEO satellites encouraged the Council to propose a strategy to develop and demonstrate a capability in the small satellite industry sector.

(c) Remote sensing applications

The development of value-added remote sensing industries was seen as a commercial opportunity for Australian industry and as an area in which the public good would be served. Included in the proposed actions were a national Earth observation education and training program to address the needs of domestic and Asia-Pacific region users.

(d) Space Industry Development Centre (SIDC) program

The objective of this program is to support and accelerate the development of commercial space products for domestic and international markets. These capabilities were seen as essential to international recognition of Australia as a space-qualified nation, able to contribute to and benefit from scientific and commercial participation in overseas space programs.

There are three such centres in operation in Australia: the Satellite Communications Research Centre at the University of South Australia, the Space Centre for Satellite Navigation at the Queensland University of Technology and the Space Centre for Microwave Technology at Griffith University in Queensland.

The largest of these centres is the Satellite Communications Research Centre. For over three years prior to the commencement of the program, the Centre had been actively engaged in research and development in conjunction with Government agencies and private companies. The approval of SIDC funding of A$500,000 (A$1.00 equals US$0.75) per annum for the SCRC had a major impact on its ability to undertake programs which contributed to industry development and public good activities. The funding arrangement was based on the following understandings:

- the SCRC would undertake space-related projects specializing in the area of satellite communications for the benefit of the Australian space industry,
- SIDC funding of up to A$500,000 per annum would be provided over five years, and

- SIDC funding would only be made available on a dollar-for-dollar basis with industrial partners who were required to provide matching funding.

Over the first four years of its operation, the Centre has undertaken research and development tasks involving significantly more funding from industry clients than that required to match the funding provided through the Australian Space Office. This has enabled the Centre to undertake a wide range of tasks in collaboration with space related industries.

In particular, 25 separate projects were undertaken in collaboration with 12 different Australian companies operating in the area of satellite communications. The total value of projects undertaken for overseas organizations was approximately A$1,600,000. At the expiration of the five year funding period, it is probable that there will be sufficient demand for its expertise for its work to continue.

The industry development benefits that have flowed to Australian companies have been significant and many of the companies which participated with SCRC in joint projects are now involved in production and export of space communications equipment.

4.3 Objective 3: Fostering Space-Related Research

This objective recognizes the importance of space-based scientific research in Australia. The Five Year Plan promotes high quality research by existing research organizations and encourages international collaboration. Examples of such research include:

- Project Endeavour, the Australian designed and manufactured ultra-violet telescope which flew on the Space Shuttle,

- the Australian designed and developed Along Track Scanning Radiometer (ATSR) space flight instrument, which is deployed aboard ESA satellites, and the Advanced Along Track Scanning Radiometer (AATSR), which is to be flown an the Envisat-1 spacecraft in 1998,

- Australia's leading research in hypersonic flight.

4.4 Objective 4: Supporting International Policy Objectives

This objective would be achieved through the development of international cooperation at both the public and private sector level. In particular space activities would be used to further the Government's foreign policy objectives of developing closer economic and technological ties with countries in the Asia-Pacific region.

4.5 Objective 5: Education

The Australian Space Council noted that the benefits derived from space activities are generally not well known. It considered that it was important to raise the overall level of industry and community awareness of Australia's space activities and capabilities. Therefore it proposed a program to improve awareness among the public and among governments, industry, academia and research institutions.

4.6 Objective 6: Project of National Significance

The centrepiece of the National Space Plan is a project to establish a launch services capability in Australia. This would require Government support for establishing or refurbishing the infrastructure for light launch facilities at Woomera (which is suitable for polar orbits) and in the longer term in the North of Australia, from which a variety of orbital inclinations and altitudes could be achieved. It also involves considering options for Australian designed and manufactured payloads and funding two initial launches. The benefits of such a project would include:

- the development of a new industrial capability,
- a means of launching Australian-built demonstration satellites,
- a means of providing public good and semi-commercial services such as environment and pollution monitoring, surveillance, meteorology and communications,
- the catalyst for an Australian satellite development program,
- the ability to participate in the new market for LEO telecommunications satellite launches,
- the prospect of securing a share of the Asian launch business,
- the participation of the private sector in the Australian launch industry.

In addition to the recent announcement concerning the Australian-Russian launch partnership, the Australian Space Council recently commissioned a feasibility study for a Low Earth Orbit remote sensing satellite (ARIES) which will provide data for mineral and environmental mapping. This satellite could be launched from Woomera on a START launch vehicle.

5. The Future

In March 1995 the Australian Government announced a review of the National Space Program by an Interdepartmental Committee on International Space. The report of this committee is not yet available and no decision on its recommendations will be made until after the forthcoming Federal election on 2 March 1996.

Until the Government's decision on the future of the National Space Program is known, the Australian space community will be apprehensive about the future. Cancellation or severe curtailment of the program is likely to have the following consequences:

- Australia's dependence on a wide range of space-derived information would continue to increase but its access to the primary technologies upon which the information is based would be impaired. Australians would become even more vulnerable to the discretion of the operators and owners of the technologies as to whether or not to continue to allow us access to space products and services. Australia's international leverage in space technology would diminish.

- The opportunity to benefit economically from significant commercial opportunities in the international space industry, particularly in the Asia-Pacific region, would be lost. In particular, it is unlikely that Australian industry would be able to participate in the future development of the constellations of LEO telecommunications satellites, through the manufacture of satellites, the development of ground station control systems or the development and manufacture of hand-held terminals and mobile terminals.

- Australian scientific research and development activity would be prejudiced.

- Australia would be deprived of the opportunity to advance its foreign policy objectives by using space as a forum for international cooperation.

The decisions taken over the next few months will critical in determining whether or not Australia enters the new millennium as a participant - or as an observer - in humankind's quest to improve life on Earth through the use of space.

Serving Humanity Using Space: A Great Idea, But Who Pays the Bill?

J. Egan, Egan International, 1701 K Street NW, Suite 205, Washington DC, 20006, USA

Abstract

Space enthusiasts, and I consider myself one, of all stripes have suggested many and varied ways in which space can serve various humanitarian purposes. Some have already been brought to fruition. Such projects as weather satellites which provide early warning of impending natural disasters, other earth observation satellites which provide us with the means to make decisions about various aspects of our environment, and communications satellites which are able to operate when terrestrial infra-structure has been crippled by war or natural disaster or where such infrastructure does not exist are a few examples. As the participants gather at this interdisciplinary, international forum to discuss this topic and, hopefully, "move forward from the discussion of the 'problems' to the formulation of 'innovative solutions'", attention must be paid to the economics of this topic as well as the technologies and programmatics needed to do the job.

Serving humanity, no matter whether this is done using space or through some other means, is normally characterized in economic terms as a public good. This means that there is not a commercial/profit motive for the performance of this function and it is, therefore, funded through some form of public/ government payments. Historically, much space activity in itself, with the notable exception of communications satellites, has been viewed as a public good receiving most of its funding from governments. Unfortunately, "times, they are a changing!" Worldwide budgets for all government services, including those characterized as humanitarian, are being slashed and space activity, humanitarian or not, is on the cutting block along with everything else.

This paper addresses the economic issues of providing services to humanity using space in this era of tight budgets. It addresses the possibilities of commercial participation in the provision of such services [much like many municipal services, which were formally government functions, are now being done under contract by the private sector] and the role international cooperation may be able to play in reducing the costs to individual governments or allowing poorer governments to benefit from the activities of those more able to pay. Both the difficulties faced in meeting the economic challenges as well as the substantial benefits which may be gained will be discussed. It is hoped that this discussion of the economic problems and possibilities will form a useful context in which past programs and new ideas can be discussed. It is also hoped that providing this context will avoid final recommendations from this Symposium proposing many wonderful technical and programmatic ideas which are impossible to carry forward or even suggest seriously because they fail to meet the basic economic hurdles now required governments.

1. Introduction

I have been consulting in the space industry for over 15 years; after that time, I must be willing to accept the fact that I - and my company - are undoubtedly part of "the industry". However, when I speak at symposia such as this or in other fora, I am invariably identified as the person with the outside point of view, the one who is making the controversial statement or asking the difficult question or, in general, deviating from the accepted point of view of industry. While I have long been a proponent of space activity, in particular of commercial space activity, and believe very strongly that space has a very important role to play in providing both economic and social benefits to humanity, I know that my topic, which focuses on who pays the bills will, once again, thrust me into the role of the "bearer of bad tidings". It is a role to which I have become accustomed. However, as I am often reminded, anyone can point out problems, the useful person is the one who can suggest solutions as well. I will attempt to do both.

For those of you who are concerned about people like me who are always being controversial, I should mention my associate, Wilfred Mellors, who, at the recent IAF Conference in Oslo, pointed out that we had made a number of suggestions on the same weather satellite issue at an IAF Conference two years prior to which apparently no one had given any notice whatsoever. Undaunted, however, we had pressed on to expound our views on the same organizational and economic issues once again. For those of you who came to this Symposium with the desire to discuss just the wonderful things which can and even should be done using space in the service of humanity and do not really want to listen to people like me who are concerned about the tedious detail of how a program actually gets the funding which it needs, I tell you this story so that you will know that you are in good company when you choose to ignore what I have to say.

2. The Space Culture and its Influence on Funding for Space Activity

There have always been visionaries who have looked into the heavens and seen great and wonderful things for the future. These have ranged from science fiction writers such as Jules Verne to more practical visionaries such as Werner von Braun. They have all caused us to wonder about the future and even caused some to work toward doing things in space. However, in the late 1950s the pace of both thought and activity regarding space changed with the arrival of a totally unexpected advocate for space - politicians. Up to that point the space advocate had been doing science and engineering work aimed at modest goals using space. With the support of the politicians, the sky, literally, was the limit and the "space race" become a battleground of the Cold War. Success in space was equated with winning the battles for technological supremacy and prestige between the superpowers. This new and totally unexpected dominance of space in this struggle got the US to the moon, built the space shuttle and Mir, and provided funding for many other programs. There is no doubt that during this time significant benefits in terms of scientific knowledge and, to some degree, practical benefit, were gained - but these were byproducts and not the real goal.

Another byproduct of this popularity of space was a culture which developed among those involved in the process. Groups of people in the aerospace/defense industry and in space advocacy groups such as the old L-5 Society and now the National Space Society developed a particular view of the world in which space has a distinct, "special" role. As Winston Churchill once said: "*We want lots of engineers in the world, but we do not want a world of engineers*". Space activity, however, became the haven of a special brand of space "engineers" which I will call, for lack of a better term, "space people". This group has dominated the thinking about the importance of space and in particular how space activities should be funded. Space has been characterized, by space people, as a place where activities can be pursued which are beyond petty, even mundane, outside concerns such as financial justification or the need to provide pragmatic, practical benefits. Things are done in space for the same reason that someone climbs a mountain - as Sir Edmund Hillary said, "*Because it is there*". It is, in their view, a place for pure exploration, pure science, and pure discovery. Space people strongly believe that governments should fund space activity because it is, by definition, good - that is, things done which involve space are good in, and of, themselves. Questioning whether or not a space activity should or should not be done, or even that they should be held to some form of a cost-benefit standard was, for these people, somehow almost sacrilegious.

2.1 "Times they are a changing! "

In the early years of the space program, when it was new, exciting and different, the space people's view was seldom challenged. However, with the end of the Cold War and changes in priorities in government programs, many politicians have begun to abandon their automatic support for all things associated with space activity. There are increased pressures from those in government for space people to justify the continued support for space programs on other than the old grounds of technology leadership and on the interesting and good things that can be done in space. The problem for space people is that times have changed but, for many of them, their view of their world has been very slow to change; it had been a firmly held view for so long that it is a difficult one to abandon.

My first message to you then is a blinding flash of the obvious. The old methods through which space based activities have been funded, that is direct government funding of space activities as part of national space programs, is a shrinking pie which, some day, may be very small indeed. This is true on a worldwide basis (with the exception of Japan and a few other countries where direct spending on space is either stable or even rising a bit). There are many reasons for this, not the least of which are the general economic conditions caused by high government deficits and severe competition for limited resources within government. However, I believe, a second, though nowhere near as significant, reason lies closer to home, with space people themselves.

The history and culture which has grown up around the space industry, at least in Western countries, is extremely insular. Space people have not only emphasized that things related to space are good in, and of, themselves, they have also developed the mystique that things related to space are technically and intellectually challenging, beyond the reach of the average lay person, and in a realm of specialization so rarefied that it is almost beyond comprehension. We have an expression in the US that "You don't have to be a rocket scientist to ...". It is used to imply that the topic under discussion is not so difficult to understand that one has to have the extraordinary mental abilities of a rocket scientist to get the point. The problem is not in that statement but in its corollary. That is in order to understand and benefit from space one has to have the extraordinary mental abilities of a rocket scientist. While this view may be very good for the rocket scientist's ego, the perception left with the general public is that space activity is the narrow province of scientists and engineers and is of limited, if any, value to them. While one can argue whether or not this is actually true, like it or not, perception is reality and this perception leaves space activity with few friends and supporters outside its own closed circle. This not a good position to be in when fighting for a piece of an ever shrinking government budget pie.

2.2 Providing Products and Services vs. Public Good From Space

It is probably a good idea to make a clear distinction between space as a provider of a product or service and space as a provider of a public good. From an economic point of view the difference between these two can be characterized by their funding source. A product or service is normally paid for by the consumer of that product or service while a public good is normally paid for by some collective form of funding such as tax revenue. The obvious example of the former is communications satellites. Communications satellites moved away from the traditional government supported space model years ago. Space people equate communications satellites with the satellite itself and its launch, that is with the space component. The users of communications satellites, such as AT&T, HBO, BT, KDD, and many others would never characterize themselves as space companies. They are communications companies or broadcast companies or, increasingly, information companies. As Ricardo de Bastos, Lockheed Martin Vice President for Business Development, said at a conference just a few weeks ago: "*It wasn't too long ago that a spacecraft was a really exciting technological marvel ... today a satellite is simply a commodity, a capital asset that has to earn its keep against competing communications technologies*" .

The communications satellite industry receives little or no government funding for its work. It does most of its own research and development and pays entirely for its own equipment and launch services. In other words it is a going concern, a functioning business. It provides a practical, useful service to a customer base which is willing to pay for that service at a level which makes it a profitable business. Like most businesses, it provides mundane services on a regular, reliable basis and competes on customer service, features and cost.

2.3 Providing Products and Services for which People are Willing to Pay

Here, I believe, is the first lesson which should be learned regarding who will pay for services from space for humanity. If the service is practical and useful to a customer and can be provided at a reasonable cost, the customers themselves will pay. This is perhaps obvious and really is an area which should not be of much concern to this Symposium because entrepreneurial industrialists will find these markets, if they exist, and will provide services if the demand exists, with little or no help from outsiders. We can see that happening with GPS ground equipment, Big and Little LEO satellites, and in many other fields.

However, I do believe that there is an important point to be made here for this Symposium's consideration. That is the need for space people to focus on the end user, the final customer, and be willing to find a direct solution to their problems using space technology. This again sounds very obvious, but such an approach has not been a characteristic of space people. For example, in the early history of remote sensing, space scientists decided what problems needed to be solved and built sensors which could solve those problems from space. Their purpose was to see if the space technology would work, and it did. Then they said, this is great; now, can anyone use these data? No one asked the potential users of remotely sensed data what problems they had or determined what practical, useful problems can be solved by building these sensors. Certainly, no one ever asked the question: can we build a system to provide needed data in an economic fashion? This was just not part of the culture for many years. Fortunately, there are many signs that this view is starting to change.

It may be difficult for many space people who want to push the limits of technology and science in space to accept that it is the mundane but useful, repetitive products and services, delivered at an acceptable cost, for which customers are actually willing to pay. This may not be what space people want to do, but it is what most people who have jobs do every day. For space activity to benefit customers directly, it must change its approach. Space people must stop saying that this is done in space, and you should recognize that it is good for you. They must rephrase the question and ask: what practical problems need to be solved and how can space technology be used to solve those problems? As I noted above, there are signs this is starting to happen (e.g. studies on using space in disaster mitigation). There just needs to be more emphasis along this line. Using this approach, space may not get the credit and notoriety that space people would like to see (just as people who make silicon chips get little credit for the information revolution happening all around us) but humanity will benefit.

It should be noted that I have use the word customer in referring to the user of the space related products and services. People often equate the word customer to individual consumers and then directly to consumer products or services such as cars or telephones. However, the word customer is a collective noun. A customer can be a corporation, government department, or a whole country as well as an individual consumer. Government emergency relief agencies, for example, can and often do purchase communications services, remotely sensed data, or even weather data from commercial suppliers. They need not own and operate these services themselves. The decision as to whether a space asset is better owned and operated by the private sector or the government should be made on a combination of economic grounds and whether the activity is a legitimate function of government. It should not be based on traditional solutions or on the desire of space people for separate capability that somehow justifies the use of space.

In my view, then, the first question which should be asked by those interested in using space to be of service to humanity is: what can we do using space for which people or organizations are willing to pay? As I have already mentioned, this rather simple sounding question will require an entirely different approach to deciding how space-related activities are funded from that currently taken by space agencies worldwide. This is not the way that most space people have addressed the problem in the past, but it is, I submit, the logical way to proceed in the future. If for no other reason than the current budget problems in most countries, it is an approach which will have significant practical benefit in the near term.

2.4 Space as a Public Good

With regard to purely public good activity, space people, believing strongly that things done in space are by definition public good, have generally expected that space-related activity would be funded from public moneys and in the public interest. Where there has been a clear and useful product from this activity, this has presented little problem. For example, despite all of the problems with cost overruns and delays which occurred with the GOES NEXT geosynchronous weather satellites program in the US, no political figure ever once seriously suggested that this program be canceled. The general public, as well as the meteorological community, have become dependent on pictures from space both in our weather forecasting and on the nightly weather reports. There is a tangible benefit and, though there is always pressure to do the job better and cheaper, there has never been any real problem with funding.

However, on other programs, such as the space station for example, there has been a continuous problem with political support and funding. This is not to say that there are not real, tangible benefits which may be derived from the space station. There may well be, but they are not as direct and tangible as those associated with weather satellites. The station is seen by many as an expensive toy for space people and, recently in the US at least, it has owed its survival more to the themes of international cooperation and relations with the former Soviet Union than to any direct benefit to the US taxpayer. It is generally believed that funding for new large programs such as the international space station will be increasingly difficult to obtain.

Space people argue that this shrinking of support and funding is because the politicians who dispense the funds have lost the "grand vision" and do not see all that can be accomplished by returning to the Moon or going to Mars. They talk of humankind's destiny to explore and of space being the final frontier. They point out that space activity is supported, according to opinion polls, by an overwhelming majority of the general population. All of this is very well and good but this grand vision misses several important points. First, the public support for space is what I call a veneer. That is, people consider that space activity is, in general and of itself, good, but when given a choice between a space activity and some more concrete activity, such as building a veterans hospital, the more concrete activities are almost always considered more important. As noted earlier, because space people have been very insular in their views and done little to demonstrate the direct benefits of space activities either to other parts of government or to the public in general, advocates for space have very few cards to play. They cannot say, except for weather satellites and in some other areas: "Look what we are doing, the public cannot live without this, your department cannot live without this - therefore, you cannot cut our budget". Unless this situation changes, space budgets will become easier and easier to cut as pressure from competing interests mounts.

Space people, in general, do not appear to see this problem. In the US, for example, there has been legislation proposed this year to provide tax incentives or holidays for investment in space. When this legislation comes up for debate, the first question which will invariably be asked is, why space? If tax incentives are to be handed out in a time of government deficits, why of all places space? Why not cancer research, environmentally beneficial projects, or a host of other areas? The space people's answer is, of course, because space is good and space is special. Unfortunately, other than space people, no one really believes this and, in my view, such tax schemes will never become reality and, quite frankly, never should. If space is going to make it, it should be able to do it on merit, just like everything else.

3. How to Spend Existing Government Funding Better

Am I saying that there is no justification for government funded space activities? No, I am not. There clearly should be, as there has always been, some money in government discretionary budgets to fund space research and technology developments. The question is: how much money, and how should it be spent? I believe that if those of you attending this Symposium wish to really see space become of benefit to humanity, and have the funds for it doing so made available, then two specific actions must be taken by space people or, perhaps better said, by the space community. These are:

- First, a re-focusing of the priorities of the space community must occur which emphasizes near term, attainable goals and, in particular, goals which will benefit people directly in a very tangible way. Talking about returning to the Moon or going to Mars is great but, in the short term, it is impractical - no one has the money. Thus, whenever these goals are mentioned by space people the message conveyed to the general public is that space people are totally out of touch with both reality and the public will - which, hopefully, is not true. In the near term, space activity should be focused in areas such as monitoring the environment, enhancing communications, providing assistance in disaster relief, and other pragmatic endeavors. If such a course of action is followed, it will allow space people to claim with justification that they are helping other government departments and, indeed, humanity and not just asking for more funding to do their own, self-serving "pet projects".

- Secondly, recognizing that space activities, like all government activities, will have to achieve more with less, ways have got to be found to do things in space in a less costly manner. I believe that the best way to accomplish this [and I recognize that what I am about to say will not go down well with all the space people who have pet projects, both big and small] is to utilize as much as possible of the available space funding to attack the biggest problem facing real progress in space -- the very high cost and unreliability of access to space. About one third of the NASA budget is spent on the space shuttle. It is old technology, it is extremely expensive to operate, and it will eventually either wear out or have another failure. This puts the NASA launch program in a real box. If it spends resources to maintain the shuttle, then there are insufficient resources to work on innovative alternative solutions to reduce the cost of getting to orbit dramatically. If it spends these resources on coming up with new solutions which will dramatically reduce the cost of getting to space, then it will have no capacity for manned space flight and no way to build or utilize the space station. It is a terrible box in which to be. Getting out of that box will require extremely courageous decisions, and, since courageous decisions are rare in political circles, it is most likely that the US, at least, will not even try to figure out how to get out of this box until it is too late.

What benefits would be derived from taking these steps, as unlikely as they may be. With regard to the first bullet point, by demonstrating the practical benefits derived from space, political and public allies would be developed which may make the funding process much easier. The existence of these allies might eventually make the discretionary funding of purely space activities such as astronomy, astrophysics, and even exploration, a bit easier as well. The danger with this course, however, is that emphasizing the practical applications of space means that space becomes simply a tool to provide a product or service and loses its "specialness" which is so important to space people. Signs of this are already showing up in the strain between operational organizations which utilize space, such as EUMETSAT and NOAA, and their respective space agencies, ESA and NASA, regarding future generations of weather satellites. This, of course, is only a problem for space people. For most others this change would be viewed as a good thing. I personally believe that space is a tool and that space activity would be far better served if it were recognized as such.

With regard to the second bullet point, we are all aware that the introduction of new commercial launch vehicles such as Ariane 5, Atlas 2AR, Delta 3, and Sea Launch, as well as the Russian and Chinese launch vehicles, is going to cause a drop in launch costs of perhaps as much as 25% from current levels. Though this is welcome, it is not sufficient to permit the development of new space based ventures, such as power generation for use on Earth, manufacturing and tourism, which have been discussed and dreamed about by space enthusiasts for many years. But, if the cost of access to space could be reduced by a factor of ten or more, there is little doubt that significantly more could be done for humanity for the same or even less funding. Therefore, I suggest that a very powerful indirect way that the participants in this Symposium can assist space to be of benefit to humanity is by strongly advocating programs which will result in significantly reducing the cost of access to space.

4. Benefits From International Cooperation

There is one final area that I would like to touch on briefly in which, I believe, benefits to both humanity and cost reduction can be gained. This is through international cooperation. The space industry worldwide, because of its nationalistic base, is full of duplication, redundancy and problems with the "not invented here" syndrome. Given the national security nature of much of space technology, especially launch vehicles, there was, perhaps, a good reason for this. However, it must be recognized that this fact leads to a significant amount of inefficiency and resultant higher costs.

There is a very successful history of international cooperation on science programs. Recently, larger programs such as the space station and polar environmental satellites are being carried out in internationally cooperative ways. In my view, if the true benefits which could be derived from international cooperation are to be realized, cooperative activity must start earlier in the process. By this I mean that an agreement should be made to conduct basic research on a jointly funded basis in areas such as launch vehicle technology where the aim is to reduce launch costs significantly (sharing of results from the RLV and FESTIP programs between the US and Europe would be a start). The basic concept behind such an international program is that getting off the planet at a low cost is a goal for humanity, not a specific country or company. Therefore, a pooling of funds and joint management of a program dedicated to this purpose makes sense.

The primary argument against this approach is the desire of one country or company for leadership in space and the belief that being part of an international program would cause them to lose a competitive advantage. This argument has successfully blocked many international efforts to date, but it is not really valid. First, for basic research, there are a number of examples of research efforts, though not international ones, in the computer industry and elsewhere where multiple, competitive companies have pooled funds to conduct basic research on a collective basis very successfully. Secondly, if it were invented, any technology used for truly inexpensive access to space would be like the technology of an internal combustion engine - it is basic, and knowledge of the technology involved does not prevent the automobile industry from being global and highly competitive. Finally, the cost of developing a really new, very low cost space vehicle will not be small. Without some kind of pooling of funds on an international basis, it is unlikely that any single country or region will be able to provide the resources needed to undertake the project. However, collectively, it may be feasible. Therefore, one can argue that there is a strong probability that cheap access to space will never happen without some form of an international pooling of resources.

5. Conclusion

If space people wish to take up a cause, I would suggest they forget returning to the Moon or going to Mars, at least in the near term. The strong believers should focus their total energy on achieving a much more tangible and mundane, though perhaps as difficult, goal - reducing the cost of access to space. If that were to happen, a large part of the question asked in the subtitle of this paper, "Who will pay the bills?", will be answered. The bills will be a lot easier to pay because they will not be as large. The funding sources in government will be easier to tap because the benefits will be more understandable and visible. It is also possible that much of what is now asked of governments will be available through commercial channels.

In other words, I believe that government money is necessary and should be used primarily to reduce the cost of access to space significantly. Once this happens, in my view, industry will be able to do much more to provide products and services of value to humanity from space without further government assistance. This will allow the available government funds to be used for space science and other programs which are in the public good. This approach requires a long-term view and, I recognize, difficult choices in the near term. It may be, however, the only way actually to fund efforts to use space to serve humanity in the future.

Preface to the Proposal Summary of the 1995/96 Master of Space Studies (MSS) Team Design Project

R. Thomas, MSS Team Design Project Faculty Chair 1995/96, International Space University

Each year in the International Space University's Master of Space Studies (MSS) program, the students engage in a team-based activity to produce a report or product of significance to the international space community. In doing so, the students develop their skills in intercultural and interdisciplinary communication, teamwork, and knowledge synthesis. This work prepares them to be especially effective in their subsequent professional careers. Team Design Project activities are also expected to provide a fresh look at developing and/or utilizing space science and technology.

The early stages of such a fresh look for the MSS class of 1995/96 are summarized in the following paper. This document was prepared by the MSS students themselves for presentation at the ISU Symposium *Space of Service to Humanity*. The paper summarizes proposed Team Design Project work on the students' chosen topic: space applications to sustainable development and environmental protection.

This year's Symposium provided a key input to the MSS students in developing their subsequent team project work. After the Symposium, the students decided to narrow their focus to desertification and deforestation. Their work concentrated upon exploring how to improve access to data in Developing Countries and to the information needed to combat these two phenomena (especially desertification), and upon developing systems for monitoring desertification. The students' analyses and recommendations will be presented in a final report to be published in August 1996 by the International Space University.

Space of Service to Humanity: Master of Space Studies 1995/96 Team Design Project Proposal Summary

The 1995/96 Team Design Project members are: **W. Balogh, A. Berinstain, S. Challis, P. Clerc, M. Davis, I.M. De Angelis, C. De Matos, L. Falvella, O. Ferrandon, G. Fujii, D. Gendre, J-C. Grenouilleau, N. Hadj Sahraoui, F. Haessig, H. Kagawa, S. Mei, A. Metzger, D. Morros, J. Mugwe, G. Pagel, G. Pandozy, J. Riedman, F. Rosa, D. Saint-Pierre, C. Stott, F. Toschi, R. Uda, L.Valignon, O. Vogt**

1. Introduction

During the 1960's the Government of Brazil launched a campaign to colonize the Amazonian region, an area of South America barely touched by humans. Before long, the rapid economic development of the region and the associated population growth produced dramatic effects on the environment.

Satellite images of the Amazonian region began to show disturbing deforestation patterns. The images demonstrated that most of the deforestation was due to clearings for cattle raising and agricultural uses and not for fuel production. The Government legislated to discourage the cutting of trees. By the end of the 1980's the deforestation rate was reduced significantly. The Brazilian Space Agency, INPE, was able to measure these changes using images from Landsat satellites. With the assistance of accurate data, the Government of Brazil is pursuing a "sustainable development" policy of allowing economic development of the Amazonian region, without endangering the environment. INPE continues to monitor the situation.

At the end of the 20th century, Earth is beset with environmental problems. The huge economic growth of the Western world following the Second World War resulted in the massive consumption of fossil fuels and other non-renewable forms of energy. The by-products of the consumption of fossil fuels have shifted the fragile equilibrium of the Earth's atmosphere. Our large cities are choked by atmospheric pollution and our natural resources are polluted by industrial waste. Industrial by-products have caused a depletion of the atmosphere's ozone layer. There is evidence to suggest that Earth is undergoing an unnatural warming. Unless global solutions are identified and implemented urgently, the ability of this planet to support humankind in the future is in question. Economic, social and ecological disaster on a large scale is a very real possibility.

Seventy-five per cent of the world's population, living in the "developing world", consumes twenty per cent of the world's resources. By the year 2025, the population of the "developing world" will have increased by two thirds, whereas no significant increase of the population of the "developed world" is expected.

Economic growth in the "developing world" will put increasing pressure on the world's environment. The rapid growth of energy consumption per capita will cause enormous "thermal pollution". The increasing population of the world will exacerbate the world's food supply problem.

To avoid potential disaster, deforestation, soil salinity and desertification must be arrested. Better land use techniques must be developed. Water resources must be properly managed. Urban pollution must be reduced. Toxic industrial by-products must be minimized. Education must be accessible to all humans. Political conflict must not be allowed to interfere with environmental solutions. A policy of sustainable development is needed to protect the Earth's environment and natural resources, while supporting the economic growth of the "developing world".

Science now gives us the potential to solve some of the world's major problems. It allows us to formulate prognostic models from which solutions can be derived. As a result of advances in scientific knowledge, we are better able to understand the impact of a proposed solution on other systems.

Space-aided technologies are scientific applications which offer unique advantages. Space transcends political boundaries. It is no longer an arena of conflict between nations and is now a forum for international cooperation. It is a symbol of humankind's quest for a better future.

Space applications provide a global coverage of the Earth. They overcome the geographical limitations of terrestrial observations and measurements. A multiplicity of wavelengths can be used to gather different information about the same object or event. They provide objective, accurate, measurable and repetitive global data which cannot be controlled by any single government.

As the Brazilian example demonstrates, space applications can be used to define the problem and to aid the solution. To implement corrective actions in the near future, decision makers must be provided with adequate information in a timely fashion. Because of the global nature of these problems, the generation and distribution of the information will need to be as wide as possible.

A conviction that space must be used to serve humanity on Earth has led the members of the Team Design Project to select the themes of environmental protection and sustainable development for this year's Team Design Project. The challenge we face is to design solutions which will help to harmonize these themes and which will make a significant contribution to the preservation of Earth for the benefit of future generations.

2. Technical Approach

The following step-by-step approach is proposed to meet the overall goals of the project:

- Definition of the proposed Program objectives. They will be defined after research into the broad areas of sustainable development and environmental protection and an analysis of past, present and planned programs addressing those issues. The selection of issues will be based on criteria developed by the project team.

- Mission Design. Existing and planned programs will be assessed to identify data, technology, social and political gaps. Based upon clearly identified needs, an approach or program, including space and ground segments, will be proposed.

- An implementation plan, including a financial plan, a training plan, a Pilot Program, a data policy and an organizational structure. This phase will also evaluate the political, economic, legal, social, and cultural impact of the Program upon possible user communities and partners.

The project will start on April 24, 1996. However, pre-project activities are planned to aid the project team to focus on a subset of possible issues related to environmental protection and sustainable development. An initial research phase and interaction with world space experts during the ISU Symposium *Space of Service to Humanity* will allow the project team to develop a comprehensive list of issues before proceeding to the main research phase.

Considerable resources will be used during the different phases of the project. The expertise of resident ISU faculty members and at ISU Affiliate campuses with expertise in environmental protection and sustainable development will help to focus the research and guide the project team. International organizations such as the United Nations Organization, non-government aid organizations and space agencies will represent direct sources of information and assistance.

3. Management

An innovative management structure is proposed, adapted to the unique characteristics of the project team. In particular, the proposed structure distributes responsibilities and authority equitably and gives all team members the opportunity to be involved in the coordination of the project as well as in the hands-on research, design and implementation.

4. Deliverables

The deliverables of the project will include:

- an assessment of the issues that are inherent to both environmental protection and sustainable development;
- a review of current and future programs aimed at tackling these issues;

- a review of the benefit of space-aided assets in terms of the these issues;
- a complete program to address the issues, including a mission analysis using space assets;
- a promotion and financial plan;
- an implementation plan including a training plan, a Pilot Program, a data policy, and organizational structure;
- a report on the process of the project including an outline of the project history, the management structure, key decisions made and lessons learned.

5. Schedule

The project will commence on April 24, 1996 and the final presentation will take place at the ISU Central Campus on July 24, 1996. The project team plans to spend a total of 7360 hours on the project, representing 3.5 person-years of effort.

Report on Panel Discussion 1: Global Issues

S. Challis, **G. Pandozy**, 1995/96 Master of Space Studies (MSS) students, International Space University

Session Chair: **J-B. Mennicken**, DARA GmbH, Cologne, Germany

The first day of the Symposium *Space of Service to Humanity* was dedicated to papers concerning Global Issues. During the course of the day, the speakers discussed many important issues concerning space and its role in serving humanity. This was reflected in the richness of the panel discussion. A lively and interesting debate was fueled by a variety of fields, including space solar-powered satellites, technology transfer and space commercialisation, as well as a variety of philosophies regarding space activities.

J-B. Mennicken began by announcing that the panel discussion would be divided into two sections. The first section would be dedicated to general questions concerning the issues already presented by the speakers, whilst the second would be exclusively for discussion on the MSS Team Design Project.

The debate opened with a question concerning the rationale behind space power satellites. A member of the audience questioned the need for such satellites to generate power, believing that solar power devices could be used as effectively on Earth. **P. Creola** (who in his paper had referred to microwave transmission of power generated from space power satellites) was quick to point out that a considerably smaller amount of sunlight per unit area could be used directly on Earth than that which could be harnessed in space. He also pointed out that a large percentage of the Earth's surface would be needed to help generate the level of power required to maintain an explosive world population, and this would have a significant impact upon other crucial land uses such as agriculture and urbanisation. In support of this view, **G. Gibbs** emphasised the potential behind the concept, stating that NASA alone was currently investigating at least three design architectures. A second member of the audience questioned the feasibility of solar power satellites by indicating the potentially high operating costs of such a system.

Choosing to respond, **J. Egan** conveyed the opinion that the major cost would be for the launch and construction of a solar power satellite; operational costs of the satellite would be considerably smaller in comparison. He also pointed out the need to assess the cost of disposal, an issue which he felt had not been discussed enough. Finally, he emphasized the need to cut the cost of launches which, he commented, *"would change our whole philosophy in space."*

G. Soffen was asked to explain the "face" on Mars. Smiling, he replied that the "face" had simply arisen in the imaginative minds of Viking engineers when they were analysing the data from the Martian Orbiter. He emphatically stated that there is no geological evidence to prove that it is anything other than a natural formation. G. Soffen then asked the space professionals present to predict how many "space people" there would be ten years from now and whether they thought the field would expand or contract. **P. Wood** responded optimistically, giving a figure three times greater than today's work force, but **G. Khozin** underlined that this forecast would not be valid for Russia in light of the current socio-economic climate. Before moving on, **J-B. Mennicken** pointed out that a clear definition of "space people" should be made before answering the question, since the term could also include people working in the media, telecommunications, and other businesses.

The next question concerned the issues and problems related to technology transfer in the space field. **V. Hood** spoke of the success of ESA training programmes for remote sensing technology demonstrations in Asia, especially for the production of rice in Indonesia and Thailand, and the hope that these programmes would be available to all nations in Asia in the future. However, she indicated that continued funding for these programmes was never certain.

M. Davis then spoke of the relationship between technology transfer and space activities in Australia. He explained that industry in Australia had initially been reluctant to participate in space activities but, with government subsidies, it had become possible to exploit both technology and intellectual property through industrial partnerships.

J. Egan had a different opinion on the role of governments in trying to stimulate technology transfer. In his view, the results had been unsuccessful, believing that governments had focused too much on the transfer of technology without considering the rationale for such activities.

Expanding the debate, an audience member shared his concerns on the apparent shift in space activities from fundamental research to commercialisation: *"The pendulum has swung too far."* He believed that the motivation to invest in research no longer exists and that Space of Service to Humanity can only be achieved with further extensive fundamental research.

J. Egan responded that, in his view, industry will only work on projects that assure a profitable return and that a government's idea of fundamental research does not match that of industry. He returned again to the issue of launch costs and the outrageous expense of space travel. He believed that space is expensive because people have decided to make space special and that things should be different.

G. Soffen was more optimistic, believing that through programmes such as the Hubble Space Telescope young people are being inspired to create a new vision for space exploration and research.

P. Creola spoke of funds as the central problem, and emphasised the need to find a balance for both practical and emotional objectives in space. He concluded: *"We have good space missions with huge budgets and good space missions with small budgets."*

An MSS student in the audience conveyed feelings similar to those of the audience member who had begun the discussion on fundamental research and commercialisation. She stressed her belief in the importance of fundamental research and her fear that dreams and visions are being abused in order to sell products.

J-B. Mennicken closed this section and began a discussion on the theme for the MSS Team Design Project, calling upon the panelists to give their opinions and advice. The feeling from the panelists was that the students' choice of focus for the project, sustainable development and environmental protection, was ideal for Space of Service to Humanity. However, students were advised to be wary of the mistakes made by past space initiatives. There was a general feeling that the students should focus their efforts on trying to convince the decision-makers of the role which space has to play in serving humanity, and that the students should consequently focus on a single fundamental issue.

The panel discussion was dominated by controversy over the importance of space activities whose economic return is neither easily defined nor immediate, together with those projects which would be difficult for industry or government to implement. Also raised were the issues of technology transfer, fundamental research and commercialisation in the space sector.

The general feeling from the panel was that the importance of both fundamental research and commercialisation in space could not be ignored, and that the world's governments and businesses have to find a balance between the two. However, there seemed to be no clear opinion on the benefits of technology transfer and the involvement of governments in space activities.

Session 2
Space of Service: Remote Sensing

Session Chairs:
P. Eymar, Corporate Director, Sales & Marketing, Aérospatiale Espace & Défense, France
T. Kuroda, Corporate Chief Engineer, NEC Corporation, Japan

Past, Present and Future Benefits of the ESA Earth Observation Programmes

P. Gilson, P. Osterman, ESTEC (European Space Research and Technology Centre), European Space Agency, Postbus 299, 2200 AG Noordwijk, The Netherlands

Abstract

Since the launch of its first Earth Observation satellite, METEOSAT-1, in 1977, the European Space Agency (ESA) has largely contributed to the availability of satellite remote sensing data for Earth sciences and operational applications for the international community. This paper will present: (1) a review of the different elements of the past ESA Earth Observation programmes, from METEOSAT-1 to ERS-2. This includes a summary of the main scientific and operational achievements resulting from these missions; (2) a description of the missions currently in preparation in ESA, in order to ensure the continuity of data availability for the next 10 years, i.e. MSG (METEOSAT Second Generation), ENVISAT-1, and EPS/METOP; (3) a description of the candidate Earth Observation mission profiles addressing user requirements beyond the year 2000. This includes the proposed Earth Explorer missions (research or demonstration missions for Earth sciences) and the proposed Earth Watch missions (pre-operational missions for emerging Earth Observation applications).

The focus will be on the benefits resulting from these missions for the scientific community and the populations of the international community, rather than on the detailed technical description of the satellite systems.

1. Background

Over the past three decades, Earth orbiting systems have grown enormously in number, complexity and utility. Many of us make daily use of space systems for long distance communications or for planning weather-dependent activities. Satellite systems providing data for weather forecasting or intercontinental telecommunication links are now sufficiently reliable and cost-effective to work in a fully operational manner. However, space technologies are far from delivering today the full range of services within their capabilities. Substantial research and development work is currently being carried out, and more will be necessary, to allow further applications to become possible in the future.

As all new technologies, remote sensing techniques evolve through a series of maturity levels such as research, demonstration, pre-operational, operational, and commercial.

In Europe, no Earth Observation space system has yet reached the true commercial stage. This is mainly due to the substantial financial resources needed for their development. Although the French SPOT system is selling pictures on a commercial basis, it has been developed using governmental funds.

The other systems have mainly been developed by European governments through international cooperation, via the 14 nation European Space Agency (ESA).

As is the case in other disciplines, e.g. launchers and telecommunications, the ESA programmes are focused on the development of the space systems up to and including the pre-operational stage. When a space system is mature enough to reach the operational stage, it is handed over to a separate organisation, such as Arianespace, EUTELSAT or EUMETSAT. Therefore, the operational activities are in general not the responsibility of ESA.

In this context, the benefits for the international community of the ESA programmes are of several types. In addition to fostering international cooperation and European space industry competitiveness, which are benefits common to all ESA programmes, the ESA Earth Observation programmes have specific benefits which can be grouped as follows:

- scientific achievements relating to our understanding of the Earth environment,
- demonstration of new techniques for future services to the populations,
- operational achievements of systems originally developed by ESA.

The purpose of this paper is to outline in broad terms the past, present and future Earth Observation programmes of ESA, focusing mainly on their benefits for humanity.

2. ESA Earth Observation Programmes Launched or in Development

The four fundamental objectives underlying the Agency's present Earth Observation programmes are:

- the study and monitoring of the Earth's climate and environment on various scales, from local or regional to global,
- the monitoring and management of the Earth's resources, both renewable and non-renewable,
- the continuation and improvement of the service provided to the worldwide operational meterological community,
- the contribution to the understanding of the structure and dynamics of the Earth's crust and interior.

Between these objectives span all four Earth Observation disciplines (i.e. land, atmosphere, ocean/ice and geodesy/geophysics).

The following paragraphs will give a short description of the missions implemented by ESA to meet these fundamental objectives. Then, the last part of this paper will review the benefits of these missions.

2.1 METEOSAT

Ever since the Television and Infrared Observation Satellite (TIROS-1) sent back its first cloud formation images in 1960, Earth Observation satellites have provided a unique view of the weather on our planet. At the suggestion of the United Nations, a complete network of meteorological satellites has been set up to improve weather forecasting capabilities. Since the end of the 1970s, this system has provided coverage of the Earth by means of satellites in polar and in geostationary orbit.

- **METEOSAT - Pre-Operational Programme.** Following the decision of the European meteorological community to enter the field of space meteorology in the early 1970s, METEOSAT-1 was launched in November 1977. Placed in geostationary orbit at 0° longitude, it provided a permanent view of most of Europe, the whole of Africa, the Middle East and the Eastern half of South America (in total over 100 countries). This became one of the geostationary satellites which consitutes the basis of the World Weather Watch of the Global Atmospheric Research Programme. METEOSAT-2 was launched in June 1981 to replace METEOSAT-1, followed by METEOSAT-3 in 1988.

- **METEOSAT Operational Programme - MOP.** As financing and operating satellites is not the role of ESA, a dedicated organisation, EUMETSAT, was set up for this purpose in 1986. Since then, ESA's responsibilities include the construction and launch of the satellites on behalf of EUMETSAT which owns and finances the operational spacecraft. The satellites were also operated by ESA until November 30, 1995, when the new EUMETSAT ground segment took over this responsibility. The first satellite of the METEOSAT Operational Programme, MOP-1 (METEOSAT-4), was launched on 6 March 1989, followed by MOP-2 (METEOSAT-5) in March 1991 and MOP-3 (METEOSAT-6) in November 1993.

- **METEOSAT Transition Programme - MTP.** The last spacecraft in the initial METEOSAT series will be METEOSAT-7, to be launched in 1996/97, this time under a EUMETSAT contract.

- **METEOSAT Second Generation - MSG.** Since the mid-1980s, ESA and EUMETSAT have initiated ambitious preparatory programmes aimed at a drastic improvement of the METEOSAT mission towards the beginning of the next century. As a consequence, the MSG system has been proposed. The second generation of METEOSAT satellites will thus provide continued and improved meteorological data, resulting in enhanced benefits for hundreds of millions of people. It will also have the capability to support climate and environment monitoring and Earth resources management. MSG-1 is currently in its development phase. Its launch is planned for the year 2000.

The METEOSAT system has a threefold mission:

- to construct accurate images of the Earth in three spectral bands (visible, water vapour and thermal infrared) at 30 minute intervals. The visible image has a spatial resolution of 2.5 km, while the other channels have a 5 km spatial resolution;
- to collect environmental and meteorological data from up to 4000 measuring stations on ships, planes or buoys and to transmit them to a central facility;
- to transmit processed images and meteorological data directly to the user stations in more than 130 countries. The 75 primary user stations are mainly meteorological offices receiving high resolution data in digital format. In addition, several thousand smaller user stations located in Europe, Africa and Asia receive data in analogous format. Since 1991, when METEOSAT-3 was positioned over the Americas at the request of the National Oceanic and Atmospheric Administration (NOAA), the US weather service, to replace an American weather satellite, the number of METEOSAT users has increased even further.

The future MSG will offer the following improvements on the original METEOSAT system:

- enhanced multispectral imagery (12 channels instead of 3);
- higher resolution imagery (1 km resolution in the visible band and 3 km in the other bands, compared to 2.5 and 5 km for MOP);
- higher temporal resolution (continuous image-taking in the various channels with a baseline repeat cycle of 15 minutes).

These new features will provide to the users improved image accuracy and products such as wind velocity vectors or surface temperature, as well as new atmospheric data such as air mass analysis by using some H_2O and CO_2 absorption channels. In addition, the capability to collect data from Data Collection Platforms and to disseminate meteorological data to primary and secondary user stations will be maintained. Finally, provisions are also made on board to carry as a passenger the "Geostationary Earth Radiation Budget" scientific payload, and to perform a search and rescue function by receiving beacon signals and relaying them to surveillance stations.

2.2 ERS

ERS-1, the first European remote sensing satellite, was launched by ESA on July 17, 1991. Its almost twin successor, ERS-2, was launched on April 21, 1995. Both satellites are currently operated in "tandem" mode. This unique opportunity should continue at least until the end of May 1996.

ESA's original objectives for the ERS mission were both scientific and economic. The three primary objectives were to:

- increase our scientific understanding of coastal zones, global ocean processes and polar regions. Major contributions to the World Climate Research Programme were foreseen, and it was anticipated that the data, used in conjunction with in situ or other satellite measurements, would enable significant advances to be made in physical oceanography, glaciology and climatology;
- develop and promote economic and commercial applications. Based on an improved knowledge of ocean parameters and sea-state conditions, surveillance and planning of coastal and marine activities worldwide was defined as a primary target for applications development;
- explore the potential of radar data for land process studies and applications.

For this purpose, it was decided to use a low altitude (800 km) polar orbit, and the following sensors were selected for ERS-1:

- a synthetic aperture radar, providing all-weather, day and night imaging of the Earth's surface and ocean surface wave spectra,
- a wind scatterometer, providing wind speed and direction at the ocean surface,
- a radar altimeter, providing accurate sea and ice surface elevation, indications of wave heights, various ice parameters, sea surface wind speed and gravity information from the sea bed,
- a scanning radiometer, providing accurate sea surface temperature, cloud top temperature, cloud cover, water vapour and land surface temperature.

The sensors of ERS-2 are identical to those on board ERS-1, except for the following additions (Reference 5):

- additional channels (visible and near IR) were added to the scanning radiometer to provide data for vegetation and land temperature studies,
- a new Global Ozone Monitoring Experiment (GOME) was included.

The number of ERS ground stations is also expanding. There are now mobile stations that allow data to be provided for areas not previously covered, such as Central and Eastern Africa.

2.3 ENVISAT

In order to capitalise on the experience gained from the ERS satellites and to provide continuity of data after these missions, ESA has undertaken the preparation of an ambitious programme called ENVISAT. ENVISAT-1 is currently in its development phase, and its launch is planned for mid-1999.

ENVISAT-1 is mainly concerned with research objectives and the flight of pre-operational instruments. Its primary objective is to make a significant contribution to environmental studies, notably in the areas of atmospheric chemistry and ocean/ice studies (including marine biology of the ocean surface).

This will provide data necessary to produce better models of the Earth's climate. As secondary objectives, ENVISAT-1 will also seek to enhance the capability to monitor and manage the Earth's resources, and will contribute to a better understanding of solid-Earth processes. Finally, ENVISAT-1 should consolidate the ERS pre- operational mission, in particular as far as the monitoring of sea ice is concerned. A mix of 10 optical and microwave instruments constitute the payload selected to achieve these mission objectives (see Table 1).

AATSR	Advanced Along-Track Scanning Radiometer (VIS + IR)
ASAR	Advanced Synthetic Aperture Radar
GOMOS	Global Ozone Monitoring by Occultation of Stars
MERIS	Medium Resolution Imaging Spectrometer (VIS + near IR)
MIPAS	Michelson Interferometer for Passive Atmospheric Sounding
RA-2	Radar Altimeter
SCIAMACHY	Scanning Imaging Absorption Spectrometer for Atmospheric Chemistry (UV + VIS + IR)

Table 1. The most important ENVISAT-1 instruments

ENVISAT will represent the European element in the International Earth Observation System. Together with ADEOS from Japan and EOS from the USA, this will provide the main information source for scientists to observe, monitor and understand Earth processes in the first part of the 21st century.

3. Future ESA Programmes

3.1 EPS/METOP

Meteorological data from satellites in polar orbit have been made freely available to the rest of the world by the US for more than 30 years, and on a fully operational basis for nearly two decades. The NOAA is responsible for the US civilian polar orbiting meterological satellite programmes.

NOAA has indicated that it will discontinue operation of one of its two polar orbiting Sun-synchronous meteorological satellites by the end of this millennium. Through several years of cooperation between NOAA and EUMETSAT and to ensure long term continuity of a full operational polar-orbiting satellite system, the EUMETSAT council decided to prepare a EUMETSAT Polar System (EPS) proposal to meet the objectives of an operational meteorological and climate monitoring mission which, together with the geostationary meteorology satellites, will provide a more complete service weather forecasting and climate monitoring.

The EPS/METOP mission objectives are tailored to meet long-term European operational requirements of meteorological and climatological data from polar orbit. EUMETSAT and ESA have together worked out a spacecraft specification that will meet the requirements in a "morning" polar orbit. The programme proposal is currently foreseen to include three satellites. EPS/METOP-1, which ESA proposes to develop for EUMETSAT, is planned to be launched not earlier than the year 2001, and together with the two follow-on missions, it is intended that uninterrupted services will be ensured up to the year 2014.

The main candidate instruments currently being considered are the following:

- **Advanced Very High Resolution Radiometer (AVHRR).** Ocean measurement (Sea Surface Temperature - SST), clouds and Earth radiation budget, land measurements (snow cover, Land Surface Temperature - LST, vegetation).

- **Multi-Frequency Imaging Microwave Radiometer (MIMR).** Ocean measurement (SST), land measurement (snow cover, LST, vegetation, soil moisture), sea ice (extent, concentration, type), cloud and Earth radiation budget (cloud liquid water), precipitation (parameterisation), water vapour content.

- **High Resolution Infrared Radiation Sounder (HIRS/3).** Atmospheric minor constituents, upper tropospheric water vapour, cirrus clouds.

- **Advanced Microwave Sounding Unit-A (AMSU-A).** Sea ice.

- **Microwave Humidity Sounder (MHS).** Clouds and Earth radiation budget, sea ice.

- **ARGOS DCS-2.** Data collection and location system.

- **Advanced Infrared Sounder (IASI).** Ocean measurement (SST), cloud and Earth radiation budget, atmospheric minor constituents, land measurement (snow cover, LST, vegetation)

- **Ozone Monitoring Instrument (OMI).** Atmospheric minor constituents (ozone content and profile).

- **Advanced Scatterometer (ASCAT).** Ocean measurements (surface wind vectors, surface stress, fluxes and air/sea interchange).

3.2 The Earth Explorer and the Earth Watch Missions

In order to meet the complex requirements from a growing community of users from different sectors (public or private, research or application), the Agency has proposed to depart from the multi-objective programmes leading to large satellites such as ENVISAT and to adopt a dual mission concept. This concept is foreseen to be developed within a new optional framework programme. Following these requirements, the ESA Ministerial Council in October 1995 approved basically two distinctive schemes, the Earth Explorer and the Earth Watch.

The four fundamental objectives underlying the Agency's present Earth Observation programmes, mentioned in the introduction of this paper, can be complemented by a fifth objective reflecting the growing importance of Earth Observation for user applications: the initiation and consolidation of the services for application communities with emerging needs for Earth Observation space data.

These fundamental objectives meet the increasing need and concern from scientific and public users, i.e.:

- the rate of environmental and climate change are likely to be increasingly influenced by mankind's activities;
- environmental and climate changes and natural disasters will have increasingly serious social and economic impacts;
- the continuous monitoring and management of Earth resources will be increasingly needed to ensure a sustainable development.

These concerns impact directly on the general well-being of mankind. In the post-2000 era, the Agency's Earth Observation programme must take full account of these concerns and the associated data requirements.

(a) The Earth Explorer Mission

A research and demonstration programme focusing on advanced understanding of different Earth system processes. The mission will support primarily scientific research, but could also address research needs of operational entities. Each mission will focus on a particular research field or regroup a limited number of research fields. The demonstration of specific new observing techniques and associated technologies could also fall under this category. Earth Explorer missions will entirely be funded by ESA, with mission durations tailored to the specific mission requirements, and including the possibility of providing instruments through announcement of opportunities. This strategy could lead to possible Earth Explorer missions approximately every three years starting in 2004.

(b) The Earth Watch Mission

A programme with a pre-operational/operational character to correspond to the needs expressed by user organisations and conducted in cooperation with these organisations. The responsibility for this type of mission will be transferred to (European) entities providing operational services. The time for this transfer will depend on the maturity of the user community and on the related status of the operational entity. To ensure such a scenario, data continuity over a period of about 10 years would be required. It is intended to have an Earth Watch mission every 5 years, starting in 2005.

These two schemes will provide the stability and continuity which is needed by the users. In support of this dual mission concept, the associated ground segment will evolve toward a homogenous framework of decentralised facilities. Together with a close cooperation between national programmes and other European entities and global initiatives, the long term goals set out in the European Policy for Earth Observation from Space are safeguarded.

In the dual mission strategy, focus will be on:

- consolidation of the achievements and investments of currently approved Earth Observation programmes by ensuring the continuation of the provision of essential data and deeper involvement of user entities;
- implementation of missions to address research as well as existing and emerging fields of applications;
- evolution of the ground segment to improve the user access to and availability of Earth Observation data and products;
- improvement and enhancement of the awareness and utilisation of Earth Observation data products including a stronger involvement of the private sector and value-added industries.

The Earth Explorer and Earth Watch missions will cross-fertilise with each other in that the former could demonstrate new technologies which later on could be used for routine operational applications. On the other hand, the Earth Watch mission could provide data to support scientific activities also.

4. Benefits

The applications of the missions described above, and the activities which will benefit the most, are summarised in Figure 1. A fuller description follows.

4.1 The Benefits of METEOSAT

The most well-known benefit of operational meteorology is the provision of improved weather forecasts. METEOSAT data play a central role and have contributed significantly to improvements in weather forecasting. Thirty years ago, weather forecasts were limited to about 24 hours ahead in areas a few hundred kilometres across. Meteorologists used in situ measurements of atmospheric parameters such as temperature, humidity, wind speed and direction, and barometric pressure to draw barometric maps and make an "educated guess" based upon linear approximations and experience. The major sources of better weather forecasts today are the improved numerical models and more powerful computers, but also the availability of meteorological data from satellites. For example, measuring cloud movements as recorded in consecutive METEOSAT images gives wind speeds at altitude; the intensity of the signal in the thermal infrared channel indicates cloud-top temperatures; the analysis of cloud structure contributes to identify underlying weather phenomena like hurricanes, thunderstorms or blizzards. A study performed by Dr J. Thornes of Birmingham University has concluded that METEOSAT data account for approximately 20% of improvements to the weather forecasts in Europe.

Applications			MPP	MOP MTP	MSG	ERS-1	ERS-2	ENVISAT	METOP	Earth Explorers	Earth Watch
Science	Climate modelling & monitoring	Oceans Ice Processes and interactions Atmosphere (Energy transfers, water cycle, ...) Land	Y - Y Y	Y - Y Y	Y - Y Y	Y Y Y Y	Y Y Y Y	Y Y Y Y	Y Y Y Y	YES (depending on the mission)	yes (depending on the mission)
	Environment modelling & monitoring	Land (Vegetation, ...) Oceans (Biomass, ...) Atmosphere (Chemistry, Ozone, ...)	y y y	y y y	y y y	Y - -	Y - Y	Y Y Y	- y y	YES (depending on the mission)	yes (depending on the mission)
	Solid Earth	Topography Crust dynamics Geoid	- - -	- - -	- - -	Y Y Y	Y Y Y	Y Y Y	- - -	YES (depending on the mission)	yes (depending on the mission)
Operational	Meteorology	Land (Weather forecast) Oceans (Ship routing, ...) Data dissemination	P P P	O O O	O O O	- DP -	- DP -	- dp -	O O -	yes (depending on the mission)	YES (depending on the mission)
	Resources Management	Land (Agriculture, forestry, cities, mapping, ...) Oceans (Fishing, off-shore, ...)	- p	- o	- o	DP DP	DP DP	dp dp	o o	yes (depending on the mission)	YES (depending on the mission)
	Search & rescue		-	-	o	-	-	-	o	-	TBD
Exploitation			1977 - 1989	1989 - 2000	2000 +	1991 - 1996	1995 - 1999	1999 - 2004	2001 +	2002 +	2004 +

Figure 1. Synopsis of ESA Earth Observation missions and their applications

Legend: UPPERCASE = primary mission; lowercase = secondary mission;
"Y" = Yes; "-" = No; "D" = demonstration; "P" = pre-operational; "O" = operational

From an economical point of view, the majority of the benefits from these forecasting improvements arises from the following four economic activities:

- **Construction**. By using weather forecasts to plan the use of resources more carefully and to schedule critical operations, such as pouring concrete, construction companies can substantially reduce their costs.

- **Transport.** Land, air and maritime transport are obviously in some way dependent on the weather. Timely knowledge of impending bad weather enables roads to be cleared of ice and snow, reducing accidents and loss of life, while minimising environmental damage caused by salt and de-icing chemicals. Routes of aircraft and ships are now adjusted in real time as weather fronts develop, and knowledge of wind speed and direction allows efficient aircraft fuelling.

- **Agriculture.** Regular spraying of fields with pesticides and herbicides is a costly operation. Success can be jeopardised by rainfall washing chemical products from the plants. In addition to the economical benefits resulting from the weather forecasts, contamination of ground water is also reduced because less chemical products are washed off the plants.

- **Energy production**. Experts estimate that today's power generation capacity would have to be increased by 2-3% if there were no weather forecasts. The anticipation of energy demand peaks related to weather changes allows us, for example, to minimise the need to liquefy and store unnecessary gas for pipeline pressurisation, or to optimise power plant operations such as reservoir filling, planning of maintenance work and cooling water supply.

A study has compared the average annual expenditure on METEOSAT with the estimated annual benefits to the economy after 10 years of the METEOSAT operational programme. It concluded that the return factor is more than 2.5, for the ESA member states' economies alone. As METEOSAT imagery is used in many developing countries and by the US weather service, the figure for the benefits on a global scale must be considerably higher. Additional benefits which are even more difficult to quantify, are the minimisation of loss of life and goods by early warning of natural disasters such as hurricanes and blizzards, as well as flood or desert locust invasions. Finally, public weather forecasts on television, radio and in the newspapers allow us all to plan our lives a little more carefully and efficiently.

4.2 The Benefits of ERS

(a) Scientific Achievements

As far as the scientific achievements of ERS-1 & 2 are concerned, a plethora of results have been published (see References 1 to 3 for examples). These scientific achievements improve our understanding of oceanic, atmospheric, sea ice, polar ice, land and Earth's crust processes, and their interactions. They contribute to the global effort undertaken to study and monitor the Earth's climate and our global environment. In particular, the radar altimeter measurements of the ocean and the polar ice sheets, accurate to a few centimetres, and the 0.5°C accurate sea surface temperature measurements of the radiometer are invaluable tools to monitor and detect ocean circulation features, polar ice balance and global warming phenomena. Similarly, the GOME instrument provides accurate measurements of the ozone content of the atmosphere, allowing us to study and monitor the effects of human activities on its depletion. ERS-1 and ERS-2 are thus playing a key role in the monitoring of our global environment, with the aim to better protect it for future generations.

(b) Experimental/Pre-Operational System

More than just a scientific tool, the ERS system was also originally intended as an experimental/pre-operational system forming the basis for Europe and Canada to move towards a fully operational system. This was justified in the longer term on the grounds of direct economic benefit resulting from the services provided. The primary industries which were originally expected to benefit from ERS included off-shore, shipping and fisheries. With more experience in using the data, the potential markets which could benefit are much broader, encompassing agribusiness, oil, gas and mineral exploration, civil engineering, and telecommunications.

A few examples of new techniques successfully demonstrated by ERS results are listed below, and further described in Reference 4. Wind scatterometer data have been shown to be capable of discriminating meteorological features with considerable extra detail compared with model background plots. These data are used for global scale numerical weather predictions on a quasi-operational basis. The SAR was able to demonstrate its potential for ship routing optimisation. On the one hand, the data on the ocean wave characteristics provided by the SAR, when it is operated in "wave mode", have been used to optimise the routing of ships. On the other hand, SAR images were used in several projects in Europe and Canada to detect open water areas within sea ice, and to monitor ice drift more accurately than with drifting buoys. These data were transmitted daily to ice-breakers on a pre-operational basis. These projects have demonstrated the capability of SAR images to contribute to ship routing during the ice season, and hence to become a key component of the ice monitoring system, e.g. for the Northern Sea Route.

ERS images were also used to demonstrate the capability of SAR to complement or improve traditional sources of data in a variety of activities such as:

- detection of water pollution such as oil slicks (under suitable weather conditions),

- land use applications (crop identification and growth monitoring; forest monitoring, in particular in tropical regions; land mapping, including digital elevation models, etc....),
- monitoring of natural disasters (e.g. flood development and recovery, or volcanic eruptions) and improvement in their modelling (earthquakes).

The ability of the ERS microwave instruments to penetrate atmospheric clouds and darkness at night have enabled images and data which are not available from optical sensors to be obtained. The demonstration of the capabilities and reliability of these techniques by ERS-1 and ERS-2 has opened the perspective for a better scientific understanding of our Earth's climate and environment, and paved the way towards new services in the future for the populations of the international community. The ERS space system will continue to provide these services until the launch of the next ESA Earth observation satellite in low Earth orbit, in 1999.

4.3 The Benefits of ENVISAT-1

With the exception of the wind scatterometer, which is not part of the ENVISAT-1 payload, the ENVISAT sensors will provide continuity and enhanced capabilities with respect to the corresponding ERS instruments (Reference 6).

In the study of the atmosphere, SCIAMACHY will be an improved version of the ERS GOME. It will be used in synergy with the new MIPAS and GOMOS instruments to determine the atmosphere's composition and allow the study of its associated chemical processes, including those affecting the ozone concentration at all altitudes, as well as "greenhouse gases".

For the study of ocean, land and ice, the microwave instruments will continue the ERS mission, with some improvements. One example is the better tolerance of the Radar Altimeter to changes in surface topography which will allow robust elevation measurements of the land surface. Other improvements are the multi-polarisation and variable look-angle/swath-width capabilities of ASAR, which are expected to improve significantly the discrimination of surface features such as vegetation.

Finally, the performance of AATSR will be almost identical to the ATSR-2 on board ERS-2. However, it will be complemented by the very flexible MERIS spectrometer which will not only study the "ocean colour", i.e. chlorophyll and other suspended matter, but also ocean pollution, coastal effects, sea ice, vegetation, agriculture, forestry, cloud characteristics, atmosperic water vapour content and aerosols.

In brief, the main benefits of ENVISAT-1 which can be predicted today are the amplification of the scientific and pre-operational achievements of ERS, with additional breakthroughs in the fields of atmospheric chemistry and land processes. But the ERS experience has shown that many more, unexpected applications will appear during the exploitation phase, which will help us to address present and future environmental and resource management issues.

4.4 The Benefits of EPS/METOP

The primary mission benefits of EPS/METOP will be achieved in the areas of operational meteorology and climate monitoring (Reference 7).

(a) Operational Meteorology

The frequent imagery from the geostationary meteorological satellites covers most of the globe, except the polar regions. METEOSAT provides images at intervals of 30 minutes; MSG will provide images every 15 minutes and primarily support short range local and regional forecasting activities. The polar satellites fly at much lower altitudes (around 850 km instead of 36000 km) and provide global imagery in much more detail but at less frequent intervals. Polar orbiting meteorological satellites complement the geostationary satellites by fulfilling additional objectives for operational meteorology, such as:

- **Global operational sounding**. Provides information on the three-dimensional fields of temperature and humidity of the atmosphere, to support Numerical Weather Prediction (NWP) models.
- **Global VIS/IR imagery.** Provides cloud imagery of forecasting applications, sea surface temperatures and global radiation budget parameters. Supports the global sounding mission through identification of cloud-free areas.
- **Global ocean surface wind vectors**. Provide wind speed and direction over the global ocean surface and supports the NWP systems by provision of these data in otherwise data-sparse regions.
- **Prediction of damaging ultraviolet radiation**. Predicting the ultraviolet radiation which may cause damage to life on Earth.
- **Forecasting of volcanic clouds.** Detection of volcanic aerosols and $S0_2$ ejected into the atmosphere by volcanic eruptions. These forecasts will enable the optimisation of route planning of modern jet aircraft.

(b) Climate Monitoring

Small climate changes will provide a challenge to survival over wide areas, through the effects of flooding, drought or pollution. The natural variation of climate over time makes it difficult to detect the long term implications of anthropological activities. Thus, it is necessary not only to monitor many parameters, but to do so over long periods of observation lasting decades or even longer.

In fact, many of the climatology requirements coincide with, or overlap with, the requirements for operational meteorology. The mission objectives for the monitoring of climate change, which are in addition to those already established for operational meteorology, are listed below:

- **Earth radiation balance**. Provides information on Earth's outgoing and incoming thermal radiation and the resulting global net radiative balance.
- **Sea ice monitoring**. Provides information on the coverage of ice and snow over the oceans for climate monitoring (in addition to ship routing in operational meteorology).
- **Ice and snow monitoring over land.** Provides information on the coverage of land surfaces by ice and snow for climate purposes (in addition to operational meteorology).
- **Global precipitable water mapping**. Provides information on the global distribution of precipitable water in support of climate monitoring (and NWP).
- **Global ozone mapping.** Provides information on the distribution of ozone in the upper atmosphere.

Secondary missions are the contribution to Earth science and operational surveillance. EPS/METOP will continue to support these areas as a natural evolution following the contribution from ERS and ENVISAT.

(c) Contribution to Earth Science

This will be ensured through support to:

- **Atmospheric research**, which could benefit from better understanding of atmospheric chemistry and radiation and energy balances.
- **Oceanic research**, which could benefit from better understanding of the general circulation of the oceans and the ocean heat transfer, both of which are known to exhibit strong seasonal variations.
- **Hydrological research.** The slow response of the climate is partly modulated through the slow turnover time of water *within* the oceans. The associated exchanges of water and energy determine the amplitude and regional pattern of the climate change.
- **Cryospheric research**, which could be stimulated through in-depth monitoring of sea ice, continental ice and through climate modelling. The global cryosphere is composed of the Antarctic and Greenland ice sheets, mountain glaciers, Antarctic ice shelves, sea ice and the snow cover. Land ice occupies about 11% of the continental surfaces. Sea ice and ice shelves occupy up to around 7% of the total oceanic area.

(d) Contribution to Operational Surveillance

In addition to the benefits to Earth science, contributions to other operational activities may spin off in a variety of fields such as:

- **Environmental monitoring.** The control of air, sea and land pollution is increasingly important if we want to preserve our planet. Oil spills and ocean pollution can be detected or followed from space by more or less direct methods, depending on the resolution of the sensor. For instance, sea surface temperature and winds can be useful tools. Further benefits could be the detection of volcanic eruptions and the smoke from forest fires. Consequences of environmental damage such as floods, fire, droughts and deforestation can also be determined through information about vegetation or soil moisture.

- **Marine applications.** Reliable sea state forecasts, apart from more immediate use for navigation safety, both commercial and for leisure, can also be used for a growing number of industrial activities. These could be the offshore search for and exploitation of oil and gas, as well as for ship routing optimisation. Fishing activities could benefit through increased safety and improved accuracy in targeting fishing areas. It has been shown that half of the world's sea food is caught in extremely limited regions where there are upwelling systems, which represent 0.17% of the world's ocean surface.

- **Renewable resource management.** The establishment of global inventories through satellite data is necessary to obtain current data to handle future tasks. The examples of benefits are numerous and include crop inventories and production, forest monitoring and management, water resource assessment, mineral resource exploration, urban planning and mapping.

In order to further enhance the economic use of a flight in polar orbit for related scientific and technical objectives, or for humanitarian purposes, the following ancillary missions benefits should also be recognised:

- **Search and rescue support**, which continues on an interim basis the communications facilities and location detection used in support of global search and rescue operations.

- **Space environment monitoring**, which provides information on the space environment, including charged particles, which is used to help explain instrument and spacecraft anomalies when operating in orbit.

4.5 The Benefits of the Earth Explorer & Earth Watch Missions

(a) Earth Explorer Missions

This programme will be implemented in successive slices covering a pre-defined number of years. Ideally, all member states would participate according to their Gross National Product, although another key of contribution might be eventually preferred according to the interest of the individual member states in fostering Earth Observation (References 8, 9).

Following close consultation with the user community, nine potential Earth Explorer missions have been identified:

- **An Earth radiation mission** to advance understanding of the Earth's radiation balance, which is of fundamental importance to the Earth's climate.
- **A precipitation mission** to observe precipitation, especially in tropical regions.
- **An atmospheric dynamics mission** to observe three-dimensional wind fields in clear air in both the troposphere and the stratosphere.
- **An atmospheric profiling mission** to observe temperature profiles in the troposphere and stratosphere for climate research.
- **An atmospheric chemistry mission** to advance understanding of the chemistry of the atmosphere, including the study of chlorine species and the hydrogen oxides.
- **A gravity field and ocean circulation mission** to derive a highly accurate global and regional model of the Earth's gravity field and its geoid.
- **A magnetometry mission** to observe the Earth's magnetic field.
- **A surface processes and interaction mission** to advance understanding of biospheric processes and their interactions with the other processes that occur in the Earth/atmosphere system.
- **A topographic mission** to observe the topography of the oceans, the land and the polar ice sheets.

These nine Earth Exploration mission candidates will be reviewed by the general Earth Observation community by the end of March 1996 and presented to the community at a consultative workshop scheduled for the end of May 1996.

(b) Earth Watch Missions

These (pre-)operational satellites will nominally comprise a series of missions which will be based on the objectives outlined in the European Earth Observation policy. The Earth Watch missions will be cooperative endeavours with complementary roles and responsibilities among the parties involved. It is expected that the implementation of the Earth Watch missions will call for significant involvement by the value-added industry for downstream ground processing, in particular the small and medium size enterprises, being complementary to the institutional and operational entities.

Considering the cooperative nature of the Earth Watch missions, the definition of the activities of the partners involved will be elaborated during the preparatory phases, taking due account of the specifics of each mission.

Two Earth Watch-type missions are currently in preparation in ESA (MSG and EPS/METOP). Looking to the post-2000 era, at least five other groups of potential Earth Watch communities have been identified, namely those interested in:

- **Coastal zones**. This covers a wide field of applications including bathometry, sea ice monitoring, oil spill monitoring and forecasting, mineral and hydrocarbon exploration, flood surveillance and prevention, inland waters, fisheries, coastal erosion and surveillance, river discharge, coastal land use, etc....
- **Ice monitoring**. This includes both operational monitoring of sea ice fields and forecasting of their evolution.
- **Land surface**. This spans a variety of users on a national and a European level, including crop forecasting, crop damage assessment, forestry, land use, cartography etc.... Some of these relate directly to the enforcement of regulatory measures.
- **Atmospheric chemistry**. The long-term need for atmospheric chemistry monitoring is only partially addressed by the ozone instrument on EPS/METOP. It is also necessary to consider monitoring halogens in the stratosphere.
- **Open oceans**. This includes ship routing, sea state monitoring and forecasting.

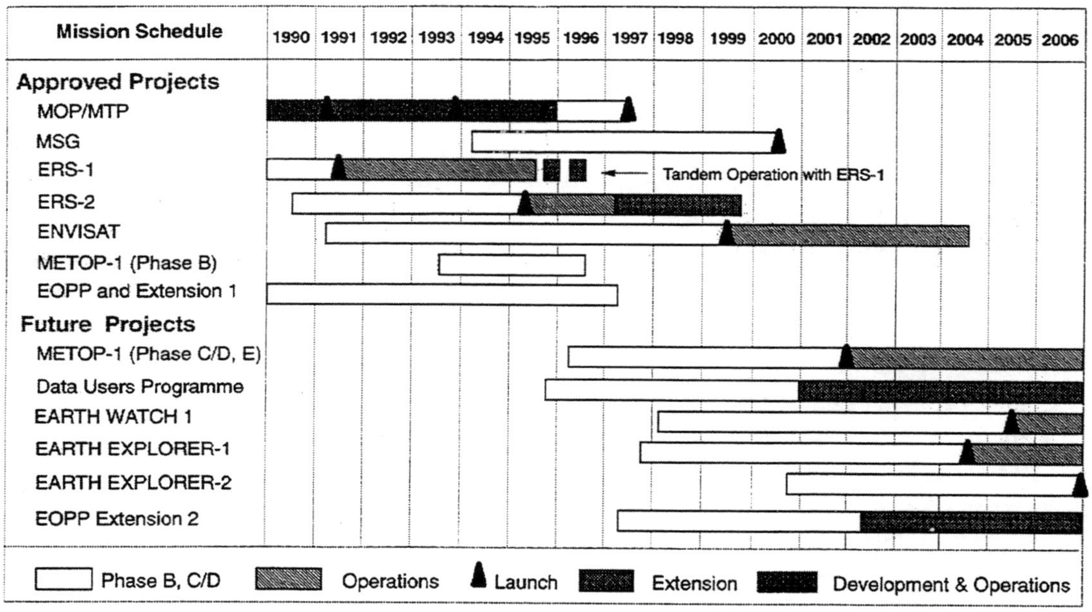

Figure 2. Observation of the Earth and its environment

To prepare the users and promote future Earth Observation programmes, the focus will be directed on:

- information and training for both on-going and future ESA missions,
- expanding data usage through monitoring of activities performed by Principal Investigators and within pilot projects for ESA missions and, in specific cases, support for the derivation of thematic information,

- harmonising user information services for better access by users to ESA missions and associated products through the various ESA and national facilities.

For these three areas, action will be directed to users both within and outside Europe, with particular attention paid to the needs of developing countries.

An indicative schedule is shown in Figure 2. The schedule describes how the planned future missions could build up and continue from the current approved missions.

5. Conclusions

Civilian Earth Observation from space is a method of acquiring physical data which allows us to achieve the global coverage and frequent sampling (in time and space) required for many scientific disciplines and daily operations.

This paper has presented the role played by ESA in developing and demonstrating remote sensing technologies for both scientific and application activities. It has also addressed the ways in which these missions are operated by ESA for scientific and pre-operational demonstration purposes, and by specialised European institutions for operational applications.

In particular, examples of scientific achievements have been presented for the study of land, ocean/ice, atmospheric and solid Earth processes. For pre-operational missions, examples of applications have been outlined, showing how the ESA programmes are paving the way towards new applications which will allow a better use of Earth's resources by mankind in the future. ESA's plans for a future dual mission concept and to ensure economically viable applications have been presented. Finally, examples of the benefits directly resulting from improved weather forecasts (made possible by METEOSAT in the past and by MSG and EPS/METOP in the future) have been demonstrated.

These efforts will significantly contribute to the key objectives of the European Policy for Earth Observation from space, namely:

- to develop our knowledge of planet Earth,
- to preserve planet Earth and its environment, and
- to manage life on Earth in a more efficient manner.

References

1. European Space Agency:*Proceedings of the First ERS-1 Symposium, Space at the Service of our Environment, 4-6 November 1992*. ESA SP-359, ESA, The Netherlands, 1992
2. European Space Agency: *Proceedings of the Second ERS-1 Symposium Space at the Service of our Environment, 11-14 October 1993*. ESA SP-361, ESA, The Netherlands, 1993
3. European Space Agency: *New Views of the Earth - Scientific Achievements of ERS-1*. ESA SP-1176/1, ESA, The Netherlands, April 1995

4. European Space Agency: *First Workshop on ERS-1 Pilot Projects, 22-24 June 1994.* ESA SP-365, ESA, The Netherlands,1994
5. European Space Agency: *ESA Bulletin n° 83.* ESA, The Netherlands, August 1995
6. European Space Agency: *ESA Bulletin n° 76.* ESA, The Netherlands, November 1993
7. EPS/METOP Mission Objectives, October 1995, OEE/DYM/M IS/5588
8. European Space Agency: *Future ESA Earth Observation Objectives and Programmes.* ESA/PB-OE(95)24, Review 4, ESA, The Netherlands, October 1995
9. Readings, C.: "Earth-Observation Missions Beyond the Year 2000;" *Earth Observation Quarterly*, December 1995

A Plan for a Global Disaster Observation Satellite System (GDOS)

T. Kuroda, S. Koizumi and **T. Orii**, NEC Corporation, 4035 Ikebe-cho, Tsuzuki-ku, Yokohama 224, Japan

Abstract

Since 1987 the authors have been studying the implementation of a World Environment and Disaster Observation Satellite System (WEDOS). WEDOS has been proposed in order to establish a satellite system whereby any point on the Earth's surface could be observed at least once per day in order to provide accurate information in regard to the occurrence of disasters and changes in the environment, and to ensure greater safety for all the peoples of the world.

Recent studies by the authors have shown that when any disaster occurs it is very important and necessary for disaster observation systems to be able to provide information as promptly as possible, during both the daytime and at night. From such a viewpoint the authors feel that WEDOS needs further improvement in regard to disaster observation. This paper describes the Global Disaster Observation Satellite System (GDOS), with specific focus on disaster observation, as an alternative version of WEDOS.

1. Great Hanshin/Awaji Earthquake

At 5:46 am on January 17, 1995 the Kobe/Awaji region of Japan experienced a large and powerful earthquake, causing great damage to the local infrastructure, such as highways, railroads, harbor facilities, etc.... The epicenter of the earthquake was located at a depth of 14 km and at a latitude 34° 36' North, longitude 135° 03' East. The magnitude of the earthquake was 7.2 on the Richter Scale; more than 5,500 people died and 171,000 buildings collapsed.

Prior to the earthquake it was thought that the emergency communication systems of the Japanese Disaster Prevention Agencies were well established, and they were expected to be able to function as required during a disaster to provide prompt and precise communication of disaster information to the appropriate organizations.

However, the earthquake destroyed almost all of the communications facilities in the central area of the Kobe/Awaji region, and disaster information was initially reported only from the surrounding areas. Information was not received from the central area until several hours after the earthquake struck, by which time the resulting fires had spread, resulting in an increased death toll.

In any natural or man-made disaster, minimizing the scale of secondary disaster depends on the success of the prevention and rescue activities carried out in the initial stages of the disaster. The larger and the wider the range of the disaster, the more difficult fast and precise notification becomes. In this respect, the suitability and advantages of the use of disaster observation satellites is obvious.

Had WEDOS been available for the Great Hanshin/Awaji Earthquake, the first satellite image would have probably been acquired at 6:00 am, only 15 minutes after the earthquake occurred. At 6:30 am, the approximate scale and status of the disaster would probably have been clarified and distributed to the relevant organizations by means of satellite image data. The provision of the data would have assisted in the initial fire fighting and rescue activities. The actions which might have been taken based on information provided by WEDOS after the earthquake occurred are based on the fortuitous fact that the time of occurrence of the earthquake coincided well with the time when the orbit of the WEDOS satellite was above the earthquake area. This was a coincidence, and would not generally be the case for most disasters.

After investigation and analysis the authors have found that it would have required 4 hours 30 minutes (at a maximum) during the daytime for WEDOS to have acquired visible images after the occurrence of the disaster, assuming a fine, clear day. Additionally, it would have required 8 hours (at a maximum) during the 24 hour period after the occurrence of the disaster for WEDOS to acquire infrared images for use in fire fighting and rescue activities.

The design preconditions of WEDOS, as illustrated above, were that the frequency of observation would be once per day for observation of any point on the Earth's surface. It can therefore be concluded that it is very difficult to acquire information within a time period of several hours or less, regardless of the time of day or the weather conditions. This leads to the proposal by the authors for a new system, referred to as the "Global Disaster Observation System" (GDOS), which would provide a viable solution for disaster observation based on the previously proposed WEDOS.

2. The GDOS Concept

The Great Hanshin/Awaji Earthquake gave the authors the chance to investigate and learn about the functioning and operation of disaster prevention and observation systems, telecommunications systems, fire defense systems and rescue systems during the occurrence of an actual disaster, and about the difficulties experienced by the relevant organizations in taking the appropriate counter-measures in the initial stage of the disaster using existing systems.

The authors propose that the requirements that need to be satisfied by an observation and communications system in the event of large scale disaster are as follows:

- Prompt provision of information on all aspects of the disaster (e.g. location of collapsed buildings and fires, in the case of an earthquake) to the relevant organizations in order that a full view of the disaster can be perceived within just a few hours.

- Observation and report of the full view of the disaster to the relevant organizations, regardless of the time of day or the weather conditions.

- Detection of the location of damaged and destroyed structures, such as highways and large buildings, including condominiums.

- Detection of fire outbreaks in residential areas.

- Efficient and continuous (non-interrupted) performance of the required functions from a location remote from and unaffected by the effects of the disaster.

Based on these requirements GDOS was conceived as a disaster observation system and was proposed (References 8, 9) as an alternative version of WEDOS, which has been proposed by the authors over the past several years (References 1-8).

2.1 Objectives of GDOS

GDOS has been proposed as a more effective system for disaster observation on a global scale than WEDOS. The main objective of GDOS in the occurrence of large scale disasters is the minimization of damage, to be accomplished worldwide as follows:

- Operation of GDOS in tandem with conventional disaster prevention systems so that a full view of any disaster over a wider area can be obtained.

- Provision of detailed disaster information to enable effective deployment of rescue and fire defence services in order to suppress potential secondary disasters to the minimum.

The main objectives of GDOS in ordinary times are for planning and preparation for the prevention of disasters, support of rescue activities and support of planning for reconstruction in disaster-affected areas, to be accomplished as follows:

- Provision of data for preparation of databases for disaster prevention purposes and zone maps of potentially hazardous and disaster-prone areas.

- Provision of data for the improvement of prediction accuracy of disasters, including earthquakes.

- Prompt detection of disasters such as forest fires and similar small scale disasters.

2.2 GDOS System Study

The establishment of the GDOS system requires consideration of the number and type of observation satellites required in each orbit, the type and capabilities of the onboard observation systems, the number and type of communications satellites and facilities, the ground system, etc.... These are proposed for the overall design of the GDOS system as follows:

(a) GDOS Desirable Features

Taking into consideration the required capabilities of the GDOS system in the occurrence of large-scale disasters and the objectives of the GDOS system as mentioned in Section 2.1, the desirable features of the GDOS system are:

- Capability to acquire a full view of the disaster-stricken area within an average time of 1 hour 30 minutes after the occurrence of the disaster (30 minutes at the earliest and 2 hours 30 minutes at the latest).
- Capability to observe the disaster-stricken area with a resolution of 5 m regardless of time of day and in any weather conditions (including cloud, rain and snow).
- Capability to observe the disaster-stricken area with a resolution of 2 m and a ground observation width of 40 km.
- Capability to observe the disaster-stricken area at periodic intervals (i.e. time intervals of every 2 hours) and for extended intervals (i.e. two times of 25 minutes for each time interval).
- Capability to detect various types of disasters promptly, such as forest fires and volcanic erruptions.
- Capability to detect vertical movements, variations and dislocations of ground areas to a resolution of several centimeters from data obtained by multiple observations.
- Capability to create a complete and detailed map of the Earth's surface and disaster prevention maps efficiently.
- Capability to ensure that GDOS observation data are useful for earthquake prediction studies.

(b) Number of Satellites, Their Orbits and Observation Frequency

The number of satellites, their orbits and observation frequencies to result in optimum observation and real time data relay to the relevant organizations worldwide are described as follows:

- **Multiple observation satellites in circular orbits and their ground tracks.** The deviation of a satellite in a Sun-synchronous subrecurrent orbit will usually be by an angle of $\phi°$ westwards if the node longitude deviates by $\phi°$ while the satellite orbits the Earth once, as shown in Figure 1.

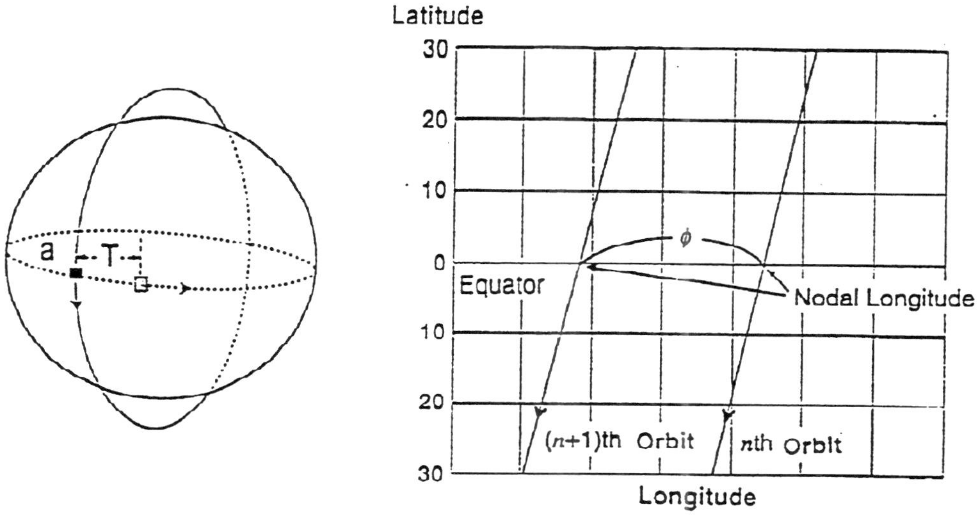

Figure l. Observation satellite on Sun-synchronous subrecurrent orbit and its ground track

If N satellites are each spaced evenly along a circular orbit, the ground track of each satellite will deviate by an angle of $(\phi/N)°$ westwards per orbit, as shown in Figure 2.

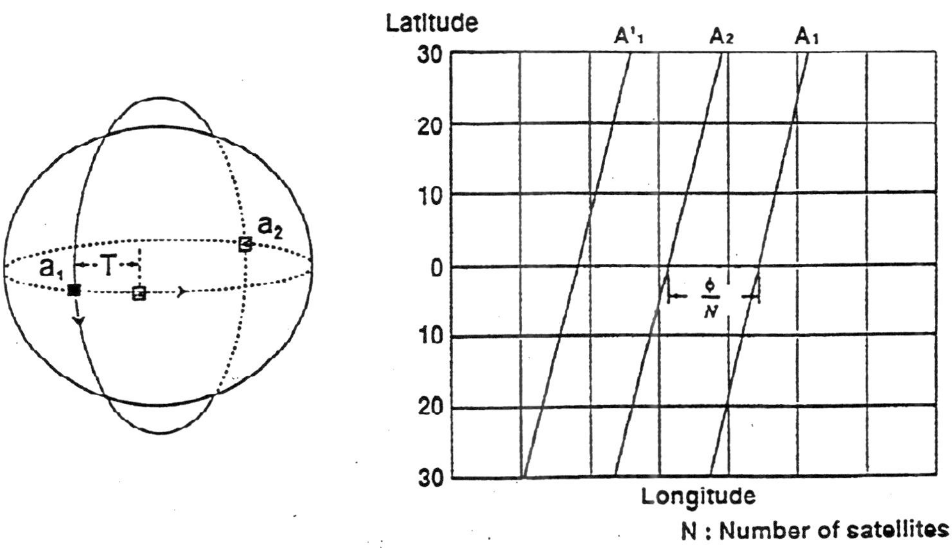

Figure 2. Multiple observation satellites on a Sun-synchronous subrecurrent orbit and their ground tracks

- **Ground tracks for multiple observation satellites on multiple Sun-synchronous subrecurrent orbits.** The analysis presented here presupposes that N observation satellites are evenly spaced on M Sun-synchronous subrecurrrent circular orbits at time intervals of τ hours. By adjustment of the phase (latitude) of satellites in adjacent orbits, ϕ/N can be divided into M satellite orbits for the time interval of τ hours, and the ground tracks as shown in Figure 3 will then deviate westwards by $\phi/NM°$. In other words, N satellites in M orbits can observe the entire surface of the Earth from North to South or vice versa in one day, with an equal phase (latitude) difference of $\phi/NM°$.

  To meet the above conditions, the phase (latitude) difference $\gamma 1$ between any two satellites on any two adjacent orbits can be obtained from the following equation:

$$\gamma 1 = \frac{360}{T}\left(\tau + \frac{T}{N \times M}\right) - 360\frac{n}{N}$$

$$= 360\left(\frac{\tau}{T} + \frac{1}{N \times M} - \frac{n}{N}\right)$$

  where n is a positive integer selected to minimize

$$\left|\frac{\tau}{T} + \frac{1}{N \times M} - \frac{n}{N}\right|$$

  The above condition can be satisfied if the phase (latitude) of satellite a(m) in orbit A is advanced by $\gamma 1$ degrees relative to satellite b(n) in orbit B, or satellite b(n) is delayed by $\gamma 1$ degrees relative to satellite a(m), where m = n+1.

(c) Optimum Number of Satellites and their Orbits to Increase Observation Frequency

The authors have chosen to investigate an observation satellite system of multiple observation satellites, having a combination of two types of Sun-synchronous subrecurrent orbits, in order that the observation frequency of the satellites is increased. Specifically, two time intervals τ between adjacent Sun-synchronous subrecurrent orbits of 1 hour and 2 hours have been selected for investigation. In this case, the number of orbital planes M (= $12/\tau$) are 12 and 6, respectively. If the total number of observation satellites (N x M) is set to be 24, the number of satellites N on each orbit will be 2 and 4, respectively.

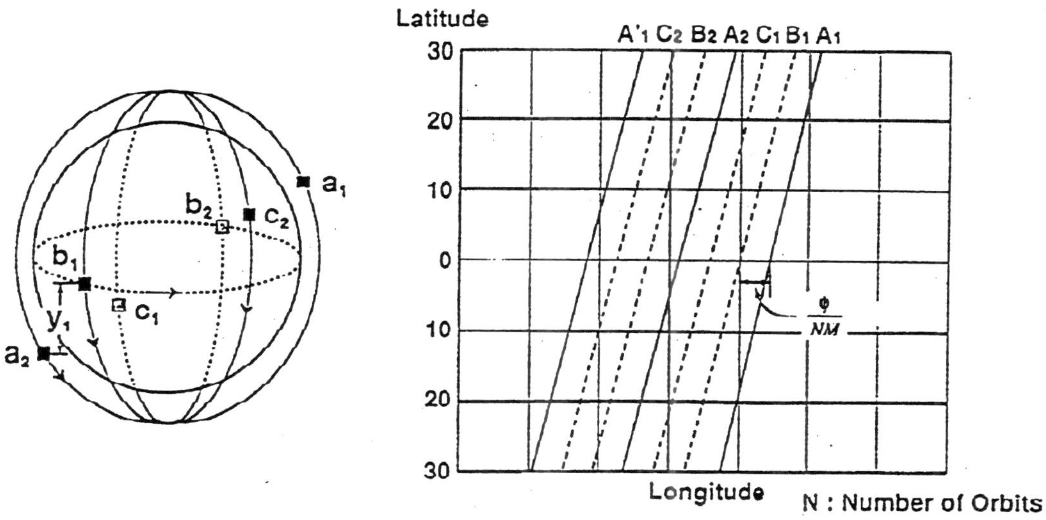

Figure 3. Multiple observation satellites on multiple Sun-synchronous subrecurrent orbits and their ground tracks

The revolution interval T of the satellite around the Earth is set to be 98.2 minutes, assuming that the altitude of the satellite is 700 km (in this instance ϕ = 24.55° and the ground track distance is 2732.88 km at the Equator and 2368.73 km at a latitude of 30°).

Case 1. Assuming that N = 2 observation satellites are evenly spaced on M=12 Sun-synchronous subrecurrrent orbits at time intervals of τ = 1 hour and revolution interval T = 98.2 minutes, then:

$$\text{Phase Difference:} \quad \gamma 1 = 360 \,.\, (60/98.2 + 1/24 - n/2)$$
$$= 360 \,.\, (0.65266 - n/2)$$

Selecting n = 1 in the above equation gives $\gamma 1$ = 54.96°. This means that if 24 observation satellites are on the 12 Sun-synchronous subrecurrent orbits (2 satellites per orbit), each having a time interval of 1 hour, the latitude of satellite b1 must be delayed relative to that of satellite a2 on the adjacent orbit by 54.98°. Figure 4 shows an example of the ground tracks in this case, for the assumed altitude of the satellite being 700 km.

In order to investigate the satellite observation frequency, it is assumed that the variable observation range of the onboard satellite sensors is 700 km on either side of the ground track for the TR (thermal infrared radiometer) and the VN (visible near-infrared radiometer), and 690 km for the SAR (synthetic aperture radar) swath width.

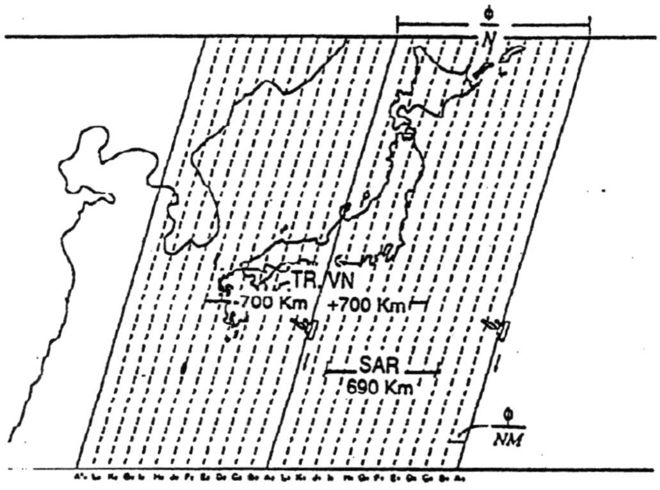

Figure 4. Case 1 - Example of ground track of satellite and variable observation area

The observation frequencies in 24 local time zones of 24 satellites having onboard TR, VN and SAR sensors are given in Table 1, for the conditions described in Case 1. Note that the observation frequencies given are the values taken at the subsatellite points, which is the most strict condition.

In this case, observation by the onboard TR sensor is performed hourly regardless of the time of day, and observation by the onboard VN sensor is possible, although during the daytime only. Reception of data from the onboard SAR sensor is considered to be alternatively available and unavailable for 6 hour time periods. For the example shown in Table 1, the observation frequency for each time zone is indicated by 0 (for no observation) or 1 (for one observation) and is determined by the location (latitude) and the phase of the subrecurrent orbit of the observation satellite.

Case 2. Assuming that N = 4 observation satellites are evenly spaced along M = 6 Sun-synchronous subrecurrrent orbits (at time intervals of τ = 2 hours) with orbital period T = 98.2 minutes, then:

Phase Difference: $\gamma 1 = 360 \,.\, (120/98.2 + 1/24 - n/4)$
$= 360 \,.\, (1.26366 - n/4)$

The value of n is selected from the relation m = n + 1 = 6. If the value of m is equal to or greater than the value of N, a multiple of N must be subtracted from m (i.e. 6 - 4 = 2 in this case). Selecting n = 5 in the above equation gives $\gamma 1$ = 4.92°. This means that if 24 observation satellites are in 6 Sun-synchronous subrecurrent orbits (4 satellites per orbit), the latitude of satellite b1 must be delayed relative to the latitude of satellite a2 on an adjacent orbit by 4.92°. Figure 5 shows an example of the ground tracks resulting in this case, where the altitude of the satellite is again assumed to be 700 km.

Local Time Zone		0	1	2	3	4	5	6	7
Sensor and number of observation times	TR	1	1	1	1	1	1	1	1
	VN	-	-	-	-	-	-	-	1
	SAR	1	1	1	1	1	1	0	0
Local Time Zone		**8**	**9**	**10**	**11**	**12**	**13**	**14**	**15**
Sensor and number of observation times	TR	1	1	1	1	1	1	1	1
	VN	1	1	1	1	1	1	1	1
	SAR	0	0	0	0	1	1	1	1
Local Time Zone		**16**	**17**	**18**	**19**	**20**	**21**	**22**	**23**
Sensor and number of observation times	TR	1	1	1	1	1	1	1	1
	VN	1	1	-	-	-	-	-	-
	SAR	1	1	0	0	0	0	0	0

Table 1. Example of satellite observation frequency and observation local time zone for the conditions as given in Case 1

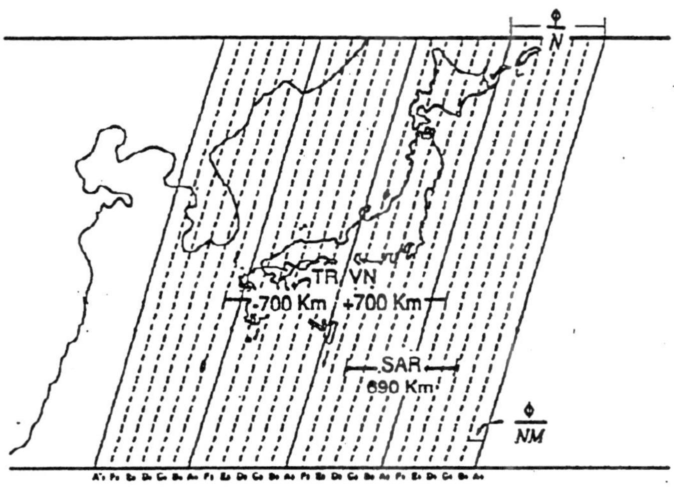

Figure 5. Case 2 - Example of ground track of satellite and variable observation area

The observation frequencies in 24 local time zones of 24 satellites having onboard TR, VN and SAR sensors are given in Table 2. Note that the observation frequencies given are the values taken at the subsatellite points, which is the most strict condition.

In this case, two hourly observations by the onboard TR sensor are performed regardless of the time of day, and two hourly observations by the onboard VN sensor are performed, although during the daytime only. In the observable time zones each sensor can perform two observations at 25 minute intervals. Reception of data from the onboard SAR sensor is possible once every 2 hours. For the example shown in Table 2, the observation frequency for each time zone is indicated by 1 (for one observation) or 2 (for two observations), and is determined by the location (latitude) and the phase of the subrecurrent orbit of the satellite.

Local Time Zone		0	1	2	3	4	5	6	7
Sensor and number of observation times	TR	2	-	2	-	2	-	2	-
	VN	-	-	-	-	-	-	-	-
	SAR	1	-	1	-	1	-	1	-
Local Time Zone		8	9	10	11	12	13	14	15
Sensor and number of observation times	TR	2	-	2	-	2	-	2	-
	VN	2	-	2	-	2	-	2	-
	SAR	1	-	1	-	1	-	1	-
Local Time Zone		16	17	18	19	20	21	22	23
Sensor and number of observation times	TR	2	-	2	-	2	-	2	-
	VN	2	-	-	-	-	-	-	-
	SAR	1	-	1	-	1	-	1	-

Table 2. Example of satellite observation frequency and observation local time zone for the conditions as given in Case 2

The results for Case 1 and Case 2 are summarized as follows:

- If the sensors onboard the satellites were selected to be TR and VN sensors only, Case 1 is preferable to Case 2 in view of the hourly observation frequency. However, even if SAR sensors were included onboard the satellites, Case 1 would have the distinct disadvantage that the unobservable time zones would last for 6 hours.

- In Case 2, two hourly observations of two times are allowed, with an interval of 25 minutes, by the onboard TR sensor regardless of the time of day and by the onboard VN sensor (during daytime); this allows for the acquisition of stereoscopic images. Two hourly observation at one time is allowed by the onboard SAR sensor, regardless of the time of day. Therefore, Case 2 is advantageous as a disaster observation system when the onboard TR, VN and SAR sensors are in use.

(c) Satellite System

The number of satellites and their orbits were investigated in the preceding section (b) in order to ascertain the best observation frequency of the disaster observation satellite system. The conclusion was that the optimum satellite constellation has 24 GDOS observation satellites, with four satellites each on six Sun-synchronous subrecurrent orbits (having a time interval of 2 hours). This constellation would additionally require six data relay satellites in geostationary orbit (the same as for the WEDOS system) which are used for tracking and control of the observation satellites, for emergency communications from the ground, and for image data distribution.

The weight of each GDOS observation satellite is expected to be approximately 2 tons, and the weight of each GDOS data relay satellite is expected to be approximately 1 ton.

Satellite type	Local node time		Satellite systems	
	Descending	Ascending	WEDOS	GDOS
Observation satellite in Sun-synchronous subrecurrent orbit	06:00	18:00	8 (c)	4 (e)
	08:00	20:00	0	4 (e)
	10:00	22:00	8 (a, b)	4 (e)
	12:00	00:00	0	4 (e)
	14:00	02:00	8 (a, b)	4 (e)
	16:00	04:00	0	4 (e)
Observation satellite in inclined orbit	-------		2 (d)	0
Data relay satellite in geostationary orbit	-------		6 + 6* *spare in orbit	6 + 6* *spare in orbit

Table 3. Comparison of type and number of satellites and satellite orbits for WEDOS and GDOS
Note: Letters in parentheses indicate the satellite type (given in more detail in Table 5).

Type of sensor	Resolution	Ground observation width	Variable angle of observation (ground distance)
VN-1 Visible Near-infrared Radiometer-1	2m	40 km	±43° (± 700 km)
VN-2 Visible Near-infrared Radiometer-2	20 m	120 km	± 43° (± 700 km)
SW Shortwave Infrared Radiometer	20 m	120 km	± 43° (± 700 km)
VT Visible Thermal Infrared Radiometer	40 m	40 km	± 43° (± 700 km)
SAR Synthetic Aperture Radar	5m	40 km	18° - 50° (690 km)

Table 4. Observation sensors onboard GDOS observation satellites and their main performance data

The constellation of the GDOS satellites and their orbits are shown in Figure 6, and the concept of the data relay satellites for inter-satellite data transmission and for data transmission to and from the ground stations is shown in Figure 7. The constellation of the WEDOS satellites and their orbits is shown in Figure 8 for reference. A comparison of the GDOS system and the WEDOS system is given in Table 3. A summary of the major types of sensors onboard the GDOS observation satellites and their main performance data is given in Table 4.

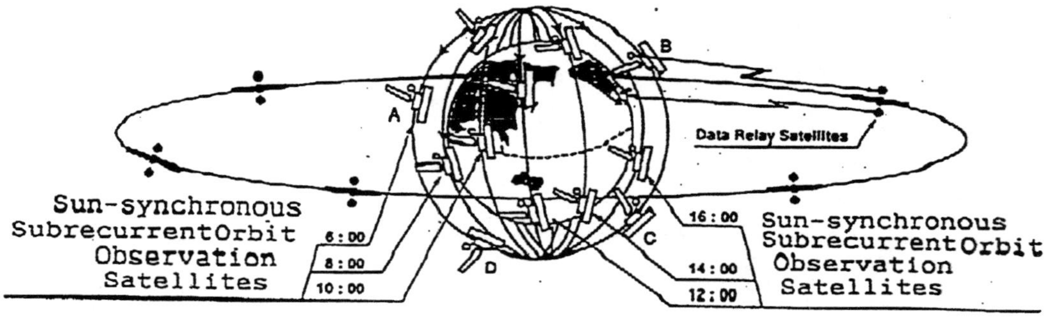

Figure 6. Constellation of GDOS satellites

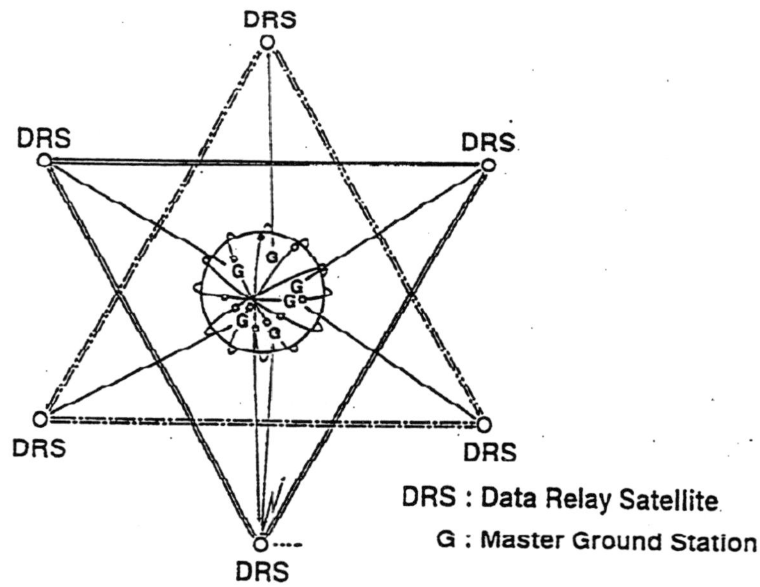

Figure 7. Concept of data transmission using data relay satellites

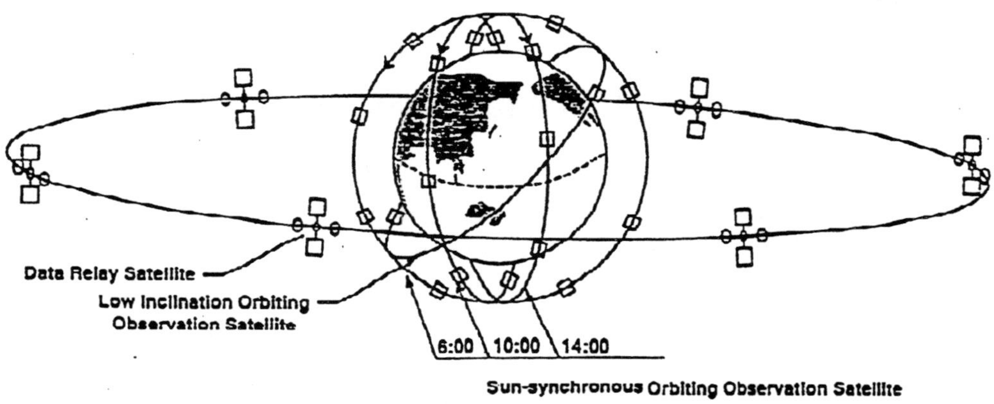

Figure 8. Constellation of WEDOS satellites

(d) Sensors

The features of the main sensors onboard the GDOS observation satellites are as follows. The TR sensor (thermal infrared radiometer) as described in Case 1 and Case 2 is classified as the SW sensor (shortwave infrared radiometer) and the VT sensor (visible and thermal infrared radiometer).

- **VN-l (Visible Near-infrared Radiometer-l).** The VN-l sensor provides high resolution stereoscopic pictures for detailed imaging of a disaster area.

- **VN-2 (Visible Near-infrared Radiometer-2).** The VN-2 sensor allows a wide area to be monitored in the event of a disaster. Databases and hazard maps for disaster prevention can be created in ordinary times, using information provided by VN-2.
- **SW (Shortwave Infrared Radiometer) and VT (Visible Thermal Infrared Radiometer).** The SW sensor and the VT sensor give early detection of forest fires and volcanic eruptions, and after detection can then provide information on their activity.

- **SAR (Synthetic Aperture Radar).** Effective monitoring of some disasters can be performed using the SAR sensor in cloudy weather and at night. Additionally, the acquisition of multiple observation data by the SAR sensor allows vertical land movements of the order of several centimeters to be investigated.

A comparison of the types of observation satellites and the sensors of the GDOS and the WEDOS systems is given in Table 5.

The observation frequency using GDOS observation sensors is twice every 2 hours for the TR sensors and the VN sensors, and once every 2 hours for the SAR sensors. If the number of GDOS observation satellites is reduced from 24 satellites to 12 satellites in order to reduce costs, in some cases observation using the SAR sensors becomes impossible for periods of 6 hours, although observations using TR sensors and VN sensors are possible every 2 hours.

To enable ordinary observation of the entire surface of the Earth and transmission of the observed data, a minimum of six data relay satellites is required, although observation of a disaster-stricken area on a local basis can be performed by only three satellites. Another reason for the necessity of the data relay satellites is that, as the onboard sensors of the GDOS satellites are directed to the disaster-stricken area immediately in the event of a disaster or the reception of disaster information from an observation satellite or notification from elsewhere, it is necessary that the operation commands be transmitted to the observation satellites via the data relay satellites. In addition, the data relay satellites have the function to relay communications in the event of a disaster and to transmit the data processed at the ground stations in the broadcasting mode.

Sensor type		Satellite type				
		WEDOS				GDOS
		a	b	c	d	e
VN-1	Visible Near-infrared Radiometer-1 (high spatial resolution)	*	*			*
VN-2	Visible Near-infrared Radiometer-2 (high spatial resolution)	*	*			*
SW	Shortwave Infrared Radiometer		*			*
VNT	Visible Thermal and Near-infrared Radiometer (high spatial resulution)	*	*			
VT	Visible Thermal Infrared Radiometer (high spatial resolution)					*
SAR	Synthetic Aperture Radar	*		*		*
IS	Infrared Sounder		*			
ML	Mie Scatter Lidar				*	
MR	Microwave Radiometer		*	*	*	
SCT	Microwave Scatterometer			*		
PR	Precipitation Radar				*	
ALT	Radar Altimeter			*		

Table 5. Comparison of the types of observation satellites and sensors of the GDOS and the WEDOS systems

(e) GDOS Ground System

The GDOS ground system consists of the Mission Management Center (MMC), the Master Ground Stations (MGS) and the User Stations.

- **Mission Management Center (one station).** The functions of the MMC are to arrange and oversee the system operation planning and management of the total system. When a disaster occurs, the MMC receives information from the related organizations and sends commands to the related MGS to ready them to perform emergency observation and command control. The MMC shall have a redundant configuration to secure its reliability, and shall be located in a remote place.

- **Master Ground Stations (6 stations).** Six MGS shall be located worldwide with approximately equal separation so that the location of each corresponds with one of the data relay satellites. The functions of the MGS are to track and control the satellites, to transmit commands to the observation satellites via the data relay satellites, to perform reception, processing and features analysis (including information showing status change) and to report such information to the relevant organizations, and to transmit processed data to be broadcast using the broadcast mode to the data relay satellites. In ordinary times the MGS monitor for the occurrence of disasters, including forest fires, and prepare databases and hazard maps useful for disaster prevention and planning.

- **User Stations.** The User Stations allow the worldwide users of the GDOS system to receive processed data broadcast from the data relay satellites and observation data from the observation satellites.

2.3 GDOS Operations

(a) Operation in the Event of a Disaster

In the event of the occurrence of a disaster, as soon as the MMC receives disaster information from the observation satellites or notification from elsewhere, commands are sent from the MMC to the MGS closest to the location of the disaster to ready it for making emergency observations of the disaster-stricken area. The MGS located closest to the disaster-stricken area transmits commands to the nearest observation satellite to point its sensors towards the disaster area and to observe the disaster area.

The operation of the GDOS system in the event of the occurrence of a disaster is shown in Figure 9 in the form of a flow chart.

(b) Operation in Ordinary (non-Emergency) Times

Normally the GDOS observation satellites continuously observe the entire surface of the Earth more than once per day to detect various environmental changes, including forest fires, as they occur, acting as a "Fire Watchtower of the Earth". The GDOS system may also be used to create various maps and databases for disaster prevention. The GDOS system can also be used to improve the accuracy of disaster prediction, including earthquake prediction, by using it in tandem with existing disaster prevention systems. At ordinary times GDOS satellite data can be directly received by existing Earth observation ground stations.

3. Conclusions

The outline of the GDOS for disaster observation has been described in this paper in comparison with the WEDOS, which has already been proposed and promoted as a worldwide environment and disaster observation system.

The GDOS observation satellites are an improved version of the Advanced Land Observing Satellite (ALOS) currently being planned by the National Space Development Agency (NASDA) of Japan. The sensors to be aboard ALOS could be utilized for the GDOS satellites after the improvement and upgrading of their performance characteristics and the expansion of their observation range. Therefore, realization of ALOS is a very important step in the verification and realization of the GDOS system.

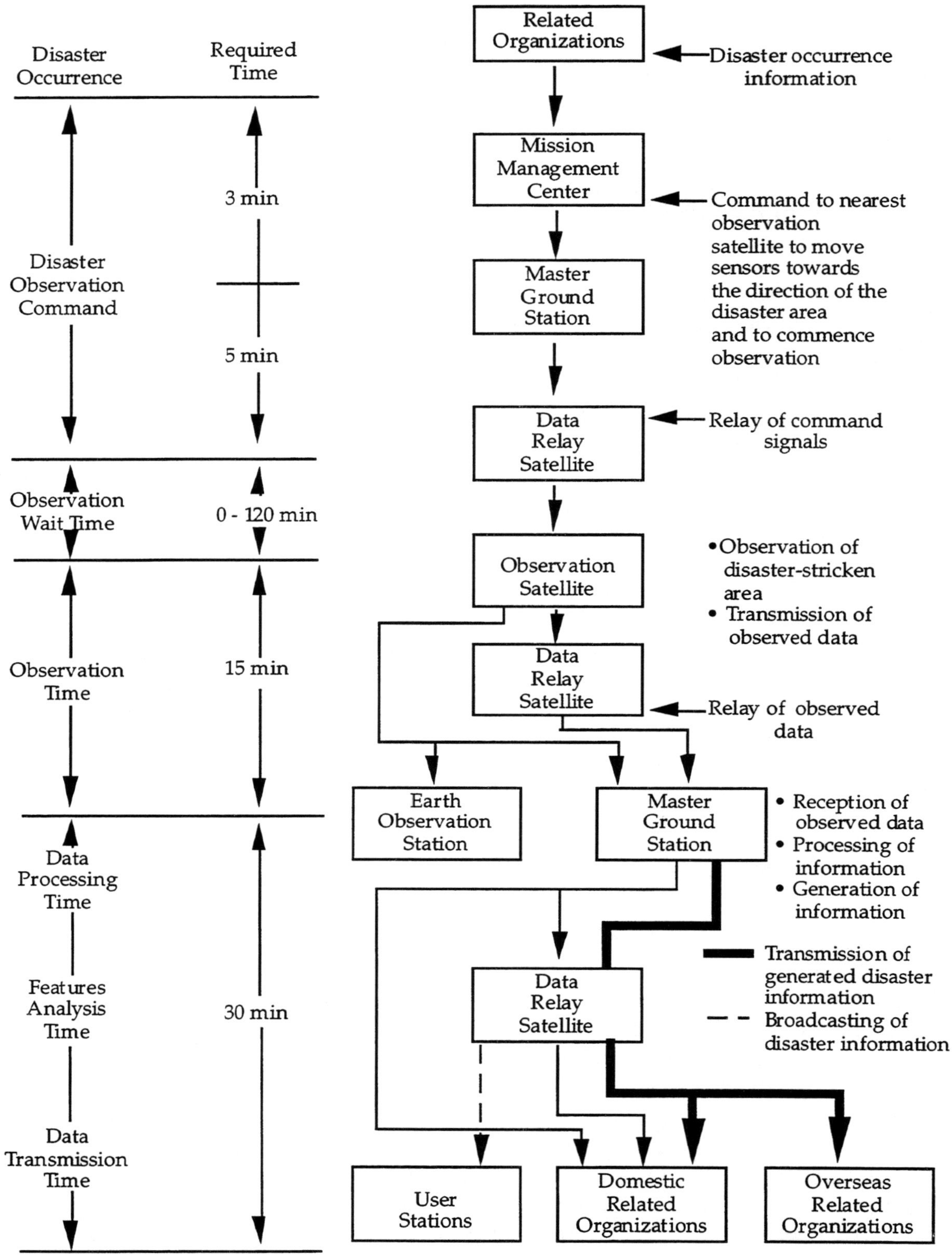

Figure 9. Example of GDOS operation in the occurrence of a disaster

The authors intend to further investigate and promote GDOS, and to make their best efforts towards a realization of this system, not only as an activity of the Society of Japanese Aerospace Companies, Inc., but also as one of the main activities of the Japan-US Cooperation in Space Project.

Acknowledgements

The authors wish to express their sincere gratitude and thanks to the members of the Society of Japanese Aerospace Companies, Inc., to the working group members of the Disaster Observation Satellite System of the Japan-US Cooperation in Space Project, and to all other people for their tireless efforts and contributions towards the investigation and promotion of the GDOS system.

References

1. Kuroda, T.: "Global Environmental Observation System." In *Proceedings of the Pacific ISY Conference, Kona, Hawaii, August 19-21, 1987*, pp. 25-32, 1987
2. Kuroda, T. and Koizumi, S.: "A Plan of the World Environment and Disaster Observation Satellite System." In *Report on the UN/ESCAP/UNDRO Workshop on the Application of Space Technologies to Combat Natural Disasters, Beijing, China, September 23-28*, p. 22. United Nations, 1991
3. Kuroda, T. and Koizumi, S.: "A Plan for the World Environment and Disaster Observation System." In *World Space Congress (COSPAR), Washington DC, August 25 - September 5, 1992*, Vol. 14, No. 3, pp. (3) 155-158. Edited by A.B. Kahle(s), COSPAR, 1992
4. Kuroda, T. and Koizumi, S.: "A Concept for the World Environment and Disaster Observation System." In *Report of the Asia-Pacific ISY Conference, Tokyo, November 16-20, 1992*, Vol. II, pp. 185-188, 1992
5. Kuroda, T. and Koizumi, S.: "A Plan for the World Environment and Disaster Observation System." In *Proceedings of the Asia-Pacific Workshop on Multilateral Cooperation in Space Technology and Application, Beijing, China, November 30-December 5, 1992*, pp.131-138, 1992
6. Kuroda, T.: "Satellite Environment Monitoring System for the 21st Century." In *Proceedings of the International Symposium on the Future of Cranes and Wetlands*, pp. 86-92. Edited by H. Higuchi and J.Minton, Wild Bird Society of Japan, 1994
7. Kuroda, T., Koshizaka, S. and Koizumi, S.: *Proposal Activities for the Concept of the World Environment and Disaster Observation System (WEDOS)*. Paper presented at the UN/ESCAP Symposium on Space Technology and Application for Sustainable Development, Ministerial Conference on Space Application for Development in the ESCAP Region, Beijing, China, September 19-21, 1994
8. Kuroda, T. and Koizumi, S.: *The Plan of the World Environment and Disaster Observation System (WEDOS) and the Global Disaster Observation System (GDOS)*. Paper presented at the UN/ESA Workshop on the Application of Space Techniques to Prevent and Combat Natural Disaster, organized in cooperation with the Government of the Republic of Zimbabwe, Harare, Zimbabwe, May 22-26, 1995
9. Kuroda, T.: *A Plan of the Global Disaster Observation Satellite System*. Paper presented at the International Symposium on Satellite Communication and Remote Sensing, SCRS'95, Xi'an, China, September 20-22, 1995

Application of Remote Sensing and Importance of Satellite Communications in a National Meteorological Services Agency

F. Bekele, National Meteorological Services Agency, PO Box 1090, Addis Ababa, Ethiopia

Abstract

The remote sensing systems and the meteorological data exchange facilities currently applied in the National Meteorological Services Agency of Ethiopia are discussed. Meteorological satellites have been shown to improve weather forecasts and early warning services of the Agency in its indispensable role as a source of data. Rain during the harvest, floods associated with heavy rainfall, the outbreak of a forest fire, and the arrival of desert locusts, etc., could be anticipated ahead of time to prevent or mitigate the associated damage. The satellite rainfall and vegetation monitoring project of the Agency with the support of the United Kingdom Overseas Development Administration and the University of Reading has contributed successfully towards these activities. The project has terminated but its activity has continued with its running costs currently being covered by part of the annual budget allocated to the Agency. There is a financial problem to replace the worn out computers, to train qualified personnel, to build an archive system, to get some spare parts, etc... However, the Agency is making every effort to continue the activity of the project.

In addition to the current application of METEOSAT and NOAA AVHRR data, the potential applications of planned future products include winds derived from cloud motions, flood monitoring, surface albedo estimation, surface solar irradiance estimation, agricultural yield forecasting, lake fisheries and determination of vertical temperature and moisture profiles. The Meteorological Data Distribution system is found to be cost-effective, efficient, fast and reliable, with meteorological data and information currently being supplied in different formats. Recently it has been upgraded to receive data from the third link-up station, Toulouse. The Data Collection Platforms and Data Reception Systems are efficient for national data collection, supporting the national meteorological data exchange network using Single Side Band radio telephones.

The contribution of a meteorological service in the national endeavor to build self-sufficiency in food production, environmental protection, sustainable development and disaster warning, relief and mitigation capacity could not be realized without the application of space technology. Therefore, by working closely with the advancement of space technology, the Agency is committed to continue the activities of previous projects and to run new ones for the benefit of the nation.

1. Introduction

Historical documents show that meteorological data collection in Ethiopia started early in the 19th century. The first meteorological institution was established in 1951 under the Civil Aviation Administration. Recurrent droughts and the growing demand for meteorological services in various sectors led to the establishment of the present day National Meteorological Services Agency (NMSA) as an independent institution in 1980. Weather-related problems like drought, flooding, forest fires, desert locust invasion, and epidemics are frequently observed in various parts of the country. NMSA, an institution established in Ethiopia to provide meteorological services for the nation on a broad scale, is contributing much in safeguarding the life and property of the nation from weather hazards, by providing weather forecasts and other advisories.

Such a service requires meteorological data collected from conventional meteorological stations and other sources. Due to the topography of Ethiopia, the required number of conventional surface meteorological stations based on the World Meteorological Organization (WMO) station network standard is estimated to exceed 3000. However, the number in use today is less than 750; and about fifty of these are used for operational activity. Out of these stations, 17 are synoptic stations which send their data to the head office every three hours, and the rest are stations reporting to the head office only at 0600 UTC every day. In the near future, the number of stations which will be incorporated in the national operational station network is expected to rise to 200. This increase in the number of stations needs a fast and cost-effective meteorological communications network.

In this paper, the stages, the achievements, the problems and future prospects of the NMSA in remote sensing and communications are discussed, with special emphasis on meteorological satellites.

2. Remote Sensing

Remote sensing is the *"science and art of obtaining information about an object, area or phenomenon through the analysis of data acquired by a device that is not in contact with the object, area or phenomenon under investigation"* (Lillesand and Kiefer, 1994). Pilot balloons, radiosondes, meteorological radars and ground receiving stations for meteorological satellites are the remote sensing devices which are in use by NMSA.

The first satellite ground receiving station was installed in Ethiopia in 1975. This station was used to receive the Automatic Picture Transmission (APT) service of the US National Oceanic and Atmospheric Administration's (NOAA) polar orbiting meteorological satellites. The first satellite reception equipment was replaced by UKWtechnik equipment in 1986 which is designed to receive meteorological data from geostationary and polar orbiting satellites. It had Primary Data User Station (PDUS), Secondary Data User Station (SDUS) and Automatic Picture Transmission (APT) systems. Maps of the cloud top height and Weather Facsimile Service were also available. Currently, NMSA gets hard copies of visible, water vapor and infrared images from METEOSAT.

Meteorological applications of remotely sensed data took another step forward when the operational satellite rainfall and vegetation monitoring project started in 1990, with the support of the United Kingdom Overseas Development Administration (ODA) and Reading University, discussed in Section 3. Since then, NMSA has begun to receive and processes real time digital data from METEOSAT and the NOAA/ AVHRR satellites.

The digital data received every 30 minutes from METEOSAT are used for operational monitoring of synoptic scale weather systems, short lived convective clouds, westward propagating storms, Arabian sea disturbances, rainfall over parts of the country yielding no conventional data, the type and area coverage of clouds, cloud top temperatures, sea surface temperatures and land surface temperatures. Therefore, satellite data in combination with conventional data are used for actual weather assessments and forecasts, for television weather presentation, for daily, monthly, and seasonal weather forecast verification, and for the preparation of agro-meteorological bulletins and various meteorological advisories.

3. The Operational Satellite Rainfall and Vegetation Monitoring Project: Achievements, Problems and Future Prospects

The aim of this project was to provide improved information on the extent, intensity, timing and impact of drought to governmental and non governmental organizations. The rainfall monitoring part of the project started in April 1990 and the vegetation monitoring part in September 1991.

3.1 Rainfall Monitoring

The METEOSAT PDUS, with a dish of 1.8 m in diameter developed by the University of Bradford, uses two IBM AT personal computers, and software installed for the purpose of rainfall estimation. This is based on the TAMSAT methodology; using METEOSAT thermal infrared data, this assumes that most of the rainfall in the tropics comes from convective clouds. There is a linear relationship between the duration of a "cold cloud" of predetermined temperature, called the Cold Cloud Duration (CCD), and the rainfall obtained from the cloud. Identification of homogenous climatic zones, selection of the best threshold temperature, calibration and validation are required to get the maximum benefit from this methodology.

When the project started, Ethiopia was classified in two homogenous climatic zones. Part of the country to the West of the 40°E longitude lies in Zone 7 and the other half of the country which lies to the East of the 40°E longitude lies in Zone 8. This classification was according to the FAO ARTHEMS system and was good to begin with. By comparing rain/no rain, cloud/no cloud classes in a contingency table, the best threshold temperature for Ethiopia was considered to be -40°C (Tadege, 1992). After determining the best threshold temperature a linear regression of decadal CCD against rainfall is made to identify the conversion parameters. The calibration is presented in the form:

Rainfall (mm) = $a_0 + a_1 \times$ CCD, with CCD in hours
Rainfall = 0 if CCD = 0
where a_0, a_1 are identified by TAMSAT for Ethiopia.

The correlation between rainfall amounts in a reference gage at the headquarters of the NMSA in Addis Ababa and neighboring gages falls off rapidly with distance (Tadege, 1992). Therefore, care must be taken when comparing conventional rainfall data (rain gage station data) with satellite rainfall estimates during validation.

3.2 Vegetation Monitoring

A High Resolution Picture Transmission receiver, two AT microcomputers and NOAA AVHRR 10 bit image software process AVHRR data to generate all possible products. The normalized difference ratio of red and infrared reflectance sensed by the AVHRR (i.e. IR-R/IR+R), known as the Normalized Difference Vegetation Index (NDVI), is used for vegetation monitoring. Channel 3 data of NOAA/AVHRR are used for forest fire monitoring by detecting hot spots.

3.3 Achievements

Decadal and monthly satellite rainfall estimate maps, decadal and monthly rainfall difference maps, and decadal and monthly composite NDVI maps are regularly provided to various governmental and non governmental organizations, along with meteorological advisories. These products are used for pest control, early warning, relief and rehabilitation, water management, food production, decision making, etc... Today, the forecast office of NMSA has a high degree of confidence in giving meteorological advisories about the occurrence of heavy rainfall, untimely rain during harvest time, etc., well ahead of time. Real time satellite rainfall and vegetation monitoring project facilities and products have contributed much for food production, environmental protection, sustainable development, disaster warning, relief and mitigation. In addition to this, the products of the project have also served to create an awareness of the usefulness of meteorology and meteorological satellites among the general public and the decision makers.

3.4 Problems

Although the project was terminated in 1994, many improvements and activities are still going on at the Satellite Data Receiving and Processing unit (SDRPU) of NMSA. The attempt to determine smaller and more homogenous zones, the planned validation of satellite rainfall estimation using a dense network of rain gage stations, the attempt to study the effect of the rainfall coming from warm stratified clouds and the effect of cirrus clouds on satellite rainfall estimates using the TAMSAT methodology are the major current activities of the SDRPU. Investigation has indicated that smaller calibration zones could improve the accuracy of the rainfall estimates derived from satellite data (Tufa, 1994).

The satellite rainfall monitoring activity has been paralyzed since September 1995 due to the encryption of High Resolution Images. Currently, NMSA receives six-hourly data at the main synoptic hours (00, 06, 12 and 18 UTC). This has become a serious hindrance for the activities of the forecast office and the SDRPU as well. In addition to this, lack of trained manpower, lack of sufficient storage media, lack of spare parts and lack of replacements for worn out equipment are the major problems of the Agency.

3.5 Future Prospects

NMSA is committed to continue the activities of the operational satellite rainfall and vegetation monitoring project. The potential applications from planned products derived from METEOSAT and NOAA/AVHRR satellites data include winds derived from observed cloud motions, flood monitoring using multi-date single band visible imagery, surface albedo estimation using METEOSAT visible images, and surface solar irradiance estimation. Prospects for using NOAA AVHRR data include the determination of vertical temperature and moisture profiles using the Tiros Operational Vertical Sounder, crop type classification, crop condition assessment and crop yield estimation. By locating the positions of transient warm pools using water surface temperature data from several AVHRR spectral bands (channels 2, 4 and 5 during the daytime and channels 3, 4 and 5 during nighttime), useful information may be able to be provided to lake fishermen.

4. Communications

The fast and reliable exchange of up to date meteorological data, which is highly important for operational meteorology, depends on the existence of reliable and efficient communication media. Single Side Band radio telephones (SSB), teleprinters, point to point microwave Message Switching Systems (MSS), Data Collection Platforms and Data Reception Systems (DCP and DRS), and Meteorological Data Dissemination (MDD) are the communication systems which are in use to exchange weather information in NMSA. However, SSB, the main national meteorological communication medium, is becoming noisy and time- consuming as the number of conventional operational meteorological stations increases.

Data exchange with neighboring countries has used teleprinters, having speeds of 50 baud (bits per second) since the mid 1960's (Tadesse, 1995). Since June 1989, this exchange has been done by point to point microwave links using an MSS with a maximum speed of 1200 baud. The increase in speed of the data exchange system of NMSA with the regional telecommunication hub has brought a relative improvement in meteorological data exchange.

DCP use a radio transmitter to provide contact with a satellite (WMO, 1982). The DCP of NMSA were installed in April 1989 at the conventional ground meteorological stations (synoptic stations) where meteorological data are collected manually by the observers. These data are received by the DCP/DRS station at the head office via the METEOSAT DCP retransmission system. All other stations equipped with a DRS can also receive the data when it is retransmitted by the METEOSAT. The introduction of the DCP and DRS system can certainly strengthen operational meteorology. However, out of the eight DCP stations installed in Ethiopia only one is currently operational. The reason for the failure of the seven DCP stations may need further investigation, but the problems may include lack of trained man power and spare parts.

A "BURL A" MDD system was installed at the headquarters of NMSA in November 1992 by the National Resources Institution with UK ODA funding. This system was upgraded in December 1995. Unlike the previous version, this new MDD system allows more interactive time to the forecaster. It is also equipped with a facility to send charts and selected data by fax.

The upgraded MDD station of NMSA receives alphanumeric and graphical meteorological products relayed by METEOSAT from the three uplink stations of Rome, Bracknell and Toulouse. It has also contributed positively to improving operational meteorology by providing various products relevant for weather forecasting and by serving as a back up for the point to point MSS link between NMSA and the regional hub, in Nairobi.

5. Conclusion

Had it not been for the contributions of satellite remote sensing products and facilities, it would have been impossible to improve various scale weather forecasts, to carry out cost-effective drought monitoring and to create a good meteorological awareness through the media. Training of meteorologists in satellite remote sensing, ensuring a wider application of satellite remote sensing, establishing an archive system for satellite data, strengthening rainfall estimations and vegetation monitoring activity, and strengthening the technical capacity required to operate and maintain the various satellite ground receiving station equipments are major activities in the future plans of NMSA.

The METEOSAT Second Generation is expected dramatically to improve the usefulness of geostationary satellite data to African users (Lafeuille, 1993). The contribution of meteorological services in the national endeavor to build self-sufficiency in food production, environmental protection, sustainable development, and a disaster warning, relief and mitigation capacity could not be realized without the application of space technology. Therefore, by working closely with the advancement of space technology, the Agency is committed to continue the activities of previous projects and to run new ones for the benefit of the nation.

References

1. Tadege, A.: *Operational Monitoring of Rainfall and Vegetation Monitoring Over Ethiopia Using Meteorological Satellites.* NMSA mimeo, Addis Abeba, Ethiopia, 1992
2. Lafeuille, J.: *EUMETSAT Services Supporting Sustainable Development.* Paper presented at the UN Regional Conference on Space Technology for Sustainable Development in Africa, Dakar, October 1993
3. Lillesand, T.M. and Kiefer, R.W.: *Remote Sensing and Image Interpretation.* Third Edition, Chapt 1, p. 1, Section 1.1. John Wiley & Sons, New York, 1994
4. Taddese, T.: *Some Points on Meteorological Communication of Ethiopia.* NMSA mimeo, Addis Abeba, Ethiopia, 1995
5. Dinku, T.: *Effect of Smaller Calibration Zones on Satellite Estimation Over Ethiopia.* Paper presented at a workshop on Estimation of Precipitation by Satellite (EPSAT), Niamey, Niger, December 1994
6. WMO: *Satellites in Meteorology, Oceanography and Hydrology.* WMO-No 585, p. 46. WMO, Geneva, Switzerland, 1982

Promoting the Use of Meteosat for Supporting Sustainable Development in Africa

J. Lafeuille, International Affairs, EUMETSAT, Am Kavalleriesand 31, 64295 Darmstadt, Germany

Abstract

The present Meteosat satellite system, although primarily designed to serve the meteorological needs of Europe, actually provides operational geostationary coverage over the whole African continent. Four major kinds of services are made available to African countries through the Meteosat System: visible and infrared cloud imagery, collection and dissemination of ground generated observation data, dissemination of meteorological forecasts elaborated in European and African centres, and products derived from the imagery, such as rainfall estimates from Cloud Cold Duration data.

The above services are considered in Africa as an essential support for a number of applications including operational meteorology, water resources management, disaster early warning and desertification monitoring. However developing countries are still far from using all the capabilities of the system, and a limited well-focused effort will help them to take a much larger benefit out of the existing and future satellites. This cooperation is expected to enable the African centres to play their full role within the world meteorological community.

In close consultation with the African users EUMETSAT has identified the following priorities: to enhance interaction with the user community at a regional level, to ensure that data and services are affordable, to support user training , and to facilitate technical access to the data

Practical achievements are reported below. The relationship established between the satellite operator - EUMETSAT - and the regional user community is highlighted as a key element in the success of these actions. Developing strong relationships with selected subregional centres also contributes to the process of local capacity building. EUMETSAT intends to work along these lines with its African partners and to coordinate its efforts with aid agencies whose objectives in this area may be complementary to those of EUMETSAT.

1. Background

The United Nations Conference on Environment and Development (UNCED) held in Rio de Janeiro in June 1992 stressed the need for international efforts to combat desertification which is thought to affect about one sixth of the world population. Sustainable land-use and water-management policies, which are key components of this effort, require extensive weather satellite data. The World Meteorological Organisation (WMO) identified several priority orientations to support this challenge:

- monitoring/assessment of drought and desertification
- application of meteorology and hydrology to development programmes, especially for land-use and water-management projects
- research on weather and climate affecting critical areas
- training and education
- support to regional centres of excellence.

EUMETSAT, the European Organisation for the Exploitation of Meteorological Satellites, is seeking to respond to those recommendations in the implementation of its programmes, in accordance with its charter. In this respect, special attention has been paid to the use of the Meteosat system in support of developing countries in Africa. This initial focus on Africa is primarily motivated by geographical location, since the African continent is at the centre of the Meteosat field of view. It is also reflecting the strong cooperation links established bilaterally between many meteorological services of EUMETSAT member states and African partners.

2. Areas of Applications of Meteosat in Africa

EUMETSAT is operating the Meteosat satellite series which ensures continuous observation in three visible and infra-red channels from the geostationary orbit at 0° longitude. This mission will be pursued in the long-term with significant improvements being added by the turn of the century through the Meteosat Second Generation (MSG). This will provide additional spectral channels, improved geometrical resolution and will deliver full disk images every 15 minutes instead of the 30 minutes for the current Meteosat system.

The Meteosat system was primarily designed to serve the meteorological needs of Europe, in particular for severe weather warning and short-term forecasting. Meteosat is also the European contribution to the space component of the World Weather Watch, and enables unique wind data and various meteorological products to the world wide community to be derived.

Meteosat is reported by African users as a very useful tool for operational meteorology and for a number of environment related applications. Due to their favourable location at the centre of the Meteosat field of view, users from the African continent are benefitting from the finest horizontal resolution of the visible and infrared imagery. Among the products derived from the imagery, the rainfall estimates based on Cold Cloud Duration (CCD) are of particular relevance to Africa. Extensive use is also made of the various communication capabilities of the Meteosat system, for the collection and dissemination of ground generated observation data, as well as for the regional dissemination of meteorological forecasts made in European and African centres.

2.1 Operational Meteorology

In the context of African developing countries, Meteosat helps to overcome the poor status of the ground based infrastructure which is far from being sufficient to fulfil the operational requirements for meteorological observation and telecommunications.

The central part of Africa is permanently affected by the Inter-Tropical Convergence Zone (ITCZ) whose yearly variation drives locally the alternation of dry and rainy seasons. Air traffic safety also requires important meteorological support which would not be possible without weather observation from space. Furthermore, difficulties in terrestrial communications increase the need to rely on automatic Data Collection Platforms (DCP) transmitting their data via satellite. The African meteorological centres tend to rely more and more on the information flow provided by the Meteorological Data Distribution (MDD) system, which broadcasts via Meteosat selected meteorological data and forecast products. Altogether, the Meteosat system is thus considered by WMO as an essential operational component of the World Weather Watch in Africa.

2.2 Rainfall Estimates and Support to Resource Management

Estimating the rainfall provides a key input for the assessment of desertification and for the management of water resources with all its implications for crop monitoring and food security. The precipitation estimates derived from CCD on a 10-day basis are therefore likely to be the main contribution of Meteosat image data in support of development projects in Africa.

As discussed by several authors, this approach is based on the correlation between the amount of rainfall and the occurence of high cloud tops below a certain temperature threshold. When applied over a large area in a period of at least ten days, this approach provides an extremely valuable input for a wide range of applications, although its performance is still limited by the rough identification of cloud type. It is currently used with minor adjustments according to the area concerned and the specific end-applications which themselves may induce particular constraints, e.g. in terms of threshold and time scale. Significant improvements are sought through combination with other sources of data, such as microwave brightness temperature. Further improvements are expected with MSG since the multispectral data should facilitate the identification of rain clouds.

A major application of this approach is the Africa Real Time Environmental Information System (ARTEMIS) set up by the Food and Agriculture Organisation (FAO). The ARTEMIS processes CCD and rainfall estimates from Meteosat image data received every hour through a Meteosat receiving station located in Rome. These data are then combined with vegetation indices generated from NOAA/AVHRR data. The ARTEMIS is used to provide vegetation monitoring reports and crop estimates to a large number of users, both in Africa through a satellite Direct Access Network for Africa as well as to various applications within FAO headquarters such as the "Emergency Center for Locust Operations" and the "Global Information and Early Warning System". This application is now well demonstrated and is expected to be transferred to African centres to be operated at a regional scale.

2.3 Disaster Early Warning

The continuous availablility of geostationary imagery makes it an important tool to support near-real time warning against natural plagues or other events such as bush fires, floods, sand storms and desert-locust invasion. It may be recalled that desert-locust invasions regularly affect the African continent and can result, for the countries concerned, in a dramatic famine whilst involving several hundred millions of US $ of emergency measures by the world community. Satellite images help to identify at a proper scale and in a timely manner the zones (with respect to temperature, moisture and vegetation) which are likely to attract the locust swarm and where operational control activities should therefore be concentrated.

2.4 Climate and Desertification Monitoring

"Strengthening the knowledge base and developing information and monitoring systems for regions prone to desertification and drought" was among the preventive measures highlighted by the UNCED in *Agenda 21*. High resolution remote sensing satellites, meteorological polar orbiters and geostationary satellites there have complementary roles.

The direct monitoring of desertification and deforestation, although mainly relying on high resolution images from remote sensing satellites such as SPOT and LANDSAT, may be usefully complemented by weather satellites such as NOAA which offer a lower resolution but a daily coverage, thus maximising the possibility of cloud free images.

The geostationary satellite information provided by Meteosat is the basis for the monthly rainfall estimates, the identification of squall lines and the migration of the ITCZ which are key aspects of climate studies relevant to Africa. The continuous coverage of the African continent by Meteosat is also providing a unique data set for the global International Satellite Cloud Climatology Project and for Earth radiation budget studies.

3. Actual Level of Utilisation

The widespread use of meteorological satellite data is facilitated by the decentralized character of the ground segment. This is an essential feature of these systems since they are primarily designed to meet a requirement for real-time access by multiple users. In practice all operational meteorological satellite systems coordinated by the Coordination Group of Meteorological Satellites have a direct broadcast capability which enable low and/or high resolution data streams to be received at affordable individual user stations. The situation is thus significantly different from that of remote sensing satellites which are normally not accessible directly by the end-user but via a ground station operator.

The current utilisation of Meteosat in Africa is illustrated by the number of user stations registered in this region (Table 1). This includes the Primary Data User Station (PDUS) receiving high resolution digital imagery, the Secondary User Stations (SDUS) receiving low resolution image formats, and the MDD receivers which are a dedicated communication capability restricted to the National Meteorological Services. The DCP are normally not understood as "user stations", but are mentioned here for completeness.

Type of user station	Function	Number registered (February 1996)
PDUS	High resolution imagery (digital)	34
SDUS	Low resolution imagery (analogue)	65
MDD	Ground observations and charts	37
DCP	Data Collection Platform	120

Table 1. Registered Meteosat user stations and DCP, in Africa

The numbers of user equipments are certainly encouraging. However, in view of the applications of Meteosat successfully demonstrated in Africa, there may still be scope for expansion. Moreover, the actual level of utilisation of the system by the present users may sometimes remain below the available capability of the system. Different kinds of limitations have be identified, related either to the availability of suitable equipment, to a lack of up-to-date information on the system, or to the availability of trained technical staff.

Furthermore, at the turn of the century the MSG and the EUMETSAT Polar System will become available. There is some concern that, unless a wide-scale effort is undertaken now, the user community will not be prepared to take full benefit from these new generations of satellite systems. Action is needed to improve the actual use of the current systems and to anticipate the transition to the new systems and products which will be available around the year 2000. In cooperation with African partners, EUMETSAT has undertaken to identify and initiate actions in response to these needs.

4. Strategy to Maximise the Usefulness

Experience suggests that a limited effort specifically oriented towards developing countries can help them to take a much larger benefit from the capabilities of the current and future meteorological satellite systems. Efforts should aim at facilitating the access to the data, ensuring that the data and products are applicable to meet the regional requirements, and ensure that the regional user community is in a position to use them.

The following strategic lines have been identified by EUMETSAT in this respect:

- to maintain a regular information flow between EUMETSAT and the African user community, in order to inform the users on the capabilities of the current and future systems and the ways to use these capabilities,

- to support the expression of regional user requirements which can help EUMETSAT to adjust its operational services to best address the regional needs,

- to ensure that data and services are affordable, through a very open data policy towards developing countries,

- to support the training of qualified technicians through a long-term cooperation with two African institutes identified by the WMO as Regional Meteorological Training Centres (RMTC), and

- to facilitate technical access to data and services, through the implementation of MDD and through encouraging the development of low-cost, standardised user stations for the current - and even more for the future - generations of satellites.

These objectives are pursued through direct interaction with the user community at the regional level. This is facilitated by the existing regional structure of the meteorological community which, whilst it does not include the whole of the Meteosat user community, certainly represents its driving force. In addition, relationships have been strengthened with selected regional entities which play the role of "focal points" and assist EUMETSAT as the permanent interface with the individual users. It is the opinion of the author that this role of focal point is not only a support to EUMETSAT to ensure the best efficiency, but also contributes to publicise the action and capacity of these centres of excellence. This cooperation with selected regional centres is a contribution to regional capacity building.

EUMETSAT is undertaking many of these actions in cooperation with the WMO, whilst maintaining coordination with other organisations through international fora such as the Committee for Earth Observation Satellites or the Space Agency Forum. Since EUMETSAT has no mandate to act as an aid agency, its support dedicated to developing countries can only involve a minor percentage of its annual budget (typically at the order of 0.3 % in 1996). This nevertheless allows it to undertake significant, well focused actions and to pursue them in the long term if relevant.

5. Practical Achievements

5.1 The EUMETSAT User Forum

An initiative was taken to convene the First EUMETSAT User Forum in Developing Countries, in cooperation with the African Centre of Meteorological Applications for Development (ACMAD), the National Meteorological Service of Niger (Météo Niger) and the AGRHYMET Regional Centre. The meeting was held in Niamey, in April 1995.

The objectives of this Forum were:

- to provide up-to-date information on EUMETSAT, its systems, programmes and policy,
- to collect users' requirements for products, data access and training,
- to outline a training support project,
- to exchange experience on regional applications, and
- to strengthen relationships with the African meteorological community.

The event, inaugurated by the Minister of Equipment and Transportation in the presence of the Minister of Research and Education, the United Nations Development Programme (UNDP) Resident Representative, and other personalities, attracted 70 participants from 30 different African countries. The audience was made up mainly of people with technical backgrounds and regional decision makers. Most MDD and High Resolution Image (HRI) users were represented, as well as four RMTC.

In its final plenary session, chaired by the President of the Regional Association 1 (Africa) of WMO, the User Forum reviewed the draft recommendations prepared by three working groups dedicated to training, MDD requirements and data policy, respectively. In the light of the amount of information exchanged during the first forum, the participants unanimously recommended to prepare a second EUMETSAT User Forum. All the recommendations from this First User Forum were then addressed by the EUMETSAT Council and its subsidiary bodies.

It was then agreed that similar events should be organised every two years in future. The Second EUMETSAT User Forum in Africa is now being prepared for December 1996, in cooperation with the WMO and regional partners.

5.2 Conditions of Access to Data

Special consideration has been given to users from developing countries when establishing the conditions of access to EUMETSAT data. It should be recalled that in the context of growing commercial value-added activities and of growing cost of observation systems and data handling facilities, the meteorological services can only justify and secure the continuity of the observation systems if the value of meteorological information is clearly recognized. A distinction is thus being made within WMO between "Basic data", which continue to be exchanged worldwide for public benefit without any kind of restriction, and "Additional data" which may be exchanged under certain conditions and possibly subject to a fee. EUMETSAT has developed a similar approach for its satellite data.

In practice, EUMETSAT is providing on a free and unrestricted basis the WEFAX data and products analogue dissemination, as well as four full disc HRI per day in each channel. Since 1995, all other data, whether HRI or MDD, are now provided in an encrypted form and access to these data is subject to a license agreement with EUMETSAT. The National Meteorological Services are granted licenses free of charge to access the MDD dissemination and three-hourly HRI for their national use. Licenses are also granted free of any charge for research and educational institutes as long as they do not carry out any commercial or operational activity. In other cases and, in particular, for commercial service providers and broadcasters, the HRI data can be accessed against a fee.

In respect to developing countries, the EUMETSAT Council has decided that National Meteorological Services from countries with a GNP per capita lower than 2000 $ should be allowed to receive all Meteosat data, for their internal use, without any charge. The threshold of 2000 $ means in practice that every country which is less wealthy than any of the EUMETSAT member states can obtain all Meteosat data free of charge for the operation of its National Meteorological Service. In addition the EUMETSAT Council reserved the possibility to consider on a case-by-case basis special licences or cooperation agreements, including provisions for the free access to data. Several cooperation agreements have been concluded; two more are currently being prepared with ACMAD and with the AGRHYMET Regional Centre.

This policy is supporting the National Meteorological Centres of developing countries to play their roles as focal points for meteorological information in their country. Further, it is helping them to bring their contribution to the world wide meteorological community.

5.3 Training

EUMETSAT has agreed to offer its long-term support to two Regional Training Centres in Africa, established in Niamey and Nairobi, in response to a request by WMO. Two workshops have been held in Niamey and Nairobi in order to review the training requirements and to define a precise training support plan. The main goal is to "train the trainers" in satellite meteorology and to stimulate and support the development of a foundation course with associated pedagogical tools.

A training course for African meteorologists is now being prepared with European and African experts. This course will address the physical basis of remote-sensing, describe the basic imaging systems from GEO and LEO satellites and put emphasis on image interpretation and its relation to synoptic scale and smaller scale weather patterns. The course will also include case studies drawn from the regional meteorological and operational context. A first three weeks training session for twenty African trainers is planned for October 1996 in Niamey, and a second one is planned for early 1997 in Nairobi. In parallel, EUMETSAT participates with WMO in a project initiated by the German cooperation agency (GTZ) to develop Computer Aided Learning modules for African meteorologists.

5.4 Meteosat Data Distribution

An additional mission of the Meteosat system is the MDD, whereby a transponder aboard Meteosat is used to rebroadcast ground-generated information received from three MDD uplinks. This has been implemented at the request of the WMO in order to provide the National Meteorological Services of Africa with an alternative - or a complement - to the Global Telecommunications System through which all National Meteorological Services can gain access to the basic meteorological data and forecasts needed for their operation.

Initially two MDD uplinks had been implemented, to broadcast observations and meteorological charts, respectively. A third channel was implemented in early 1995 in order to provide an additional capacity for meteorological information of special relevance to Africa. The MDD mission offers an easy and reliable way to disseminate specific products elaborated either by EUMETSAT or by African entities. Actually the new channel currently broadcasts some products generated by African centres such as ACMAD in Niamey and the Drought Monitoring Centre in Nairobi.

The African user community has contributed to the expression of user requirements and to the experimentation of the MDD and now participates in the evaluation of the dissemination programmes. In the MSG time frame the Low Resolution Image Transmission channel will replace both the current WEFAX dissemination and the MDD information flow. This should offer further opportunities to disseminate other specifically African products.

6. Conclusions

The relationship established between the satellite operator - EUMETSAT - and the regional user community is the key element for the success of promoting the use of Meteosat in Africa. Developing strong relationships with selected subregional centres is considered very helpful and contributes to the process of local capacity building which is a key factor for sustainable development. The practical achievements described are expected to help the African users community to strengthen its regional capacities and optimize the benefit received from satellite systems.

It is the wish of EUMETSAT to continue to work along these lines and to strengthen its cooperation with regional users' communities in developing countries. EUMETSAT is also ready to coordinate its efforts with donors who may have complementary objectives in this area, EUMETSAT assisting as far as possible through its expertise and experience in the cooperative projects that may be established between donors and regional entities.

Commercial High-Resolution Imagery From Space as a New Information Service to Humanity: A Video Address

J. Neer, Space Imaging, Inc., 9351 Grant Street, Suite 500, Thornton, CO 80229-0939, USA

I am honored to be on the same program as Dr. Arthur C. Clarke and to address this prestigious institution. Few have Dr. Clarke's vision of the potential of space to serve mankind. In 1945, as World War II was ending and the Cold War beginning, Dr. Clarke envisaged a world connected by geosynchronous communications satellites. As soon as World War II ended, the Race to Space began. The United States and the former Soviet Union entered the Space Age in 1957. Concurrent with this dawn of the Space Age was the foundation of both global communication and reconnaissance. The world has witnessed a remarkable transformation over the past four decades. National boundaries have disappeared and the physical and political walls separating one country from the other have come down. Thanks to visionaries such as Dr. Clarke and industrial pioneers such as Dr. Harold Rosen, individuals are now able to hear their neighbor's heartbeat in real time, no matter where in the world they may live. From yesterday's satellites, which weighed less than 40 kg and communicated with Earth antennae 30 m in diameter, to today's 4000 kg high-power satellites which communicate with hand-held telephones, advances in space communications technology have resulted in greater service to humanity by promoting the open exchange of information among all inhabitants of a shrinking world.

Less notable, but equally dramatic, was the pre-dawn epic of the age of transparency. In the late 1950s, the United States initiated the Discovery program, later known as Corona. Only recently declassified, the development of Corona and successive, high-resolution space reconnaissance systems have served vital democratic and global security interests. As Dr. Clarke notes in his book *How the World Was One,*

"A stable peace will never be possible without mutual trust. Yet trust cannot be blind. Ignorance and the fear it generates can be dispelled only by accurate and timely information."

Dr. Clarke suggests that a satellite with a resolution of, say, one meter could be realized without the cooperation of the nuclear powers and its results made available to everyone. Dr. Clarke's vision of 1991 will become reality late next year, when Space Imaging, Inc. will offer the world commercial, high-resolution 1 m black and white or panchromatic imagery and simultaneously collected 4 m, four-band blue, green, red and near-infrared imagery. Operating in a Sun-synchronous, near-polar Low Earth Orbit at an altitude of 680 km, Space Imaging satellites will collect digital imagery of the Earth at a rate of about 1500 sq km/minute. Using a stellar referencing and the Global Positioning System along with a high-speed ground processing system and selective ground control points, 1 m pixel patches can be geospatially fixed to Earth coordinates with less than 2 m positional error. This high-resolution imagery will redefine the space imagery and Geographic Information Systems industry, and truly open the age of transparency for all the world to access commercially, free from governmental and institutional constraints. The global economic viability of high-resolution space imagery is a clear demonstration that space may constitute a cost-effective base for services to humanity.

We will soon be able to see the global village from a high vantage point in space. This new, commercial use of space will dissolve for ever the artificial and opaque walls that separate us. As our knowledge grows through more accurate, more timely and ubiquitous Earth information, the world will truly move closer to the reality of a global village. Routine access to valuable information will no longer be restricted to the world's superpowers. All nations will be able to benefit from a clearer view of their land and regional geography, and will better value their interrelationships with other peoples of the world. If we, as a global village, are to continue our collective social, economic and intellectual development, we must do so in an environment of mutual trust and understanding. The economic by-products of decades of political posturing and the subsequent build-up of advanced defence technologies have forever changed our world. The economic growth and global security benefits arising from space communication are manifesting themselves in new services to humanity. As we prepare to enter the next century, high-resolution commercial space imagery will solidify global security, enhance our understanding of the world, the impact which we have on it and the economic benefits to be derived from it. This new information about our Earth will dramatically extend humanity's knowledge base. The dawn of the age of transparency is just over the horizon. Prepare for it!

INPE's Space Activities: Its Way of Putting Space of Service to Humanity

M. Lopes de Oliveira e Souza, Division of Space Mechanics and Control, National Institute for Space Research (INPE), CP 515, São José Dos Campos, São Paulo, Brazil

Abstract

This key-note address summarizes INPE's Space Activities, and demonstrates its way of putting Space of Service to Humanity. To do so, this paper mentions the past and present of INPE's work, detailing the activities and applications of its main areas: Space and Atmospheric Sciences, Meteorology, Remote Sensing, Oceanography, Space Engineering and Technology, and others. Finally, this work presents some initiatives tied to the future of INPE, Brazil's National Institute for Space Research, and to its efforts to put Space of Service to Humanity.

1. Past and Present of INPE

INPE was founded on August 3, 1961 with a small staff in the city of S. José dos Campos, in the state of São Paulo as Brazil recognized the early space achievements abroad, and aimed to make them relevant to Brazil. It began Space and Atmospheric Sciences research via international collaboration with NASA and other agencies, and with a Space and Atmospheric Sciences Graduate Course. Today, INPE has about 1300 persons working in 9 facilities, in the areas of Space and Atmospheric Sciences (Astrophysics, Spatial Geophysics, Aeronomy, etc.) Space Applications (of Remote Sensing, of Meteorology, of Telecommunications, etc.), Space Engineering and Technology (including Integration and Tests, Tracking and Control, etc.), and related areas such as Computer Science, Materials Science, etc. These also support 7 Graduate Courses in Space Geophysics, Astrophysics, Meteorology, Remote Sensing, Space Mechanics and Control, Combustion, and Applied Computation.

2. Space and Atmospheric Science Activities and Applications

In Space and Atmospheric Sciences, INPE has classic objectives and areas of study. Here we highlight the monitoring and the understanding of changes in global atmospheric composition, of the changing greenhouse effect, of the Antarctic ozone hole, and of some special phenomena within and near Brazil (such as magnetic anomalies). These have been carried out for a long time (18 years) via many monitoring stations installed in other states (such as Eusébio at Ceará, S.Luís at Maranhão) for balloon borne experiments, and at the Brazilian Antarctic Station "Comandante Ferraz"(62.1°S; 58.4°W) of Program Antarctica maintained by the Brazilian Navy, CNPq and INPE. Results have been obtained comparing the normal ozone layer (September 15, 1992) with the ozone hole condition (October 3, 1992), and as a three-dimensional view of the Antarctic ozone hole at the times.

To monitor changes on the Magnetic Equator and their effects on the equatorial electrojet, INPE jointly organized and participated in the Guará Campaign with NASA and other respected institutions. This was the largest campaign to date (50 scientists, > 300 engineers and technicians), launching 20 meteorological rockets and 13 sounding rockets to apogees ranging from 65 to 800 km, during August-October, 1995, from the Alcantara launch site. The corresponding measurements are being presented to specialist congresses and in the scientific literature.

To monitor station positions and the motion of the Earth, INPE installed and operates a VLBI station with NOAA/NOS/GL and other institutions. With an antenna of 14.2 m diameter in Eusébio at Ceará, this started operating on August 21-22, 1993 by finding its coordinates in the WGS-84 system (3° 52'40.2966" S; 38° 25'33.0906" W; altitude 23.0820 m). It is part of the international network of geodetic VLBI stations and contributes to the International Earth Rotation Service, IERS, in related applications (such as cartography, navigation and precise geodesy) and research programs (on anomalies in the Earth's rotation, geodynamics, motions of tectonic plates, and quantitative evaluations of global changes).

3. Meteorology Activities and Applications

It is known that the Index measuring the success of weather forecasts is lower and shorter (I < 60% in 2 days) when using subjective methods than when using numerical methods in the northern hemisphere (I > 60% in 8 days) and even in the southern hemisphere (I > 60% in 6 days). This fact and the variety of climates in Brazil motivated the Brazilian Government to envisage a partnership with the states to develop Regional Application Centers for climate, weather and hydrological monitoring and prediction. The first such center was established between INPE and "Fundação Cearense de Meteorologia e Recursos Hídricos-FUNCEME"of the State of Ceará, the driest state. The system has its heart in INPE's Center for Weather Predicition and Climate Studies, CPTEC, inaugurated in November 1994. Its main computers, including the largest supercomputer available in Brazil, are being used to produce forecasts for many users, including the daily newspapers, and to study special phenomena such as the "El Niño", its consequences in temperature and precipitation anomaly patterns, and the corresponding damages. The INPE-FUNCEME partnership has helped to increase grain production and to reduce losses, despite the extreme variations in the precipitation around its mean value.

4. Remote Sensing Activities and Applications

Noting the advances of Earth observation techniques, in 1967 Brazil decided to learn more about its territory, by acting in two complementary ways: 1) creating the Project RADAM, to cover its 8.5 million km^2 with radar and satellite image catalogs in 10 years; and 2) creating a team and facilities at INPE to learn and to use the new techniques of remote sensing. INPE began receiving remote sensing data and images with the first satellite of the LANDSAT series in 1973, just after the USA and Canada. From then until 1983 many pilot projects were developed to demonstrate and to spread the techniques through governmental and private users. Many associate users' centers have been set up since 1983-1984. Today, INPE supports the recording, processing and distribution of remote sensing satellite products (LANDSAT5 and SPOT3, and maybe ERS1 and RADARSAT in the future) through a system which mainly uses monitors, CD-ROMs, etc. instead of paper images. It includes 12 private companies and more than 100 of those users' centers with demands that place Brazil as the third largest user of those products in the world. This has permitted government agencies, scientists and staff to monitor, for example, remote Amazon regions, urbanization, patterns of deforestation, irrigated agriculture, etc. It has also led to studies of burning in the Amazon region, confirming data from other satellites, and established the change in deforestation rates. The corresponding data are available through a Data Bank System that is being improved for easy access by the community worldwide.

5. Oceanography Activities and Applications

INPE is engaged in the growing use of space techniques to study the South Atlantic Ocean, its currents, its exchanges of mass and heat and its contributions to global and local weather and climate. One example is the Project Oceanic Circulation of Western Region of the South Atlantic Ocean, COROAS, that is part of the World Ocean Circulation Experiment, WOCE. This includes: 1) mapping of the sea surface temperature from data of the AVHRR-NOAA sensor; 2) launching and tracking (through the ARGOS system), 10 buoys whose data were compared and analysed with those from the scientific ship "Professor Besnard" of the Oceanographic Institute of the University of São Paulo; 3) the installation of new recording equipment and the preservation of the old on the North East coast of Brazil, to help the fisheries.

6. Space Engineering and Technology Activities and Applications

After developing a solid activity in Space and Atmospheric Sciences since the 1960's, and in Space Applications since the 1970's, INPE with other governmental agencies decided to go into Space Engineering and Technology activities for many reasons, including the need to build our own data collection systems and remote sensing satellites after learning how to use other such equipment. This began in 1979 with the approval of the Program Brazilian Complete Space Mission, MECB, initially to build four satellites (two Data Collecting Satellites, SCD1,2, and two Remote Sensing Satellites, SSR1,2), one Satellite Launcher Vehicle, VLS, and one Launcher Center at Alcantara, CLA. Whilst the characteristics of the SCD1 are modest, it was a big challenge, from the construction and tests to the assembly for launching, using the Pegasus system, following a nonconventional but successful mission profile on February 9, 1993, with collaboration from NASA. Since then the SCD1 has worked well, being monitored and controlled by INPE's stations and outperforming its intended 6 months lifetime by several times. Its success has motivated the completion and launch of the SCD2 by the end of 1996, and even the approval of a better version named SCD3. Today, the Data Collecting System built up with the SCDs has around 250 Data Collecting Platforms (DCPs) installed throughout Brazil, including 200 just bought for the National Department of Water and Electrical Energy, DNAEE, to monitor rain, river and dam regimes nationwide. Forty new DCPs are planned to be installed in other countries of South America.

The SSR1,2 will have almost equatorial orbits to cover the important areas of Brazil (+5°; -15°, especially the Amazon region), and to complement existing sytems. They are in development now, with moderate characteristics. Their medium size and complexity provided the background for a more ambitious binational program, the "China Brazil Earth Resources Satellites, CBERS1,2". The technical characteristics of these satellites, their spectral bands and their use are more complex than those of the SSR1,2, but reachable. Today, construction proceeds towards a launch date by the end of 1997.

One by-product of this program is the opportunity to launch our first Scientific Applications Satellite, SACI1, as a piggyback to the CBERS. This is being built at INPE to carry 6 experiments of the scientific community. This experience is stimulating us to propose microsatellites and small satellites in cooperation with other countries, especially France and Argentina.

7. Other Activities and Applications

INPE has Associate Laboratories of Plasma, of Sensors and Materials, of Computational and Applied Mathematics, and other facilities to support the previously mentioned activities and to develop related fields of research.

8. The Future of INPE

Besides proceeding with the above mentioned programs, INPE is engaging in other initiatives to promote the peaceful use of outer space. Among them, we highlight Brazil's participation in the US "Mission to Planet Earth" through the construction and use of cameras and associated equipments to fly in the Space Shuttle. This puts Brazil in a select club of countries involved in such interdisciplinary investigations. A second initiative is that INPE was elected (in September 1994 in Mexico City) to host the Inter-American Institute For Global Change Research, whose headquarters is inaugurated in February 1996. A third initiative is that INPE was chosen to co-host (with a Mexican institution) every other 4 years the headquarters of the Center for Space Science and Technology Education for Latin America and the Caribbean, an initiative of the UN Office of Outer Space Affairs. The initial activities of this center are planned for 1997.

9. Conclusions

Since its foundation, INPE has made every effort to develop internal space activities, to use them to meet Brazilian needs, and to cooperate with other institutions and countries of the Earth. This is INPE's way of putting Space of Service to Humanity, that will proceed into the future.

References

1. *INPE. Atividades do Instituto de Pesquisas Espaciais: 1987.* INPE, Brazil, 1988
2. *INPE. Relatório de Atividades de 1988-1989.* INPE, Brazil, 1990
3. *INPE. Relatório de Atividades de 1990-1991.* INPE, Brazil, 1992
4. *INPE. Relatório de Atividades de 1992-1993.* INPE, Brazil, 1994
5. *INPE. Relatório de Atividades de 1994,* INPE, Brazil, 1995
6. *INPE. Primeiro Satélite Brasileiro de Coleta de Dados SCD-1.* INPE, Brazil, 1992
7. *INPE. The Amazonia Program.* INPE, Brazil, 1993
8. *Pathways to Space: 30 Years of INPE.* Editora Contexto, Brazil, 1991
9. Novo, E.M.L.M. et al.: *20 Anos de Sensoriamento Remoto no INPE.* INPE, Brazil, 1991

Environmental Management in Developing Countries: Remote Sensing Case Studies

V. Singhroy, Canada Centre for Remote Sensing, 588 Booth Street, Ottawa K1A 0Y7, Canada

Abstract

One of the most serious threats to sustainable economic growth in developing nations is the increasing degradation of their natural ecosystems and the concurrent deterioration in the quality of life for their people. This paper presents two case studies on the utility of remote sensing as an environmental assessment and protection tool in two developing countries. These case studies were selected to represent the two major climatic regions where the majority of the world's poor live. Guyana is a small tropical country on the northern coast of South America where 80 percent of the country's revenues are used to service its debt. Jordan is a small arid to semi-arid country in the Middle East where the search for drinking water is directly linked to its development. In both these countries, as in many other developing countries, earth observation data are necessary for sustainable development.

In Guyana, all of the productive agricultural lands and 90% of the population of Guyana live in the narrow and fertile coastal plain (500 x 20 km) which is below sea level. Recently, the coastal dikes constructed for sea defence and improving irrigation of agricultural areas have been eroded and broken in various areas, resulting in severe flooding. Large sections of the sea wall are collapsing and approximately 2/3 of the coastline needs permanent structures or replacement. It is estimated that future large scale flooding could destroy almost U.S. $1 billion of economic activity in the coastal areas. The failure of the coastal defences is the result of a combination of factors. These include the depletion of the mangrove forest buffer caused by wood cutting for fuel and choking by sling mud; the increase of wave activities possibly due to a gradual rise in the mean sea level; and the lack of maintenance of the engineering structures (dikes and earth dams).

This case study discusses the use of combined synthetic aperture radar (SAR) and TM images for coastal zone mapping, for determining the priority areas of coastal repair and for the assessment of flood damage of the agricultural areas. The study has shown that parts of the Guyana coastline have changed by from a few meters to half of a kilometre over the past twenty years. This has serious implications for sea defence and commercial agriculture. Areas of erosion and accretion, mangrove forest depletion and agricultural land uses were identified.

In Jordan, there is an urgent need to manage and utilise the water resources in order to comply with the immediate and projected water demand in the country. All the known aquifers have been explored. Water harvesting is a national priority as water supplies are being depleted by intensive pumping. The main concern is to optimise the utilisation of the ground water resources and to protect them against depletion and pollution. For this reason, it is necessary to map in detail the hydrogeology of the aquifers. This second case study reports on the use of C-HH SAR and SAR/TM integration for providing information on the surface hydrogeological characteristics and land use practices of these sensitive watershed areas. The SAR image was used to map wadis (their surface distribution, channel morphology, and surface roughness of the wadi sediments); hydrogeological characteristics of rock units - limestone, basalts and alluvium; fractures (used to target exploratory wells); and land practices necessary for environmental protection of the aquifers. Although the SAR image itself provided very useful information, a combination of SAR and TM provides a more suitable image to map the surface characteristics of the aquifers.

These two case studies demonstrate the need to use appropriate remote sensing techniques to assist in environmental protection programmes in the developing world.

1. Coastal Zone Mapping : Guyana Case Study

1.1 Introduction

The need for a sustainable approach to sea defence and coastal land development projects has been recognised by the Government of Guyana and a number of international donor agencies, namely the Caribbean Development Bank, the Inter-American Development Bank, the World Bank, and the European Economic Community. The Guyana case study demonstrates the role of remote sensing images in environmental management in a developing country. The study assisted in current sea defence and coastal land development projects, and provided a guide for the interpretation of SAR images for tropical coastal zone mapping. SAR and TM synergistic images were used to map the coastal geomorphology in order to locate and monitor the areas of coastal erosion and accretion. Recent remote sensing image maps provided a means of updating land cover information obtained from the 1979 air photographic coverage.

The coastal zone of Guyana is approximately 530 km in length and varies from 12 to 77 km in width. Ninety percent of the population (700,000) live on this narrow coastal plain. Elevation varies from 0.5 m below to 2.4 m above mean sea level. This coastal plain, which occupies only 7.5 percent of the total land area of Guyana, is the main agricultural region of the country, with sugar and rice being the principal agricultural exports. These crops require drainage and irrigation to maintain their agricultural productivity. Sea defence, the construction and maintenance of drainage, irrigation canals and other land development projects are therefore vital to the economic growth of Guyana. Sixty percent of the coastline is protected with some form of sea defence (earthen dams, concrete dikes, and boulder slopes). Recently, the increasing frequency and intensity of coastal storms has caused considerable damage to the sea defence structures, resulting in serious flood damage to residential areas and crops. Over the past twenty years, there has been little maintenance and rehabilitation of the sea defences. The sea walls have exceeded their design life time and are at the verge of collapse due to material deterioration and ongoing coastal erosion. In addition, any rise in mean sea level from global warming will virtually destroy all economic activity in the coastal areas. For these reasons, detailed land mapping is necessary , and shoreline changes need to be monitored.

Aerial photography at various scales has been used extensively for the various land development and mapping programmes. Coastal Guyana has a fairly good series of historical air photos. Partial coverage has been available since 1940, and the most recent 1:20,000 black and white scale photography of the coastal belt was acquired between 1979 and 1982. For twenty years (1972-92), there were only two LANDSAT images of coastal Guyana with less than 20% cloud cover. The first was the LANDSAT MSS (80 m) scene of October 1977, and the second was a September 1992 LANDSAT TM (30 m) scene. Using the LANDSAT MSS data, Singhroy and Bruce (1983) produced a regional coastal zone map describing the landcover/land use, surficial geology, geomorphology, and coastal water classes. With the acquisition of airborne synthetic aperture radar data by the Canada Centre for Remote Sensing in June 1992, it is important that the SAR images be interpreted so as to determine their uses in coastal zone mapping. This interpretation also provides a guide for the use of orbital SAR images now available from RADARSAT, ERS, and JERS. In addition, there is a need to develop a long-term monitoring system to provide early warning for areas at risk from coastal erosion. Air photography and field mapping of the coast have shown changes that occurred along the coast. However, this system does not provide an ongoing method to predict coastal erosion (Pastakia, 1991).

1.2 Study Area - Guyana

The study area extends for 100 km of coastline between the Demerara and Berbice rivers, which is one fifth of the entire Guyanese coastline. Georgetown, the capital city, has a population of approximately 350,000 and is located on the mouth of the Demerara river. This area is drained by the Demerara, Berbice, Abary, Mahaica and Mahaicony rivers. With an average of 80 inches of annual rainfall coupled with the tidal influences, the major Demerara and Berbice rivers frequently overflow, creating extensive natural levees. Currently, drainage and irrigation development projects have considerably reduced this overbank flooding.

The geology, geomorphology, and agricultural land use of the study area are described in several publications (Bleackley 1957, Daniel 1986, Singhroy and Bruce 1983, and others). Topographically, the whole coastal zone has been mapped in detail at 1:50,000 by the British Ordinance Survey. The study area 100 x 50 km is below sea level, with a gradual rise of 1 to 4 m in areas covered by beach deposits. Approximately 30 kilometres from the coastline, the land rises to a height of 10 m in the area of white sand.

The surficial sediments include mud flats, fluvio-littoral sand ridges and aprons, alluvial silt and clay, and pegasse (organic) accumulations. These deposits are underlain by recent soft marine fossiliferous clay of the Demerara Formation. Further inland within the coastal zone, interbedded Pleistocene silty clay, sand of the Coropina Formation, and Pliocene coarse sand (White Sand Series) are found (Bleackley, 1957).

The original vegetation was mixed tropical forest, most of which has been cleared for agriculture and settlement. The Wallaba forest occupies the inland white sand plateau. Mangrove forests are found on the tidal flats, and pegasse swamps occupy the inland depressions adjacent to the white sand areas. The major commercial agricultural crops include sugar, rice and coconuts. Secondary agricultural activities include tropical fruits, market vegetables and livestock.

1.3 Methodology and Image Interpretation

Airborne C-VV 40 km, wide-swath SAR image, with a resolution of 10 x 20 m SAR, was acquired in June 1992 by the Canada Centre for Remote Sensing. To facilitate visual interpretation, the SAR images were corrected from slant to ground range, and image stripes resulting from the SAR antenna pattern were removed. A 5 x 5 Gamma filter was used to reduce image speckle. Registration of the SAR image to 1:50,000 topographic maps (1972) enables us to estimate changes to the coastline, due to erosion or deposition. The coastline boundary was digitised from the 1972 topographic maps, and superimposed on the 1992 SAR image. The SAR image was combined with the only cloud-free TM image of the area, obtained in September 1992. This image combination (Figure 1) enables the interpretation of the coastal geomorphology, provided by the SAR, and the land use provided by the TM.

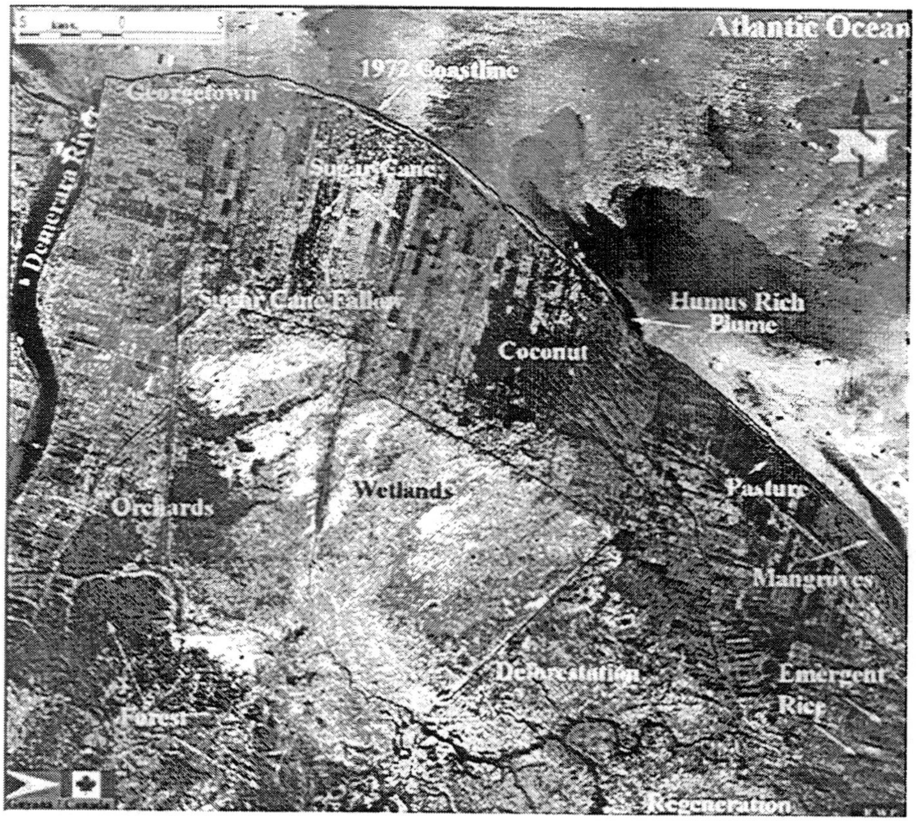

Figure 1. SAR & TM composite - coastal Guyana

- **Coastal geomorphology**

Over four years (1990-1994), there were several cases where the dikes and dams had been eroded and broken by coastal storms, resulting in severe flooding. For instance, in 1990, there were 49 breaches to the coastal structures. Assessing the risk and damage from the coastal flooding requires detailed geomorphological and land use information, which does not exist in Guyana at this time. Coastal erosion and shoreline recession have been recorded over the past 200 years at rates varying between 5-20 m a year (Cambers et al., 1994). This is true for specific recordings. However, regional estimates from remote sensing techniques shows a larger change. The comparison of the 1992 SAR images and the 1972 topographic maps has shown that parts of the coast have been subjected to severe erosion and accretion and some parts remain unchanged. Our estimate shows that the shoreline has retreated to a maximum of a half a kilometre in some areas (Figure 1) over the past twenty years. This amounts to about 25 m a year of coastal retreat. The two main erosion areas are the coastline from Georgetown to Mahaica, and the stretch from the Abary to Berbice rivers. The most stable stretch of shoreline is between the Mahaica and Abary rivers. The eastern bank of the Berbice river from Crab Island to Albion is experiencing considerable deposition and growth of mangroves. Mudflat deposition and mangrove growth has extended the shoreline by an average of half a kilometre.

Geomorphological mapping of the coastal plain of Guyana and erosion and depositional processes is essential to sea defence. The high spatial resolution of the SAR image allows the detection of small coastal erosion and depositional features. The identification of seven series of beach ridges/cheniers provide an understanding of the depositional history of the coastal plain. The offshore bars currently being formed and the identification of the small and large scale coastal erosional areas provide information on the dynamic coastal processes.

Coastal Features and Land Cover	Image Interpretation Notes
	C-VV Airborne SAR
A. Deposition	
1. Beach ridges & chenier	• Parallel, linear features mainly vegetated by coconut
2. Mudflat	• Smooth surface, dark tone
3. Spit	• Curved & straight bar associated with shoreline mangrove • Straight shoreline with mangrove, light tone, medium texture
B. Coastal erosion	
1. Jagged erosional forms	• 10 to 500 m of erosional forms mainly identified from its morphology
2. Large concave shoreline	• 10-20 km curved shoreline
C. Land cover	
Coconut	• Associated ridges and sand aprons and chenier
Rice	• Small agricultural field patterns. Dark tone flooded - early growth light tone - mature and harvested
Pasture	• Generally dark tone; due to short grass and wet ground
Sugar cane	• Large plantations associated with irrigation canals

Table 1. Visual interpretation of SAR images

- **Coastal land use**

There is a need to provide accurate and reliable land use information, given the importance of agriculture to the Guyanese economy. The SAR images have shown considerable promise in this cloudy region in providing information on agricultural land use. Visual interpretation of the SAR images is shown (Table 1).

Field work, using 1:20,000 airphotos and the SAR images, was conducted to confirm the accuracy of the visual interpretation. Table 1provides a summary of the coastal features and related land cover on the coastal plain that are recognised on the airborne radar images. However, the combined SAR and TM images (Figure 1) provided a more accurate separation of land use of this region. All the major land use practices (rice, sugar and coconut plantations, pastures, and wetlands) were identified. These images are being used to update current land use mapping in the coastal areas.

The interpretation of the synergy images also shows areas where the mangrove forests are totally destroyed by coastal erosion. Mangroves are the constituent plants of tropical intertidal forest. Communities of these salt-tolerant (halophytic) plants are dominated by trees and shrubs, which help to protect the coast from erosion. Their root structure promotes the accretion of silt and mud. Once, the entire coastline of Guyana was fringed with mangrove forest. Presently, the areas of severe coastal erosion, east of Georgetown (Figure 1) have no mangrove forest.

The visible TM bands (Figure 1) show offshore circulation patterns and water colour. The darker areas near river mouths show the extent of the humus rich plumes. The muddy and clearer waters are related to turbulence and bottom sediments in the near shore areas (Singhroy and Bruce, 1993).

2. Water Harvesting - Jordan Case Study

2.1 Introduction

Hydrogeological mapping is fundamental to the economic development of Jordan. Pumping of groundwater is already in danger of exceeding levels at which the resource can be renewed, and water quality is declining (World Bank, 1995). Groundwater is the major water resource in many areas of the country. It is comprised of both renewable and non-renewable resources. The estimated renewable groundwater is 300 million cubic metres per year (MCM/y), and non-renewable is 100 MCM/y (Ministry of Water and Irrigation and UNDP, 1992). Recommendations by the Water Authority of Jordan (WAJ) to meet the increased needs of water in the country are 1) to build dams and reservoirs, 2) to explore new aquifers, 3) to treat and reuse water, and 4) to improve the irrigation pattern (Bajjalli, 1994). Recently, the Government has been encouraging production of low-value cereals and other agriculture produces to increase its exports. This also creates a need to optimise the use of water and land resources and to control the exploitation of groundwater. There is no aquifer system which does not appear to have been tapped by exploratory wells. The distribution of groundwater in Jordan is fairly well known. The main problem is to optimise the utilisation of groundwater resources and protect them against depletion and pollution. Although the above studies have shown that there has been considerable work done for the search for drinking water in Jordan (Bajjalli 1994, GTZ 1977, etc.), the use of remote sensing techniques is limited. With the availability of RADARSAT and other orbital SAR data (ERS, JERS), it is necessary to evaluate the capability of SAR images to provide data for water harvesting. This paper provides examples of SAR images to assist in optimising the groundwater resources in Jordan.

2.2 Methodology and Image Interpretation

Airborne SAR images were acquired by the Canada Centre for Remote Sensing in December 1994. Acquisition parameters include C-HH nadir mode, 7 looks, with incidence varying from 0 to 72°, and a resolution of 4.8 m x 6.1 m. SAR data were corrected for the antenna- pattern. Speckle has been reduced using the Gamma-MAP filter before the image was converted from slant range to ground range.

Surficial drainage was obtained by applying an unsupervised classification to a synergy image created from the fusion of a C-VV airborne SAR with a LANDSAT TM image. The two datasets were integrated using the IHS integration method (Cliche et al., 1985). The three eigenvectors of the TM image were transferred to the IHS space; then, the intensity channel was replaced by the SAR image, the saturation channel was replaced by a uniform DN value of 65 while the hue was being modulated by the TM image. Finally, the image was transferred back into the RGB space, resulting in a new synergy image (Figure 2).

Jordan is divided into three hydrogeological systems that are used for groundwater exploration (GTZ, 1977): the shallow aquifer consists of Tertiary and Quaternary rocks; the Upper Cretaceous Limestone aquifer; and the deep sandstone aquifer. Remote sensing techniques are used in terms of the overall water management, in particular, the pollution of wells of the shallow aquifers, and in targeting exploratory wells.

The shallow aquifer is the most hydrologically significant in Jordan, consisting of fluvial, lacustrine, eolian sediments and basalts. It consists of groundwater of very good quality and, in some regions, extremely high permeability makes it possible to extract large quantities of water. The three main natural discharge zones are the upper Yarmouk river basin, the Zarqa basin, and the Azraq basin. This study focuses on the Azraq basin (Figure 2), where more than 550 water wells have been drilled, about 350 of which supply water for local agriculture.

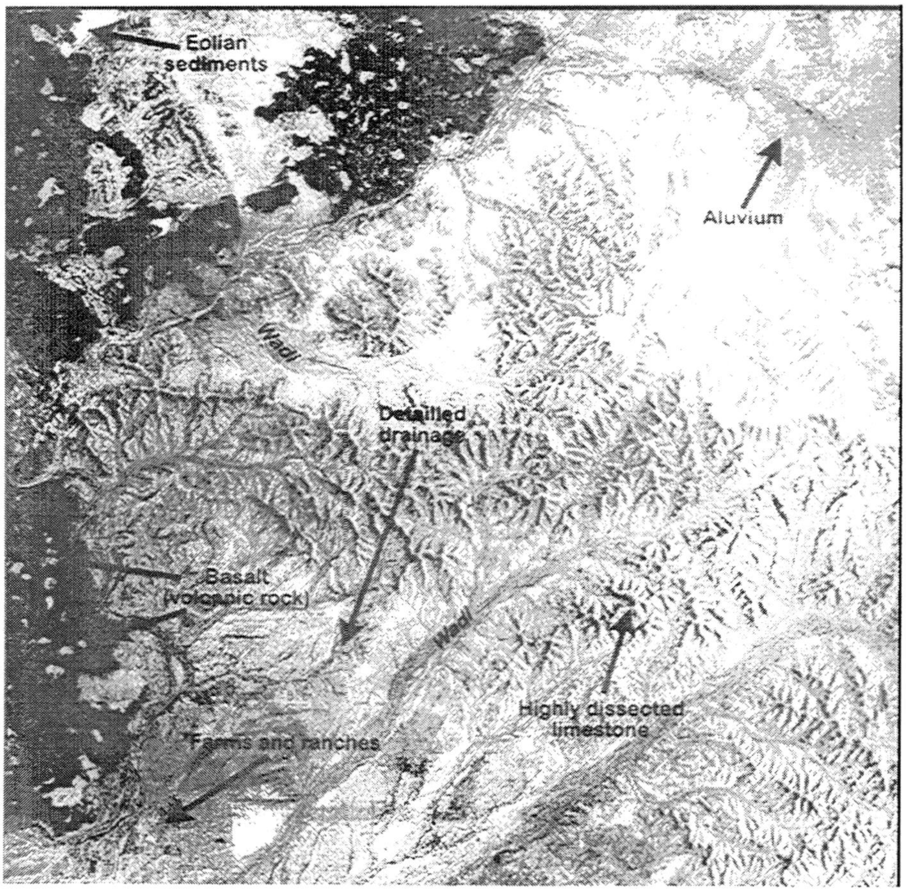

Figure 2. SAR & VIR composite - Azraq, Jordan

- **Wadis**

The management of wadis (water courses) is fundamental to the national water strategy of Jordan. Indirect recharge by wadi floods is considered to be the main source of groundwater in the shallow aquifer. The detailed mapping of the distribution of wadis and their channel characteristics (morphology, roughness and sediment composition) is necessary for water harvesting. Most of these wadis are cultivated, others are dammed to retain the water from flash floods to recharge some aquifers artificially. Their management, therefore, is linked to the overall environmental protection and to recharging the shallow aquifers. Because of the presence of temporary surface water, fertile alluvial soils and sometimes shallow groundwater, wadis are usually productive agricultural areas. Careful management is required since these dry unconsolidated river beds are easily eroded during flash floods.

Figure 2 shows the channel characteristics and wadi distribution in the Azraq area. The ploughed surface of the wadi is rougher and therefore appears lightly toned, whereas the smoother, unploughed areas of the wadi are darker. The SAR backscatter is sensitive to the sediment types within the wadis. Coarse gravels and boulders or a rough channel surface produce a light tone, whereas fine sediment, which has a smooth surface, gives a dark tone. These differences in surface roughness seen on the SAR image representing fine and coarse wadi sediments, as well as ploughed and unploughed areas, are not easily seen on the TM image. The TM reflectance of the wadi sediments is the same, despite their compositional and roughness characteristics, and is therefore less suitable than SAR to map wadi sediments.

These rough wadis with large boulders that are confined within a defined valley are suitable candidates for the construction of small dams to retain the water from flash floods. The size of the boulders, i.e. the channel roughness, indicates the water capacity of the wadi. The creation of temporary reservoirs allows the artificial recharge of the upper aquifer. Artificial recharge may be defined as the practice of increasing by artificial means the amount of water that enters a ground water (Walton, 1970). The purpose of artificial recharge is to conserve runoff from flash floods, supplement the quantity of ground water, reduce salt-water intrusion, and to store water to reduce the cost of pumping and piping. Thus, valuable water that would otherwise be largely wasted in this arid region is conserved by storage in the ground and is available to meet needed water requirements. The SAR/TM synergy images (Figure 2) are being used to locate dams in the area. The location of sites for artificial recharge is a fundamental part of the water harvesting programme in Jordan. A more comprehensive approach using GIS techniques for dam location has been reported by Waynakh (1993, 1995).

- **Hydrogeological units**

Hydrogeological maps record the status of groundwater at a point in time by defining the geology, the water, and the areas of recharge and discharge. The mapping of hydrogeological units of the various hydraulic systems is needed for groundwater storage potential. The SAR/TM images are suitable for delineating mudflats and playa, wadi sediments and alluvium, basalt, limestone, and sandstone units based on their characteristic textural and geomorphic expressions as shown in Table 2.

Mudflats and playa are characterised by a smooth surface consisting of silts and fine sand, and appear as dark areas on the SAR images. In these shallow depressions, water accumulates during flooding and fine silts and clays are deposited to form mudflats. These areas of high water potential are fed by numerous dendritic drainage patterns seen in Figure 2. The capacity of these playa to produce water is based on the feeder drainage system. Therefore, accurate large- scale drainage maps are significant for accessing their productive capacity and for the environmental protection and management of these playa.

Wadi sediments and alluvium are characterised by both coarse and fine sediments; they therefore appear as dark and light tone areas on the SAR images. The deposits are essentially boulders with unsorted gravel, sand and silt derived from various local lithological units. However, they are better delineated on the SAR/TM composite shown in Figure 2.

Hydrogeological Units	SAR-Geometric-Characteristics	Hydraulic Complex
Mudflats and Playa High water potential, mainly silt and fine sand	Smooth flat surface, low backscatter -dark tone C-VV more useful to delineate drainage density on playa surface than C-HH.	Shallow Aquifer
Wadi sediments-Alluvium High water potential, mainly sand gravel and boulders, direct recharge by wadi floods	Smooth surface-dark tone on fine sediments. Rough surface - light tone on coarse sediments.	Shallow Aquifer
Basalts Very high water potential high permeability, groundwater discharge at contact zone as springs and base flow.	Rough surface - light tone due to the angular and subangular basalt stone pavement with average diameter of 25cm. Dark depressions are mudflats in the basalt.	Shallow Aquifer
Limestone (chert and sandy) Medium water potential high groundwater recharge rates	Highly dissected surfaces and undulating hills Cherty limestones have a rough surface-light tone	Upper Cretaceous hydraulic complex
Sandstone Low-medium water potential Low ground water recharge rates	Honeycomb structure characterised by a series of lineaments, steep slopes	Deep Sandstone Aquifer

Table 2.

The basalt is highly permeable and stores large quantities of drinking water. Recharge into the ground water comes directly from precipitation. The ground water generally does not discharge from the basalt outcrop areas, but is transferred to a contact zone to adjacent aquifers. The basalt areas are easily outlined on the SAR images, and form a gently undulating surface of low relief. It has a characteristically rough surface, and therefore appears as a light tone on the SAR images. The strong backscatter is attributed to the presence of angular to subangular rock pieces with an average diameter of 25 cm and a 60% stone cover.

Two distinctive types of basalt were distinguished using an IHS, SAR/TM combination (Abdelhamid, 1995). These include the thick massive flows and rough surfaces of the alkali olivine basalt series, and the relatively smoother surfaces of the Phyric basalt.

The limestone units (Figure 2) are characterised by highly dissected distinctive fern-like dendritic drainage patterns. The chert-rich limestone is lightly toned due to the higher SAR backscatter of the angular chert pavement. The sandy limestone is more dissected than the chert-rich limestone. This aquifer system has a medium water potential, and is recharged from direct precipitation and from the basalt.

- **Fracture systems**

Nearly all the ground water resources of Jordan are found in hard-rock aquifers where the water moves in fractures, fissures and joints. Because of the low porosity of these aquifers, the velocity of the groundwater flow is high. Consequently, the ground water resources are very vulnerable to pollution caused by recycled irrigation water and waste water from human settlements and industry. The accurate delineation of surface fractures is needed, not only to target groundwater exploratory wells, but also to prevent pollution from the above land use practices in the vicinity of the fractures. The limestone and the sandstone aquifers are particularly vulnerable. In Figure 2, fracture systems correspond to the NE-SW trend of the wadis.

- **Hydrological land use**

The agricultural sector contributes about 7% to Jordan's GDP and provides about 10% of employment. The overriding constraint is the limited resource base, especially water. Irrigation efficiencies are low (40% in some areas) and cropping intensities could be improved, especially in the Jordan Valley, which has high potential. Effectively, the sector's prospects rest on its ability to adapt - to use water with increasing economy, to produce for new markets, and to switch to new, higher-value products. There is substantial scope for improvements in productivity and profitability within existing resource limitations but exceptional measures are needed to bring about these structural changes in Jordanian agriculture.

The classification of hydrological land use is necessary to conduct sustainable land management practices, in particular erosion control of the fragile desert landscape. From the SAR/TM images, field boundaries, olive plantation and areas of extensive gully erosion on weak silty and sandy soils were easily identified. Harahsheh (1995) integrated SAR images with TM, and produced an improved agriculture classification of agricultural land use. These major management practices and degraded areas, when superimposed on producing aquifers, can produce vulnerability maps for improved irrigation practices.

3. Conclusion

This paper has demonstrated that the combination of SAR and TM images has provided information for environmental management. In coastal Guyana, geomorphological and land use information interpreted from synergy images is useful for sea defence and coastal zone mapping. In Jordan, the interpretation of synergy images has provided additional information for improved hydrogeological mapping in terms of wadi mapping, surface characteristics of hydrogeological units, aquifer fracture distribution, and the land use above sensitive aquifers. When the textural and geomorphologic information from the SAR data are combined with the spectral, landcover information of the TM images, the resultant image maps are very useful for environmental monitoring and management.

References

1. Abdelhamid, G.: "Mapping of basalts and sediments in Azraq area, northeast of Jordan, using radar airborne images", *Proceedings, GlobeSAR Seminar, Amman, Jordan*, pp 24-35. Geomatics Canada, 1995
2. Bajjali, W.T.: *Recharge and Regional Circulation of Thermal Groundwater in Northern Jordan*, Unpublished Ph.D thesis, pp. 257. University of Ottawa, Canada, 1994
3. Bender, F.: *Geology of Jordan*, pp. 196. Gebruder Borntraeger, Berlin, Stuttgart, 1974
4. Bleakley, D.: "1957 Observations of the geomorphology and geological history of the coastal plain of British Guyana", *Geological Survey of Guyana*, pp. 21. Georgetown, Guyana, 1957
5. Cambers, G., Dharry, A.E.T., and Russell, C.E.: *Environmental impact analysis emergency sea defence programe of Guyana*, Vol. 1, pp. 53. Government of Guyana and Inter-American Development Bank, 1994
6. Cliche, G., Bonn, F., and Theillet, P.: "Integration of the SPOT panchromatic channel into its multispectral mode for image sharpness enhancement", *Photogrammetric Engineering & Remote Sensing, 51(3)*, pp. 311-316, 1985
7. Daniel, R.K.: *Aerial Photography and Land Development Projects in Guyana in Remote Sensing and Tropical Land Management*, (Edited by J. Eden and T. Parry), pp. 37-53. John Wiley and Sons, London, 1986
8. "GTZ: Agrar and Hydrotechnik & GTZ", *National Water Master Plan of Jordan* (8 volumes). Essen, Hanover, 1977
9. Harahsheh, H.: "Agricultural landuse classification from enhanced SAR images, Madaba and Azraq areas", *Proceedings, GlobeSAR Seminar, Amman, Jordan*, pp. 83-99. Geomatics Canada, 1995
10. Ministry of Water and Irrigation and UNDP: "Water resources policies, planning and management project", *Groundwater Resources, Jor/87/003*, Amman, Jordan, 1992
11. Pastakia, C.M.R.: *A Preliminary Study of the Mangroves of Guyana*, The European Community Article□B/946/89, pp. 32, 1991
12. Singhroy, V. and Bruce, B.: "Coastal zone mapping of Guyana using digital Landsat data", *Proceedings of the 17th International Symposium on Remote Sensing of the Environment*, pp. 659-679. ERIM, Ann Arbor, Michigan, USA, 1983
13. Walton, W.C.: *Groundwater Resource Evaluation*, pp. 664. McGraw-Hill Book Company, New York, 1970
14. Waynakh, I.M.: "Application of remote sensing techniques in the natural resources management in arid and semi-arid environments of Jordan", *Proceedings ITC*, pp. 13. Enschede, Netherlands, 1993

15. Waynakh, I.M.: "The use of synthetic aperture radar for mapping groundwater recharge zone in Madaba area, Jordan", *Proceedings, GlobeSAR Seminar, Amman, Jordan,* pp 188-203. Geomatics Canada, 1995
16. World Bank: *Agriculture Sector Adjustment Loan, Project 5JORPA072.* Ministry of Planning, Jordan, 1994

Role of Remote Sensing and GIS for Environmentally Sound and Sustainable Development

S. Murai, Institute of Industrial Science, University of Tokyo, 7-22-1 Roppongi, Minatoku, Tokyo 106 Japan (Formerly STAR Program, Asian Institute of Technologies)

Abstract

The paper reviews the role of space technology applications, particularly with a focus on remote sensing and Global Information Systems (GIS) for environmentally sound and sustainable development in Asia and the Pacific region. Though Asia is progressing rapidly in terms of economics and industry, it has a lot of serious environmental problems, such as deforestation, desertification, land degradation, etc., as well as population over-increase, industrial pollution, over-development, etc.... The author tries to introduce an eco-engineering approach to give a solution to these problems, using remote sensing and GIS. Some examples of the approach will be demonstrated.

1. Introduction

In September 1994, an epoch-making meeting, the "Ministerial Conference on Space Technology Application for Environmentally Sound and Sustainable Development in Asia and the Pacific Region", was held in Beijing, China, with the participation of ministers and sub-ministers from more than 30 countries in the region. In the conference, the "Beijing Declaration" was agreed, as follows:

- There is global concern about the depletion of natural resources and environmental degradation in Asia and the Pacific region.
- The application of space technology is beneficial, and essential for sustainable development.
- Space technologies, including remote sensing, Global Information Systems (GIS), Global Positioning Systems (GPS), space telecommunications, etc., are cost-effective and accessible.
- The application of space technology should be encouraged and promoted, based on regional cooperation and coordination.
- A regional Space Applications Program should be launched in Asia and the Pacific region.

According to the above agreement, many Asian countries have started to develop and support space technology applications for environmentally sound and sustainable development. GIS are booming all around Asia and the Pacific region since they link all aspects of remote sensing, particularly for natural resources management. However the current status on geo-spatial data access in

Asia and the Pacific region is far from the "open data policy" that is now common in the developed countries of Europe and North America.

2. Environmental Problems in Asia

According to the source of UNEP EAP-AP (United Nations Environmental Program - Environment Assessment Program - Asia & Pacific), the following problems are the most serious. The numbers in brackets show the number of countries (out of 25 Asian countries) which consider the problem to be very serious.

- **Deforestation** (16). Commercial logging (currently in Malaysia, Indonesia, Papua New Guinea, etc., and previously in Philippines, Thailand, etc.), shifting cultivation (Myanmar, Laos, Thailand, Sri Lanka, etc.), mountain overdevelopment (India, Nepal etc.), shrimp farming in mangrove areas (Thailand, Bangladesh, Vietnam, etc.) are major causes.

- **Pollution and inland water** (16). Most of the developing countries in Asia suffer from uncontrolled industrial effluents, mismanaged urban waste water, soil and mud sedimentation, etc....

- **Land degradation** (14). Land degradation usually results from deforestation. The most serious land degradation is from saline soil (North East Thailand, many states in India, etc.), soil erosion (Indian Himalayas, Nepal, Northern Thailand, etc.) and over-grazing (Inner Mongolia, Pakistan, etc.).

- **Health hazards** (13). Health hazards result from improper drinking water, malaria, malnutrition, etc....

- **Floods** (13). Many rivers in Asia such as the Bhuma Putra (Bangladesh), Ganges (India), Chao Phraya (Thailand), Yang Zhu (China), flood frequently.

- **Marine water pollution** (11),
- **Air pollution** (11),
- **Destruction of bio-diversity** (9),
- **Low agriculture production** (9),
- **Drought** (9),
- **Cyclone** (9).

These environmental problems are related to an imbalance of terrestrial and/or marine eco-systems due to human activities, particularly over-population and over-development, which can be monitored and managed by remote sensing and GIS techniques.

3. Requirements of Geo-Spatial Information

In Asia and the Pacific region, most applications of remote sensing and GIS will be as follows:

- Hydrological studies for watershed management,
- Flood prediction/flood mapping,
- Vegetation mapping/forest mapping,
- Crop production forecasting,
- Map updating,
- Forest fire monitoring,
- Land use mapping/land use change study,
- Coastal zone dynamics study.

The scale requirement for the above applications depends on the level as shown in Table 1.

Level	Scale	Resolution
Global	1:1 M	500 m - 1 km
National	1:250 K	20 - 50 m
Operational	1:50 K	5 - 10 m
Local (urban)	1:10 K	1 - 3 m

Table 1. Requirement of scale and resolution

It should be noted that maps with scales larger than 1:50,000 are classified, with military control. This prevents the timely and appropriate planning and management for environmentally sound and sustainable development of natural resources operationally. This is the main reason, as well as the recommendation that space remote sensing or satellite images with high resolution capability (1- 5 m ground resolution equivalent to 1:10,000-50,000 scale map) should be highly utilized in Asia and the Pacific region, based on an "Open Sky Policy".

The frequency requirement for map updating used to be about five years in the past. Now it has shortened to 1-3 years because of the drastic environmental changes occurring in Asia and the Pacific region in only a few decades.

The accuracy requirement is, of course, the higher the better, but 80-90% accuracy would be widely accepted at an operational level in Asia. The dynamic accuracy is considered to be more important than the static accuracy because we should trace and track the dynamic change of environmental conditions.

4. Role of Remote Sensing and GIS

There have been many discussions on how to integrate remote sensing and GIS for natural resources management. It is a fact that many GIS projects are going on in Asia and the Pacific region without satellite remote sensing. However, most published topographic maps and thematic maps of Asia and the Pacific region were produced in the 1970's and 1980's (sometimes even in 1950's and 1960's), with a mixture of new and old versions. This makes GIS information out of date, inaccurate, inconsistent and unreliable, resulting in improper policy making and mismanagement. For example, the available topographic map of 1:50,000 around the Asian Institute of Technology (AIT), in Pathumthani, Thailand, was produced in 1969; it does not include the AIT campus. Satellite imagery is the only reliable source of information for GIS applications to land use planning in this area.

It is the author's understanding that remote sensing is a powerful tool to monitor environmental changes which are induced by human impacts on the environment, as shown in Figure 1. These changes are regarded as "physical dimensions" which can be dealt with by "Earth System Science", proposed by NASA in 1983 to study the relation between environmental impacts and the changes.

On the other hand, GIS is a sort of decision support system for policy making, planning and management in terms of the human response, as shown in Figure 1 too. Human response is the main subject of social/political science, to study and analyze the influence of human driving forces (sometime human dreams) such as population growth, affiliation, economic growth, technology innovation, food supply, modern life, etc. which is the so-called "human dimension".

Existing science and technology support the human driving forces in the form of the impact on the environment, resulting from construction, agricultural development, deforestation, shrimp farming, urbanization, industrialization, etc....

However, there has been no technology to feed back from the "physical dimension" (monitoring of environmental changes by remote sensing) to the "human dimension" (human response supported by GIS). This feedback technology has been named "eco-engineering", or "global eco-engineering" in the case of a global level, by the author in 1994 at the 8th Toyota Conference. This global eco-engineering approach is the answer to how to make the Earth sustainable, even survivable, through the integration of remote sensing information and GIS.

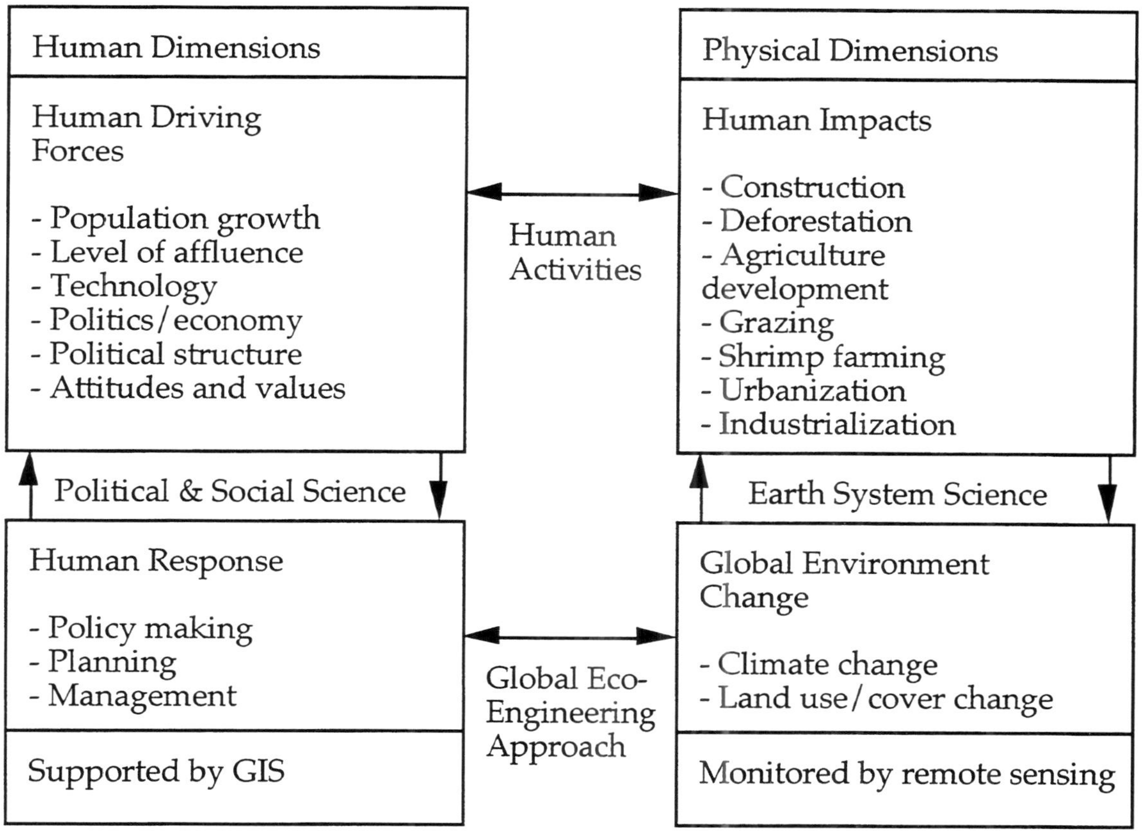

Figure 1. Concept of global eco-engineering approach

5. Case Studies of Global Eco-Engineering

Three case studies have been demonstrated by the Global Engineering Group funded by the Toyota Corporation in 1991-1994 (for which the author was the coordinator).

5.1 Human Impacts on Vegetation Changes

Actual vegetation maps produced from the NOAA Global Vegetation Index (GVI) data were compared with the potential vegetation map estimated and modeled from the virgin conditions, with the use of rainfall and temperature information.

The difference between the actual and potential vegetation maps reflects the human impacts on deforestation, land degradation or desertification. From the result of this study, human activities destroyed the forest land area from 49.3% in the original condition to 34.0% at present, increased the grassland from 28.7% to 33.9% and also increased the desert/semi desert area from 22.0% to 32.1%. According to these vegetation changes, a total amount of 12.4 G tons carbon/year was lost in the fixation of carbon dioxide through processes of vegetative photosynthesis. This information would be a sort of limit for deforestation with respect to the total balance of carbon dioxide.

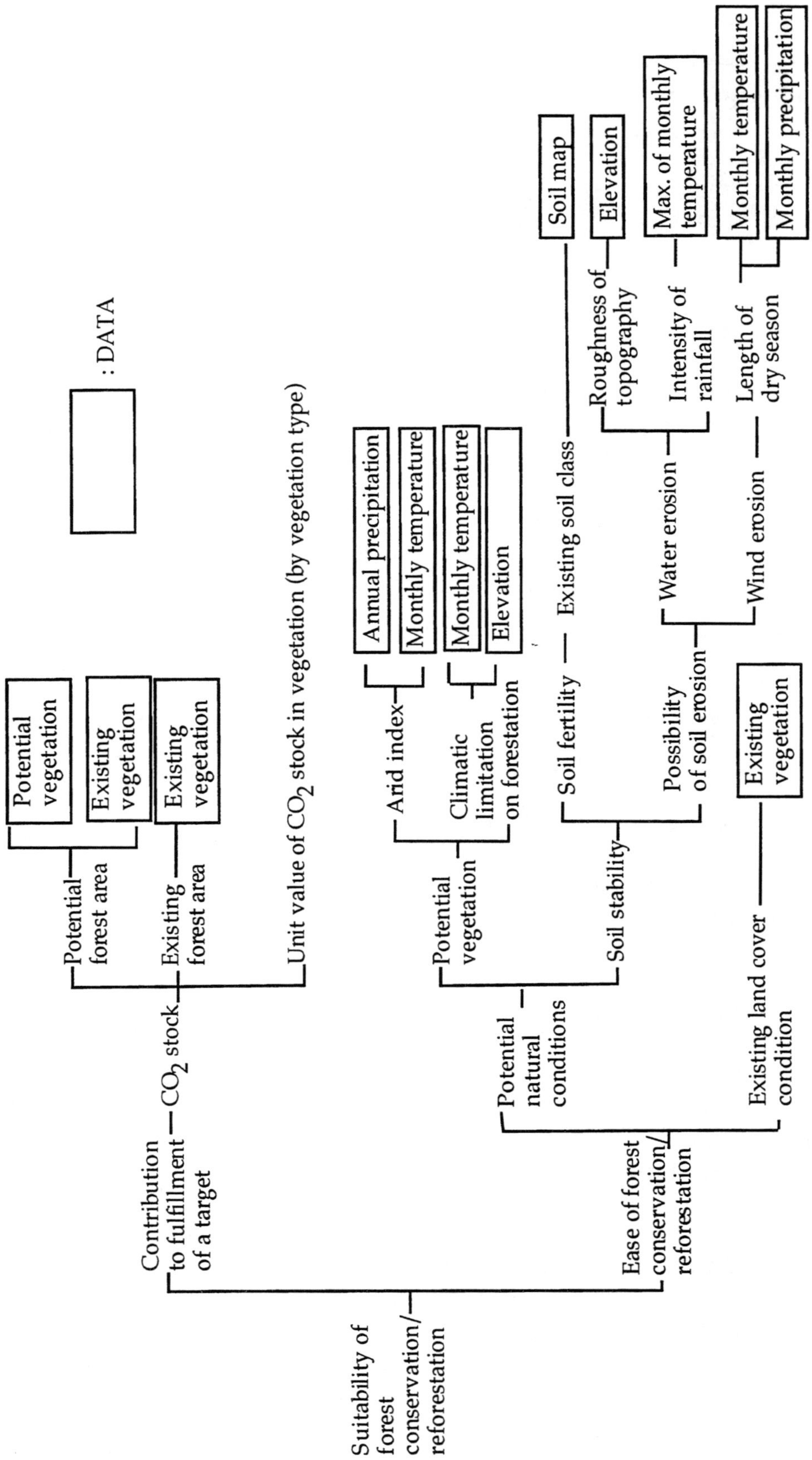

Figure 2. Procedure of suitability evaluation for forest consevation/reforestation

5.2 Global Carrying Capacity

From the actual vegetation map and other geospatial data (digital elevation model, soil, temperature, rainfall, GNP etc.), the potential crop production was estimated under the assumption that no more forest - but only grassland - shall be converted into agricultural land, and the population depends on crop production. Then, the supportable population was estimated on a per country base and summed to give the total supportable population in the world for the three cases.

(a) If we eat as we eat now in each country, the total supportable population is 7.34 billion.
(b) If we eat the world average, the supportable population is 8.97 billion.
(c) If we eat as Americans eat, the supportable population is 4.01 billion.

From this estimation, it is concluded that the sustainable population or carrying capacity of the Earth will not be more than 10 billion.

5.3 Global Forest Conservation Strategy

Those forests growing in ecologically fragile areas, with poor and erodable soil, steep slopes, wet land, etc., should be conserved because they would not recover if the forests were destroyed. Forests are also indispensable to maintain the water balance (underwater, evaporation, soil moisture, etc.) reasonably. Various geospatial data can be input to a GIS model to maximize the effect of forest conservation. Figure 2 is the model that the author's group proposed for global forest conservation strategy. According to this model, a strategy for forest conservation was produced. However, there should be further debate as to how to keep the best balance between forest conservation and agricultural development for the future survival of mankind in consideration of the carbon dioxide balance, as mentioned before. We should establish some sort of limit to the uncontrolled population increase and deforestation accordingly.

6. Conclusions

Remote sensing and GIS can play an important role for environmentally sound and sustainable development if they are recognized for their role in the linkage of human and physical dimensions. Eco-engineering, or global eco-engineering, will be a new concept to feed back from the physical to the human dimensions in defining the optimum solution for mankind.

References

1. *Beijing Declaration on Space Technology Applications for Environmentally Sound and Sustainable Development in Asia and the Pacific,* ST/ESCAP/1451. United Nations, Beijing, China, 1994
2. Konecny, G.: *Basic Considerations for the Implementation of Spatially Based Information Systems*. Paper presented at the LIESMAS Colloquim, Wuhan, China, October 1994

3. Deekshatulu, B.L.: *Indian Remote Sensing Programme and National Information Needs*. Paper presented at the Workshop and Conference on International Mapping from Space, Hannover, Germany, September 1993
4. US Government: *Coordinating Geographic Data Acquisition and Access: The National Spatial Data Infrastructure,* Executive Order 12906. The White House, Washington, DC, April 1994
5. Murai, S.: *Development of Global Eco-engineering using Remote Sensing and Geo-information Systems, Toward Global Planning of Sustainable Use of the Earth*. Elsevier, Amsterdam, The Netherlands, 1995

Limitations to the Use of Remote Sensing Technology in Kenya

B.M. Mati, Dept. of Agricultural Engineering, Jomo Kenyatta University of Agriculture and Technology, PO Box 62000, Nairobi, Kenya

Abstract

The use of remote sensing technology is becoming an increasingly necessary tool for planning development projects, even in developing countries such as Kenya. There is a need to use quick, accurate and convenient methods to gather information over the ever-changing environment. Remote sensing in this region can be used for surveying, vegetation monitoring in the rangelands, weather assessment and forecasting, water resources management, forest fire detection, early warning activities (drought, food, floods), migration patterns of wildlife, detection of pollution, planning of infrastructure, agriculture, demographic changes, and all aspects of land use planning.

Despite the great demand for remote sensing in Kenya, its use has been hampered by problems ranging from financial constraints, gaps in technology, lack of enough trained personnel, socio-political and economic constraints. These constraints need to be resolved if advances in space technology applications for peaceful activities are to benefit the people. This paper deals with identifying these constraints, their causes, and possible resolution.

1. Introduction to Kenya

Kenya, in Eastern Africa, borders Ethiopia and Sudan to the North, Uganda to the West, Tanzania to the South, Somalia and the Indian Ocean to the East. The total land area of Kenya is 582,646 km^2, of which 11,230 km^2 is water surface, mostly lakes. The country is dissected by the Great Rift Valley from North to South, and most of the lakes are found within it. Altitudes range from 0 m at the coast to 5,199 m above sea level at the top of Mt. Kenya. Other notable mountains and highlands include Mt. Elgon (4,310 m), the Nyandarua Range, Cherangani Hills and Mau Escarpment. These highlands are drained by many rivers and streams, the largest being the River Tana in the East, Athi in the South East, the Nzoia in the West and the Ewaso Ng'iro in the North.

Kenya has varied climatic conditions ranging from the arid eastern region, receiving 308 mm of rain per annum, to the cool highland areas surrounding the major mountain areas, where 1300 - 2200 mm are recorded (Jaetzold and Schmidt, 1983). As Kenya lies on the Equator, temperatures are affected mainly by altitude. Annual mean maxima range from 24° to 35° C, while annual mean minimum temperatures range from 7° to 20° C. Agriculture is the main economic activity in Kenya, employing over 75% of the population. Only about 17% of the total land area is suitable for rain-fed agriculture. In recent years, the rapid population increase has resulted in an expansion of agricultural activities onto dry marginal lands and steep catchment areas. As 85% of the population in Kenya lives in the rural areas (Central Bureau of Statistics, 1993), competition for land between crop production, livestock and wildlife has become evident, causing increased land degradation. For the proper allocation and utilization of resources to the various development activities, a reliable and accurate system of land resource assessment and monitoring, such as by remote sensing techniques, is necessary. Therefore, remote sensing has become increasingly important in Kenya.

2. Background

Remote sensing techniques for Earth resources monitoring, such as aerial photography, have been used in Kenya from as early as the 1930's (Tiffen et al., 1994). However, the use of satellite imagery became prevalent in the 1970's when the Government mandated the then KREMU (Kenya Rangeland Ecological Monitoring Unit) to conduct systematic rangeland monitoring and to develop a data bank as inputs to the planning and management of renewable resources in Kenya. This was done by acquisition of LANDSAT imagery and by systematic reconnaissance flights (Andere, 1986). Since then, other types of satellite imagery have been used such as NOAA-AVHRR (Lamprey et al., 1987), LANDSAT TM and SPOT (Delsol, 1989). The Regional Centre for Resource Surveys and Mapping (RCRSM) was also set up around the same time to cater for the regional market in the areas of remote sensing and mapping. Aerial photography has continued to be used to complement other methods of remote sensing and sometimes to supplement them. In addition, Kenya has a relatively good coverage of topographic maps providing physiography and drainage. The whole country has been mapped at 1:250,000 scale, while maps at 1:50,000 are available for many areas. However, these maps need to be updated from time to time.

3. Applications of Remote Sensing

Remote sensing has increasingly been used in Kenya as advances in the technology make it more easily accessible to scientists, researchers and other users. Its applications range from surveying, weather assessment and forecasting, water resources management, vegetation monitoring, migration patterns of wildlife, early warning activities such as drought, flood and crop yield forecasting, livestock populations and distribution, detection of pollution, planning of infrastructure, demographic trends and all aspects of land resource planning.

Some of the main activities that have been achieved in Kenya using remote sensing are as follows: LANDSAT MSS imagery and aerial photography are used to estimate forest cover (Doute et al., 1981). Estimations of crop yield by green biomass assessment using LANDSAT, SPOT and aerial photography (Peden et al., 1986) have produced good vegetation indices that can be used for future planning. As tourism and wildlife is a major foreign exchange earner in Kenya, monitoring wildlife numbers in the national reserves has been done mostly by aerial photography (Loth, 1988). Land classification, its status and soil erosion potential have been performed using LANDSAT imagery and ground truth surveys (Wahome, 1984) with reliable results. Remote sensing has also been used for drought monitoring, to assess land use dynamics, desertification, and the availability of natural resources such as water, grasses in the rangelands, and livestock management in the arid and semi-arid areas of Kenya (Isavwa, 1989). Apart from Earth resource monitoring, there are satellite receiving stations for communications, weather assessment and forecasting in Kenya.

4. Constraints in the Use of Remote Sensing Technology

4.1 Technical Constraints

Though not common in Kenya, the use of NOAA-AVHRR imagery is hampered by low spatial resolution and, when used for vegetation monitoring (Lamprey et al., 1987), it has limitations in explaining detailed changes in a wide range of agricultural and rangeland vegetation types. LANDSAT TM and SPOT give a better resolution, but they are usually unavailable for some areas, and also too expensive. The LANDSAT imagery has too small a scale to provide detailed erosion features such as rill erosion in agricultural land. The LANDSAT information may not be available on a real time basis and the quality of the image is sometimes poor due to cloudiness, especially in the rainy season. The problem of cloudiness is especially manifest at the coast where, in one study (Doute et al., 1981), there were almost no cloud-free scenes in an eight year period. The use of other remote sensing methods that penetrate cloud cover, such as radar imagery, is still limited in Kenya.

When using LANDSAT MSS, it is usually difficult to differentiate clearly between proper forest and other vegetation types such as thick bush or coffee plantations, especially in the rainy season images (Gibbon and Mather, 1986). It is also hard to delineate plantations from indigenous forest and even with digital classification this is not easily possible. When conducting rangeland monitoring using remote sensing, it is usually difficult to distinguish between various types of fodder grasses on satellite imagery (Agatsiva, 1989). This can be done on aerial photographs and by field checks, but this makes the exercise slow and expensive.

In estimating crop yields by remote sensing, the time lags for acquiring the imagery preclude its use for short-term work. In many farming areas of Kenya, field sizes are small, some being less than 0.5 hectares, so that the resolution of satellite sensors is inadequate. During the crop growing season, the maize growing areas are often under cloud cover. It is difficult to define the gradient between maize and forest in some districts (Ottichilo and Peden, 1986).

The use of remote sensing (usually by aerial photography) in monitoring wildlife migration patterns and their populations is limited by the presence of bushes and shrubs, under which the animals may hide (Loth, 1988). Such animals can be easily overlooked especially when they occur in small numbers, producing errors in the final estimate. In dealing with wildlife, several scenes are required over the same area and also over a longer temporal scale to cater for annual/seasonal migration patterns. This can be expensive and time-consuming.

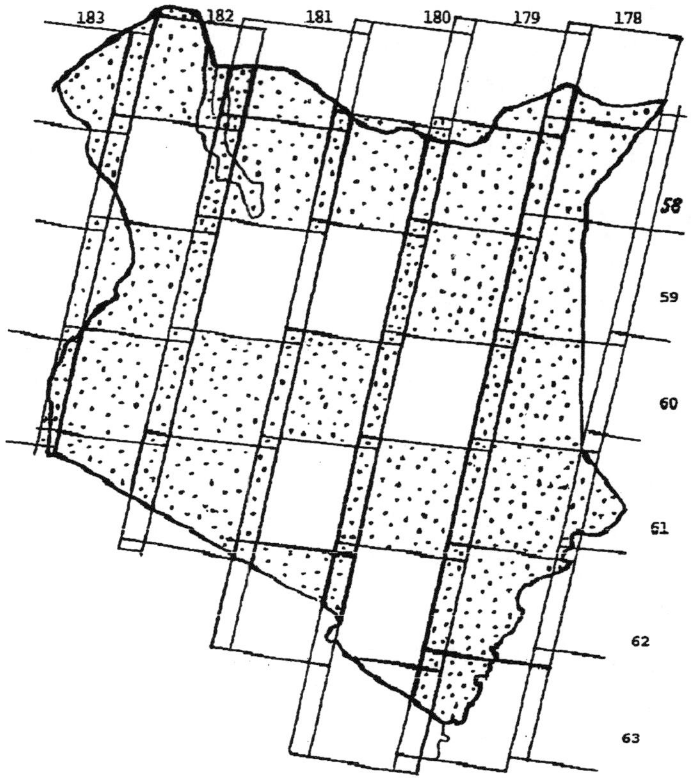

Figure 1. LANDSAT MSS color composite prints

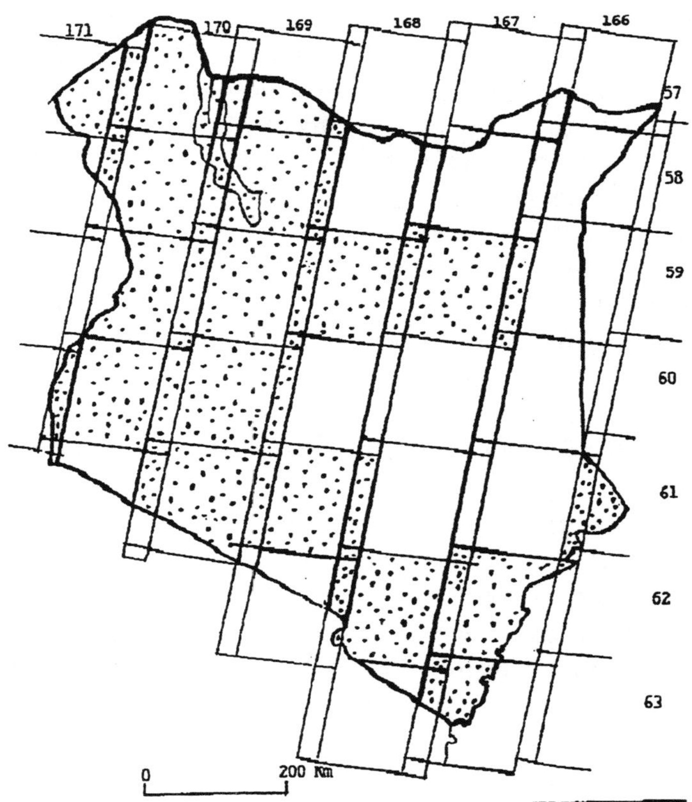

Figure 2. LANDSAT TM color composite prints

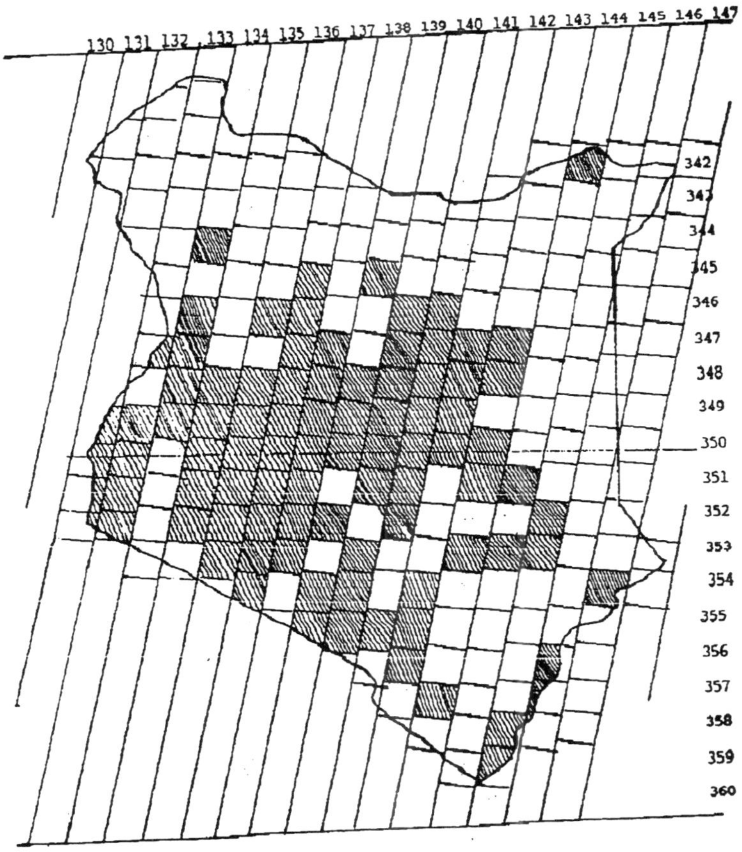

Figure 3. Location of existing SPOT images (1986 - 1991)

Remote sensing for land resource assessment and monitoring activities such as land use, soil erosion and land degradation, drought and desertification and demographic dynamics, has been hampered by lack of suitable imagery covering some of the most susceptible areas. For instance, such facilities as the LANDSAT MSS composite prints are available for isolated areas (see Figure 1), while LANDSAT TM prints cover only parts of central and western Kenya (see Figure 2). The SPOT colour composite prints (see Figure 3) are only available for certain parts of the country, leaving out the North Eastern and the coastal regions (Maurel, 1987; Beukema, 1994). Another problem is the lack of updated imagery and remote sensing data. This is because there is no ground receiving station in Kenya for Earth resources satellite data, and also the lack of adequate processing facilities for satellite data and imagery.

4.2 Socio-Economic Constraints

Socio-economic constraints that hamper technological advancement in a developing country such as Kenya contribute to the slow rate of adoption of remote sensing technology. One of the main problems is financial constraints, which limit the capability to purchase equipment and materials necessary for use in remote sensing activities. There is also a lack of funds for the direct importation of satellite imagery and digital data, which are often too expensive for would-be users (Peden et al., 1986). This is compounded by the fact that currently, Kenya does not receive satellite data regularly as there is no receiving station. The operation of a ground receiving station has been limited by poor consumer demand, mostly due to shortage of foreign exchange in the region, and low consumer awareness.

The shortage of suitable trained manpower to operate and analyse data is still prevalent (Otieno, 1982), as training in this field is still lagging behind other forms of technical training opportunities. Due to the lack of adequate remote sensing training facilitities locally, many of the technical personnel have been trained abroad and at great costs. The technical support facilities in the available institutions are in need of up-to-date equipment and materials for effective capacity building in this sector.

Remote sensing is still a relatively new technology. Therefore, decision makers need to be convinced to clearly perceive its usefulness. Sometimes there are conflicts of interest in setting priorities, where remote sensing is pitted against other conventional methods of doing things, with the result that the available resources are allocated to these alternatives. In addition, there are data acquisition problems (Isavwa, 1989) associated with the high costs and the bureaucratic red tape, which slows down the efficiency of project implementation while also frustrating the consumers of remote sensing information.

The high cost of satellite imagery and the difficulties encountered in acquiring satellite data are a result of the absence of indigenous capabilities for direct satellite reception. Because of the low level of economic development activities, there is little serious consumer demand for natural resource and environmental information (Catizzone and Fanta, 1993). There are also limitations in the flow of information (De Man, 1982) and problems of communication and poor infrastructure, especially within the arid and semi-arid areas.

5. Conclusions and Recommendations

Remote sensing, like any other rapidly evolving scientific technology, is expected to become cheaper, easily accessible and more user-friendly as it continues to advance. This will have a positive effect on countries such as Kenya, whose economies limit the use of this technology. For this to be realized within a reasonable time scale, the constraints identified in each case must be addressed and resolved through a multi-disciplinary approach.

- To alleviate the financial constraints, the Government should provide stable and reliable funding for long-term remote sensing activities. Financial assistance from external sources, including the donor community would be a welcome source of extra funding especially in the short run. Provisions for commercialization of remote sensing activities or the privatization of the public sectors currently dealing with it should be considered, with a view to making it a competitive income-generating enterprise like any other sector of the economy. The cost of remote sensing technology for use in Third World countries should be subsidized for researchers and other scientists who cannot sincerely afford it.

- Legal support from the administrative bodies, including a reduction in the bureaucratic red tape, is required for the smooth running of operations. Improvements in communication facilities both within the country and with the international community are necessary so as to reduce the time lags during data acquisition.

- The training of professionals in remote sensing at technician and management levels should be given priority as a first step in revitalizing this industry. This also includes equipping the affiliated training institutions with the relevant materials and equipments, in order to maintain high standards in the products. In-service training and retraining of existing staff should be maintained in order to keep up with new developments. The target consumers should also be sensitized by proper marketing strategies, information and education.

- Quality controls are necessary to promote high quality products and services to consumers. This can be achieved by the provision of good quality data receiving and processing equipment, provision of supporting infrastructure and proper training of the staff.

- It is necessary to diversify the range of remote sensing products available on the market, so as to reduce the technological deficiencies already being experienced in certain areas. This should also improve consumer demand, and encourage wider applications of remote sensing technology.

- Planning of remote sensing activities should be given priority at the national level with some form of commitment and social purpose. It is necessary to integrate the relevant types of socio-economic aspects with information on natural resources. Information flow should be organized from the national level down to the target groups and users.

With improvements in the financial support, training, information flow, legal, technical and administrative support, it should be possible to operate a satellite receiving station, afford good quality equipment and facilities, provide high quality products and, therefore, solve the technical constraints associated with the use of remote sensing technology.

Acknowledgements

I wish to acknowledge the support of the Department of Resource Surveys and Remote Sensing (DRSRS), Ministry of Planning and National Development, and the Regional Centre for Resource Surveys and Mapping (RCRSM), Nairobi, for availing the materials and facilities for the preparation of this paper. I would also like to thank the United Nations Office for Outer Space Affairs, Vienna, Austria and the Natural Resource Monitoring, Modelling and Management Program of Nanyuki, Kenya for their financial support.

References

1. Agatsiva, J.L.: "Sampling strategies in forestry surveys: implications of choice." In *Research Results Dissemination in Support of the District Focus Strategy* , pp. 25-27. Department of Resource Surveys and Remote Sensing, Ministry of Planning and National Development, Nairobi, Kenya, 1989
2. Andere, D.K.: "Rangeland monitoring in Kenya. The Mara ecosystem case study." In *Proceedings of the 20th International Symposium on Remote Sensing of the Environment , Nairobi, Kenya,* pp. 325-345. Environmental Research Institute of Michigan, Ann Arbor, Michigan, USA, 1986
3. Beukema, H.: *Overview of Remote Sensing Images in Kenya*. UNEP/DC-PAC, United Nations, DRSRS Ministry of Planning and National Development, Nairobi, Kenya, 1994
4. Catizzone, M. and Fanta, A.: *Needs and Prospects of Remote Sensing in Developing Countries.* European Commission Report EUR 15888 EN, Brussels, Belgium, 1993
5. Central Bureau of Statistics: *Economic Survey.* Ministry of Planning and National Development, Government Printer, Nairobi, Kenya, 1993
6. Delsol, J.P.: *Optimal Period for Maize Yield and Production Forecast Using Remote Sensing Techniques.* DRSRS, Nairobi, Kenya, 1989
7. De Man, W.H.: "Short notes on data collection and information supply for development." In *Proceedings of the 4th International Seminar on Remote Sensing Decision Making,* pp. 69-72. ITC Publications, Nairobi, Kenya, 1982
8. Doute, R., Ochanda, N. and Epp, H.: *Forest Cover Mapping in Kenya Using Remote Sensing Techniques.* Ministry of Environment and Natural Resources, KREMU, Nairobi, Kenya, 1981
9. Gibbon, H. and Mather, P.: "The use of remote sensing techniques for land use assessment and planning at district level in Kenya. A case study of Meru District." In *Proceedings of the 20th International Symposium on Remote Sensing of the Environment , Nairobi, Kenya,* pp 489-498. Environmental Research Institute of Michigan, Ann Arbor, Michigan, USA, 1986
10. Isavwa, L.A.: "Remote sensing transfer in Eastern and Southern Africa." In *Proceedings of the International Symposium on the State-of-the-Art of Remote Sensing in Biosphere Studies,* edited by P. Curran and G.M. Foody. UNEP 12.1-12.8, Moscow, Russia, 1989
11. Jaetzold, R. and Schmidt, H.: *Farm Management Handbook of Kenya.* Ministry of Agriculture, Nairobi, Kenya, 1983
12. Lamprey, R.H. and Leeuw, P.N.: *UNEP/ILCA NOAA-AVHRR Calibration Project.* A report presented to the Global Environment Systems, UNEP, Nairobi, Kenya, 1987
13. Loth P.E.: "Vegetation and landscape of Kora National Reserve." *ITC Journal* , Vol. 2, pp. 133-148. ITC, Enschede, The Netherlands, 1988
14. Maurel, P.: *Inventory of Available Satellite Data in KREMU.* Working Document No. 1, Ministry of Planning and National Development, Nairobi, Kenya, 1987
15. Otieno, J.O.: "The application of remote sensing technology to the process of natural resource planning in Kenya. Future needs and potentials." In *Proceedings of the 4th International Seminar, Remote Sensing Decision Making,* edited by J.M.M. van den Broek and H. Epp, pp. 41-50. ITC, Enschede, The Netherlands, 1982
16. Ottichilo, W.K and Peden, D.G.: "The use of remote sensing in maize production estimation in Kenya." In *Proceedings of the 20th International Symposium on Remote Sensing of the Environment, Nairobi, Kenya,* pp. 319-324. Environmental Research Institute of Michigan, Ann Arbor, Michigan, USA, 1986

17. Peden, D.G., Ottichilo, W.K. and Mwendwa, H.: "Estimating pre-harvest production of maize in Kenya using large-scale aerial photography and radiometry." *International Journal of Remote Sensing* , Vol. 7, No. 12, pp. 1669-1677. ITC, Enschede, The Netherlands,, 1986

18. Tiffen, M., Mortimore, M. and Gichuki, F.N.: *More People, Less Erosion: Environmental Recovery in Kenya.* ACTS press, Nairobi, Kenya, 1994

19. Wahome, E.K.: *Soil Erosion Classification and Assessment Using LANDSAT Imagery: A Case Study in the Baringo District, Kenya.* Technical Report No. 105. KREMU, Kenya, 1984

A Proposal on Improved Precipitation Estimates by Satellite and Conventional Methods for Agricultural Planning in Nigeria

A.A. Okhimamhe, Remote Sensing Unit, Department of Geography, Federal University of Technology, Minna Niger State, Nigeria

Abstract

Climatic aberrations of the past two decades resulting in crop failure, famine and starvation in the Sudan-Sahelian zone of Africa have made in depth climate-related studies on precipitation necessary. However, the major problem in the sub-region now is how to achieve sustainable development through self-sufficiency in food production.

There is therefore an urgent need to develop and establish an acceptable technique of quantifying precipitation on a scale commensurate with the problem. In an attempt to improve on past studies, a combination of satellite and conventional data could be used in estimating rainfall over the Sudan-Sahelian belt of Nigeria. Specifically, the research objectives include an improved technique of estimating precipitation from convective clouds during the summer monsoon. The information obtained will be applied to address some real indicators of drought with a view towards mitigating its worst effects. It will also serve as input to land use and management techniques which will serve as control measures against flood, erosion and desertification.

1. Introduction

Climatic aberrations of the past two decades resulting in crop failure, famine and starvation in the Sudan-Sahelian zone of Africa have made in depth climate-related studies on precipitation expedient in the sub-region. These disastrous events which include recurrent drought, flash floods, erosion and desertification have far-reaching effects on agricultural production. Generally, in Africa, particularly the Sudan-Sahelian belt, the major problem now is how to achieve sustainable development through self-sufficiency in food production (among other socio-economic development). For instance, in Nigeria, agricultural production was the most important aspect of the economy before independence in 1960's. The highly populated West African country has suffered in local food production partly as a result of dwindling of total land available for agriculture, the rapid growth in population and the inadequate application of climatic parameters as input in planning.

Although over 36% of land is devoted to agricultural activities in Nigeria, all available evidence shows that hunger and malnutrition are on the increase (Atinwo, 1989). There is a serious gap between population growth and the demand for food and the rate of food production is estimated at between 1% and 1.5% per annum; the growth rate of the demand for food is estimated at more than 3% per annum. A World Bank report indicates that 13.7 million Nigerians are under-nourished (World Bank, 1981). Unfortunately, simplistic attitudes to these climatic aberrations led to an emphasis on short-term solutions. For example, attempts to curb the losses incurred in terms of food shortages include relief measures, importation of selected grains, and the construction of dams for water conservation and mechanization of agriculture as a way of bridging the 'gap' betweeen low rainfall expectations and sustained food production. Other attempts have included the importation of fertilizers and the initiation of agricultural development schemes, such as the National Accelerated Production Project (NAFPP), the Operation Feed the Nation (OFN), and the Green Revolution Programme (GRP).

However, the true solution of the problem of climatic aberrations lies in carrying out real estimates of the most important factor limiting crop production - precipitation as rain. Although it is often said that successive generations of farmers have learned to live with the limitations of their local precipitation conditions through trial and error methods, famines resulting from climatic aberrations have shown that these have to be relegated to the background and replaced by well-defined research based on sound scientific knowledge. Agricultural projects have failed where planners have given insufficient recognition to the variability of precipitation which is a major determinant of crop growth, and hence food production. Proper agricultural planning should, therefore, involve quantifying precipitation in such a way that the results could easily be adapted to particular goals and in forms that can easily be understood and applied to these needs.

In view of this, there is an urgent need to develop and establish an acceptable technique of quantifying precipitation on a scale commensurate with the problem and based on the existing knowledge of its complex control, that is the general circulation of the atmosphere, disturbances and local factors. Numerous studies have been conducted on the structure of weather-producing systems in West Africa, but these have been mainly subjective and qualitative. For instance, the anomalous patterns of precipitation over West Africa resulting to climatic aberrations, especially between the 1970's and 1990's, have aroused much interest. Several explanations have been given ranging from the failure of the monsoon and tropospheric synoptic-scale features to increasing albedo effects due to denuded surface vegetation which will result in lower thermal heating needed for convection (Charney et al., 1975; Adefolalu, 1984).

Attempts have been made to study the role of the monsoon, especially in the supply of moisture and in enhancing regional instability. Features studied have included the Inter-Tropical Discontinuity (ITD)/Monsoon Trough (MT) migratory motions in relation to precipitation and its seasonality, the roles of the Tropical Easterly Jet and African Easterly Jets (TEJ/AEJ), synoptic wave types and characteristic distributions of rainfall, and the energetics and predictability of rain-producing systems.

Specifically, Archibold and Hamilton (1945) and Adejokun (1978) studied the seasonal characteristics of the monsoon, and further delineated the West African region into 4 major weather zones. Adefolalu (1982, 1983) confirmed these features in studies carried out during WAMEX. The occurrence of precipitation which is the most important phenomenon was linked with each weather zone (Aspiliden and Adefolalu, 1976). In studies relating to TEJ/AEJ, all convective activities resulting in precipitation during the summer monsoon are restricted to the area bounded by the AEJ to the North, and the TEJ to the South in West Africa. However, the relationships established between the formation intensity of specific weather events (e.g. rain, showers, etc.) to the synoptic features need amplification. Furthermore, satellite images and conventional data have been used to determine the location of convective activity (Reed et al., 1977) within the African Easterly waves while squall lines are characterized by cold cloud and heavy thunderstorms, with intense rainfall.

2. Estimation of Precipitation from Satellite Data

The insufficient surface-based conventional instruments do not provide quality data for research. The need to fill the "gaps" created if acceptable estimates of precipitation are to be made cannot be ignored. Otherwise, fundamental misconceptions will occur. Hence, there is no alternative to using all observational data platforms - particularly in space or on balloons - to monitor the environment.

Several techniques of estimating rainfall from satellite data during the past two decades have been developed (Barrett and D'Souza, 1986; Barrett et al., 1986). These techniques are indirect methods of precipitation estimation. For example, Cold Cloud Durations (CCD) are inferred from measured cloud characteristics in the visible and thermal infra-red parts of the spectrum. These are then related to the amount of rainfall. Topographic or latitudinal effects, and deviations of the seasonal amount of rainfall from its average value, are corrections that are sometimes used to improve these estimates.

Some of these "rainfall estimate" models have been previously discussed. For instance, Snijders (1991) compared the precision of ADMIT, PERMIT, TAMSAT, EARS and SDUS[1] applied to West Africa using July 1986 data and found that no significant differences exist between them, with the exception of SDUS. Carn et al. (1989) compared the TAMSAT method developed at the University of Reading (Milford and Dugdale, 1986) which used the surface temperature method and the correlation of rainfall with latitude where TAMSAT performed the best. However, several of the methods had either over- or underestimated values, when compared with rain guage data. It is therefore necessary to conduct further research aimed at improving the existing techniques, several of which can be applied to the Sudan Sahelian belt (Snijders, 1991; Carn et al., 1989).

---

1 ADMIT: Agricultural Drought Monitoring Integrative Technique
BIAS: Bristol/NOAA Inter Active Scheme
PERMIT: Polar-Orbiter Effective Rainfall Monitoring Integrative Technique
TAMSAT: Tropical Agricultural Meteorology Using Satellites
SDUS: Meteosat Secondary Data Users Station method

3. Aim and Objectives

The main aim of this paper is to propose the use of a statistical technique to estimate rainfall from satellite, conventional upper air (radiosonde) and surface-based data. The contributions of rain-producing systems over West Africa such as line-squalls and thunderstorms will be investigated. This is in an attempt to further exploit and contribute to existing knowledge of the role of precipitation and its influence on crop production through such climatic aberrations as drought, floods, erosion and desertification, with a view towards defining strategies that will aid and improve agricultural planning in Nigeria.

Specific objectives to be addressed include:

(a) Estimating rainfall from convective clouds during the summer monsoon in the Sudan-Sahelian belt of Nigeria, using a model which relates surface equivalent potential temperature to precipitable water.

(b) Applying the information obtained from (a) to address some real indicators of drought with a view towards mitigating a drought's worst effects and serving as input to land use and land management techniques which will serve as control measures against flood, erosion and desertification, and hence increased crop production.

(c) Defining strategies that will improve crop production, land use and land management practises based on the information derived from (a) and (b).

It is expected that the results will contribute to the information needs relevant to planning of crop-production in the Sudan-Sahelian belt of Nigeria. With time, the availability of funds and reliable crop production and yield data, further studies can be conducted using these expected results in a Geographic Information System and with other applied research, all aimed at boosting food production.

This proposal was designed for the Sudan-Sahelian belt of West Africa, particularly Nigeria. About 15% (140,000 km^2) of the nation's landmass is located in this climatically fragile belt which is prone to droughts, flooding and desertification (UN, 1977). Over 28 million people and more than 58 million livestock inhabit the area (Madu, 1988). Human, animal and crop losses run into billions of dollars as a result of these climatic aberrations.

4. Analysis Method

The equivalent potential temperature, θ_e, is defined as the potential temperature, θ, of a parcel of air when its water mixing ratio is zero. It is obtained from the first law of thermodynamics. Theoretically, θ_e is obtained when a sample of moist air is lifted adiabatically from its initial state until all its water vapour is condensed out. As a result, the latent heat is converted to sensible heat which is used to warm the air. It is then brought to the 1000 hPa level, dry-adiabatically; the equivalent potential temperature, θ_e, is the temperature attained at this point. Previous studies by Adefolalu (1972) and Oduro-Afriye (1989, 1992) have shown that θ_e is capable of predicting the convective instability which is prevalent in West Africa. This can lead to the development of cumulus clouds into cumulonimbus clouds. When the energy of these systems is sufficiently intense, they break through the characteristic trade wind inversions of the atmosphere over West Africa. As a result, high cumulonimbus towers develop (Adedokun, 1981) and thunderstorms and widespread rain occur.

The station most suitable for testing the proposed precipitation/rainfall estimation technique is Kano (12°N, 8°E) because of its location in the Sudan-Sahelian belt of Nigeria and the availability of upper air (radiosonde) data. Daily surface and specific humidity values will be obtained and the corresponding dew point temperatures derived and used to obtain equivalent potential temperature, θ_e, from a table prepared by Adefolalu (1972).

An attempt will be made to use the model developed, relating convecting cloud activity and precipitable water, to improve on precipitation estimation using meteorological satellite data. The research will adopt the log-linear modelling approach with total precipitable water as the dependent variable.

5. Potential Users of the Results

The results obtained will serve as input in the determination of better cropping zones, early warning systems in crop production planning, and studies on precipitation effectiveness. These will serve as important contributions to the formulation of well-articulated and ingenious strategies that could, in the long term,

(a) reduce the effects of drought and desertification and
(b) control flood and erosion.

It is expected that a user committee could work towards increasing food-production for the growing population of the Sudan-Sahelian belt of West Africa, in general, and in Nigeria in particular. Members could include: the Food and Agriculture Organization (FAO), Federal Ministries of Agriculture, Water Resources and Rural Development, i.e. Departments of Planning, Research and Statistics, etc., Agricultural and Rural Development, Water Resources and River Basin Development Authorities, etc., Agricultural Agencies and Communities, i.e. Agricultural Research Institutes, National Agricultural Land Development Authority, National Livestock Development Committee, etc., other related Agencies such as the National Committee on Ecological Problems, National Committee on Drought and Desertification, etc., and scientists and researchers.

6. Conclusions

The main aim of this paper has been to propose a technique for estimating rainfall using available data sets such as METEOSAT imagery, appropriate literature materials and computer facilities. Although archived WAMEX (Summer 1979) data exist, the assistance of such organizations as the World Meteorological Organization and the European Space Agency will be required to gain full access to them. This will, however, determine the suitability of the proposed technique in estimating the rainfall over the Sudan-Sahelian zone of Nigeria, in particular, and for West Africa in general.

References

1. Adedokun, J.A.: "Potential instability and precipitation occurrence within an inter-tropical discontinuity environment." *Arch Meteorol Geophys Bioklimatol Ser A, 30,* pp. 69-86. Springer-Verlag, Heidelberg, Germany, 1981
2. Aefolalu, D.O.: On mean equivalent potential temperature of the tropical atmosphere and the "little dry season" of West Africa." *Niger Meteorol Serv Q Mag,* pp. 15 - 40. Meteorological Department, Oshodi, Lagos, 1972
3. Adefolalu, D.O.: "Desertification in the Sahel." In *Natural Resources in Tropical Countries,* edited by O.J. Bee, pp. 402 - 438. Commonwealth Foundation of Singapore University Press, Singapore, 1982
4. Adefolalu, D.O.: "Monsoon onset in West Africa: applications of satellite imagery." *Arch Met Geophys Bokl Ser B, 32,* pp. 219-230. Springer-Verlag, Heidelberg, Germany, 1983
5. Adefolalu, D.O.: *Precipitation Trends in Relation to Water Management in Nigeria.* Paper presented at the Practical Agrometeorological Conference, Nigerian Geographical Association, University of Nigeria, Nsukka, April 1984
6. Adejokun, J.A.: "West African precipitation and dominant atmospheric mechanics." *Arch Met Geophys Biokl Ser A, 27,* pp. 289-310. Springer-Verlag, Heidelberg, Germany, 1978
7. Archibold, J.N. and Hamilton, R.A.: "Meteorology of Nigeria and adjacent territory." *Quart J Roy Met Soc 3,* pp. 231 -265. Royal Meteorological Society, Bracknell, United Kingdom, 1945
8. Atinwo, T.: "Feeding the World: The Nigerian Situation." *Daily Sketch,* p. 10, June 6, 1989
9. Aspiliden, C.I.L. and Adefolalu, D.O.: "The mean troposphere over West Africa." *J Appl Met, 16,* pp. 705 - 716. American Meteorological Society, Boston, USA, 1976
10. Barrett, E.C. and D'Souza, G.: "An objective technique for broad-scale rainfall monitoring by geostationary satellites." *Proceedings of the 20th International Symposium on Remote Sensing of the Environment, 4-10 December 1986, Nairobi, Kenya,*Vol. III, pp. 955-964. Environmental Research Institute, Michigan, Ann Arbor, USA, 1986
11. Barrett, E.C., Power, C.H., Beaumont, M.J. and Richards, T.S.: *Cloud Cover and Rainfall in the Western Sahel. Monitoring of Pastureland Ecosystems in the Saheland Mapping of Cloud Cover and Rainfall.* Final Report to the Commission of European Communities Development Directorate (DG VIII). NERC, Wallingford, United Kindom, 1986
12. Carn, M.B., Dagorne, B., Guillot, B. and Lahuec, J.P.: "Estimations des pluies par satellite en temps reel en Afrique Sahelo-Soudanienne. Essai d' utilisation d'une calibration issue du champ de température maximum de surface."*Veille Climatique Satellitaire, 28,* pp. 47 - 54. 1989
13. Charney, J., Stone, P.H. and Quirk, W.J.: "Drought in the Sahara: a bio-geophysical feedback mechanism." *Science, 189,* pp. 434 - 435. American Association for the Advancement of Science, Washington DC, USA, 1975
14. Milford, J.R. and Dugdale, G.: *Rainfall Mapping Over West Africa in 1986.* Final Report to the Food and Agriculture Organization. Department of Meteorology, University of Reading, United Kingdom, 1986
15. Oduro-Afriyie, K.: "On the mean monthly equivalent potential temperature and rainfall in West Africa." *Theor Appl Climatol, 39,* pp. 188 - 193. 1989

16. Oduro - Afriyie, K.: "Estimating precipitable water over West Africa from surface equivalent potential temperature and related potential instability indices." *Atmos Res, 28,* pp. 11-20. Elsevier Science, Amsterdam, The Netherlands, 1992
17. Reed, R.J., Norquist, D.C. and Recker, E.E.: "The structure and properties of African wave disturbances as observed during phase II of GATE." *Mon Wea Rev, 103,* pp. 317-33. American Meteorological Society, Boston, USA, 1977
18. Snijders, F.L.: "Rainfall monitoring based on Meteosat data. A comparative study applied to the Western Sahel." *Int Journal of R Sensing,* Tailor and Francis Ltd., United Kingdom, 1991
19. UNEP: *United Nations Conference on Desertification - Status of Desertification in Hot Arid Regions. Maps and Explanatory Notes 74/31.* UNEP, Nairobi, Kenya, 1977.
20. World Bank: *Nigeria: Basic Economic Report.* World Bank, Washington DC, USA, 1981

Report on Panel Discussion 2: Remote Sensing

D. St. Pierre, L. Valignon, 1995/96 Master of Space Studies (MSS) students, International Space University

Session Chair: **H. Chasia,** International Telecommunication Union, Geneva, Switzerland

The theme of the day was remote sensing (RS). The papers presented reviewed the different applications of RS (meteorological services, disaster warning, resource management) both from the user's and the provider's points of view. The role of RS in sustainable development and the emerging commercial applications were discussed.

Here is a summary of the discussions that took place among members of the panel:

1. Access to Data for Developing Countries/User-Provider Links

- In order to both promote RS and help sustainable development, RS data should be available at the lowest possible cost, especially to developing countries.

- Some steps have already been taken in that direction. For example, EUMETSAT data are readily available to any developing country or research organization, and INPE allows dissemination of data free of charge or at a low price.

- Users should be stimulated to use RS but also trained to process and use efficiently the data obtained.

- In order to improve public awareness, it is important to provide continuous information and to have the right infrastructure to follow on. (The point was raised that some of the poorest countries do not even have the necessary infrastructure to take advantage of RS.) It is typically the role of the operations community (public sector) to make this type of long-term commitment, while space agencies should demonstrate new technologies in a pre-operational phase.

2. The Relation Between Space Scientists/Engineers and Decision Makers (Politicians)

- In the past, Earth's evolution was driven exclusively by physics and natural forces; now, human activities are a significant force. RS provides us with a tool to monitor the effects of our activities.

- The panelists debated whether such a monitoring tool can influence human activities and help us to work toward the preservation of Earth's resources.

- Technology alone cannot solve problems, but is dependent upon political will. RS is a tool for diagnosis which can help in resource management and training of scientists and engineers, but the information gleaned has to be brought to the attention of politicians in their capacity as decision makers. It has been proposed that future RS programs should take this aspect into consideration during the preliminary phase of mission conception.

- Companies are used to selling their projects, to lobbying politicians. The media could help to bridge the gap between space scientists/engineers and the decision makers. Today's scientists and engineers have a duty to communicate with the general public in a language which all can understand.

3. Military RS Data

- Some panelists recommended the use of military data in order to support developing countries.

- Two objections were made:

  - No direct links exist between scientists and those involved in defence; politicians are in between and may not facilitate dialog.

  - It could be difficult (if not impossible) to use these data for civilian endeavors.

Session 3
Space of Service: Telecommunications

Session Chair:
H. Chasia, Deputy Secretary General, International Telecommunication Union

New Strategies for Telecommunications Development

R. Jakhu, International Space University

Abstract

Telecommunications essentially involves technical activity carried out, primarily by technical persons, within a framework of national policy and law, in order to achieve national social, economic and cultural goals. Efficient and economical means of telecommunications are indispensable for the overall development of nations. Satellite telecommunications are implemented and used within the national regulatory and policy frameworks for telecommunications in general. New policies and strategies are being designed to develop telecommunications. In this context, privatization and competition in the telecommunications industry are increasing and have proved to be so beneficial that to question them requires exceptional and considerable argument. First, since the availability of sufficient financial resources is generally limited, new means for generating revenue must be sought actively. In this context, levying charges for the use, or the auctioning, of radio frequencies for all communication purposes should be considered since it has proved to be a good revenue generator for some countries which have followed this course. Secondly, all attempts to attract foreign investment in the telecommunications industry must be based on a good balance of the interests of foreign service providers and the host countries, and must not inhibit the introduction of more efficient technologies and new management approaches. Thirdly, effective privatization and competition must be accompanied by regulation. The introduction of privatization and competition in each country or region must be evolutionary in nature and must be handled with great care. The approach should be unique according to the particular situation in a given country or region.

1. Introduction

The theme for this Symposium is *Space of Service to Humanity*. Three quarters of humanity lives in the so-called developing countries. Thus a space activity which does not significantly and meaningfully improve the lives of peoples in these countries is not, in my view, an activity of service to humanity.

It is a well established fact that efficient and economical means of telecommunications, which carry information to where it is needed, are indispensable for the overall development of nations. Practically useful means of telecommunications are under-developed and insufficient in almost all developing countries as compared with the developed countries. For example, the city of Tokyo has about the same number of telephones as in whole of the African continent. Moreover, there is a great disparity of telephone density between urban and rural communities in most developing countries. If access to relevant information is the key to sustainable development, is there not a paradox at the outset?

Satellites have proved to be an excellent means of telecommunications because they can provide efficient services on cost-effective bases and within a relatively short period of time. However, satellites are only one means of telecommunications, and their technology and systems have converged with the terrestrial infrastructure. In considering the use of satellites for telecommunications, the value of using other means should also be taken into account, and considered as an integrated element of the global and national telecommunications systems.

Recognizing the importance of telecommunications for general development, various efforts are being made to develop telecommunications. Many efforts are hindered by the lack of the necessary financial resources. Such resources could be national and international. Since money is extremely limited domestically in most developing countries, and since limited resources are often dedicated to more urgent needs such as food, these countries are attempting to attract foreign investment in the telecommunications industry.

Although the numerous positive indirect benefits of an implementation of even a basic telecommunications infrastructure have been largely documented (e.g. powerful repercussions of such an infrastructure on agriculture, health, education and governance sectors), one essential question remains key in any debate on the useful strategy for telecommunications development: what are the direct returns for national and international investors as compared with other investment opportunities?

2. National Capital

As noted earlier, the availability of sufficient financial resources at a national level is extremely limited. However, although investment in telecommunications infrastructure is considerably less than undertaking road, power or waterworks installations, it should remain a priority as a catalyst for social and economic development. It is therefore important to identify new investment sources which will enable a developing country to facilitate the implementation of the appropriate infrastructure.

The electromagnetic spectrum is a limited natural resource that supports the transmission of information. This resource enables carriers to generate substantial profit from the telecommunications business. It could therefore be "leased" subject to the charges levied for the use of radio frequencies, as has been done in the United States of America.

The US example provides an interesting suggestion for the levying of such charges on the electromagnetic spectrum within a country. Through the Omnibus Budget Reconciliation Act of 1993, the US Congress authorized the Federal Communications Commission (FCC) to conduct auctions, based on competitive bidding, to award licenses for the use of the electromagnetic spectrum. This auction process replaces the lotteries previously used to assign frequencies. Frequencies are still allocated and allotted by the FCC - thus the FCC first decides for which types of services each block of frequencies is to be used (although service definitions appear to be wider and more flexible than before) and, for certain specific services, where they should be used. Following this procedure, the auction process is then used to decide which service provider will be licensed to use certain individual frequencies.

The first auction took place on 25 July 1995. This auction attributed licenses in the narrow-band Personal Communications Service which is expected to be used for advanced paging and data transmission services. These licenses have been issued for a period of 10 years. The first three auctions raised almost US $10 billion for the US Government. On the day of the first auction, the FCC Chairman, Reed Hundt, stated that "*...while today we might be taking one small step toward reducing the national debt - we are making a giant leap toward increasing competition in communications and expanding our economy*." The official rationale behind the auctions, apart from raising money, appears to be that of creating competition for licenses which will lead to a more efficient management of the spectrum.

This example demonstrates how a similar policy and practice could generate new revenue for developing countries. These funds could, in turn, through partnership and/or more efficient regulatory control and enforcement, provide the appropriate negotiating leveraging abilities for a government when concluding agreements with local or international investors.

Here, I may reiterate my idea which was first expressed in 1983 that the use of radio frequencies and orbital positions should be subjected to some form of international tax (Reference 1). The revenue generated could be used for the development of telecommunications internationally.

3. Foreign Investment

In order to attract foreign investment, certain developing countries have recently started adopting new strategies in the form of privatization of their state-owned telecommunications companies and providing favorable conditions to foreign investors. A short period of such development has shown a rapid growth of foreign private investment.

However, in order to invest their money some of the foreign companies have insisted and secured such favorable rights from the host countries that one wonders whether the development of the telecommunications industry will, in fact, contribute to the development of these countries. For example, certain concession or licensing agreements in some countries stipulate that a foreign company, which acquires the assets of a state-owned enterprise and provides a substantial investment in order to provide telecommunications services, is given a complete and comprehensive monopoly for 20 to 30 years on all forms of communications, including terminal equipment and directory publications. In certain jurisdictions, this enables foreign investors to maintain up to 100 % financial control and allow no local financial participation. (It may be noted that, because of national security reasons, countries like the US, Canada and Japan do not allow the foreign ownership of their telecommunications industries to exceed twenty percent). Furthermore, the foreign investors pay very little taxes - sometimes no taxes at all - and may charge any rate for the service without any effective scrutiny or control by local public authorities. These investors are also entitled to take all their profits in hard convertible currencies out of the exploited territory, and are under no obligation to provide services to less profitable areas with thin traffic.

Data from the countries which have given monopoly rights to foreign private companies show that the rates or charges for telecommunications have not significantly decreased. Since the rates for telecommunications services remain high, the possibility of the development of business in general and of initiating new ventures remains limited.

Furthermore, these countries cannot, on their own, join international consortia, like IRIDIUM, without the consent and participation of the companies holding monopoly rights. They cannot develop telecommunication services in the rural areas with thin telecommunications traffic. In fact, in some countries, they cannot even adopt new policies and regulatory reforms without the possibility of violating their agreements. Indeed, these agreements are not very different from those signed by the colonized countries with the foreign trading companies of Europe during the fifteenth and sixteenth centuries.

This situation is not generally known because of the private and confidential nature of concession and licensing agreements between governments and corporate entities. Seemingly complex, these agreements are often negotiated and signed by a single individual representing a small group of interests on behalf of the host country without any public scrutiny or long term regulatory consideration.

Since relevant policies and laws are non-existent or totally out-dated and inadequate in most of the developing countries, concession agreements with the host government in fact become the law governing telecommunications matters. Therefore, these countries are not in a position to (a) take advantage of their rights under such agreements, (b) introduce any policy and regulatory reforms without violating their obligations under these agreements, and (c) introduce new technical means for efficient and economic telecommunications services. According to certain agreements, it may be impossible for a country to install a small dish antenna without the permission of a foreign service provider. These countries would thus remain totally at the mercy of foreign investors for extended periods, and the revolution in telecommunications technology, including the added-value of satellite services, would not improve the lot of the peoples of countries which have committed themselves to foreign companies.

In order to avoid such situations, I suggest the adoption of national polices which stipulate that the introduction of privatization must be accompanied by national regulation which provides a framework for competition and reinvestment in the national telecommunications sector, particularly in thin traffic areas. The establishment of a national framework on telecommunications (and broadcast) policy and law would enable governments to better achieve national social, economic and cultural goals. In other words, all telecommunications must be implemented within the national regulatory and policy frameworks for telecommunications in general in order to improve the general welfare of the society.

Developing countries should establish a regulatory framework to which all foreign investment agreements must comply. Such regulatory frameworks should contain provisions dealing with competition, fair and equitable deals, monitoring and accountability of the foreign investors. Ideally, these countries should seek advice on legal and financial matters before signing an agreement with a foreign investor and try to negotiate Build-Operate-Transfer (BOT) type of agreements with the foreign investors.

However, in practice, it may be difficult for many developing countries to define and implement such a regulatory framework since, as mentioned earlier, mostly such agreements are signed by single individuals under various forms of pressure without the benefit of much information from all sectors of the government and the general public. In these cases, one may ask what it is possible to do in assisting a developing country which has already signed a binding agreement?

A possible solution for such a country lies in its adherence to an international trade agreement, similar to the GATTS and NAFTA models, which insists on the provision of competitive services within the territories of member States. In undertaking this process, it would be imperative to exchange information with other national administrations facing similar challenges as to compensatory damages which can or should be negotiated with the foreign telecommunications companies. Another idea would be to declare a portion of their territory as free trade zone, and establish teleports to attract investment in telecommunications facilities.

4. Appropriate Regulatory Framework: Deregulation vs. Re-regulation

Privatization and competition in the telecommunications industry are becoming so common that even to question them needs exceptional and considerable arguments. It is important to emphasize that privatization and competition must go hand-in-hand with national regulation. In this trilogy, lack of one will defeat the purpose of the other two. The introduction and strengthening of competition have also been expressed as the adoption of "deregulatory policies". However, the term "deregulation" has become one of the most misunderstood terms in general, and in the telecommunications industry in particular. "Deregulation" does not mean lack or lessening of regulations and laws. In fact, it means re-regulation and competition.

"Deregulation" must be understood as "competition" rather than "less regulation". The complexity and volume of regulations dramatically increase with the introduction of "deregulation". There are numerous examples. For instance, in Canada telecommunications have traditionally been provided by government and private monopolies. With the adoption of the policy of competition, the first national Telecommunications Act was passed in 1994. It is expected that the CRTC - Canada's regulatory body - will be issuing hundreds, if not thousands, of new regulations and policy directives pursuant to the Act.

Laws and regulations are indispensable as rules of the road for players in the telecommunications industry, especially when they are numerous and are of both national and foreign origin. In any democracy, it is the rule of law, rather than personal dictates of a leader or a private individual or company, that governs national activities. On the other hand, the lack of appropriate laws and governmental policies, or the existence of dubious laws and policies, creates chaos and uncertainty which are counter-productive for development in general and investment and business in particular.

The introduction of privatization, competition and regulation in the developing countries must be handled with great care. A delicate balance must be achieved between national interest and private sector prerequisites for investment. On the one hand, because every country presents a specifically unique context, the approach by each should be tailored locally for the benefit of the local population, with both short and long term considerations in perspective. No specific and single model has one hundred percent relevance to all countries. On the other hand, any attempt that will be made in order to introduce privatization, competition and regulation must carefully take into consideration the appropriate interests of the telecommunications operators that have invested or will invest millions of dollars (or other currencies). Otherwise, to attract foreign investment would be very difficult, if not impossible.

The objectives of new regulatory policies to be adopted by the developing countries should be to (a) clearly establish the authority of a government over all aspects of telecommunications, (b) clarify the relevant law and policy-making procedures, (c) clearly define the delegated decision-making functions of the concerned government officials, (d) update existing rules and elaborate new ones in response to the emerging context, and (e) clearly establish a policy and regulatory framework within which any concession or license may be awarded to a national and/or foreign private investor, with specific reference to the condition and costs of the said concession. In sum, any legislation which is adopted must essentially be based on a good and fair balance of interests of the service providers, the regulators (the governments) and the public.

Since telecommunications policy and law expertise is almost non-existent in various parts of the world, it is suggested that, in addition to technical training centers for the use of telecommunications technology, a form of regional Institutes for Telecommunications Policy and Regulation should be established. These Institutes should be independent bodies where government officials could learn and develop the appropriate tools to cope with the challenges posed by, and to take advantage of the benefits offered by, the information and telecommunications revolution. The principal tasks of these institutions should include (a) teaching and training, (b) holding conferences and seminars while providing a forum for regulators, (c) carrying out research and publications, and (d) providing general and specific consultation, in the field of telecommunications policy and regulation.

5. Conclusions and Recommendations

First, at a national level, and particularly in developing countries, privatization and competition in the telecommunications sector must go hand-in-hand with the creation of clear national telecommunications policies and regulatory frameworks.

Secondly, all attempts to attract foreign investment in the telecommunications industry must be based on a good and fair balance of the interests of foreign service providers and the host countries. Neither the national regulatory framework nor the concession agreements should inhibit the introduction of more efficient technologies and new management approaches.

Thirdly, new means for generating revenue must be sought actively. In this context, levying predetermined and/or auction charges for the use of radio frequencies for all communication purposes (telecommunications and broadcast) should be considered since it has proved to be a good revenue generator for certain countries which have followed this course.

Fourthly, if privatization is implemented and additional funds are raised through levying charges, they should be dedicated to the implementation and improvement of the telecommunications infrastructure through various related activities (applied R&D, education, international representation and participation in treaty negotiations, etc.).

Finally, information and experience sharing systems, perhaps through regional Institutes for Telecommunications Policy and Regulation, should be created to help government authorities to make optimal decisions for their short and long term telecommunications development strategies.

Acknowledgement
The author expresses his appreciation to M. Berdnikoff for suggesting some editorial changes to improve the text of this paper .

References

1. Jakhu, R.: *Legal Regime of the Geostationary Orbit.* DCL Thesis, McGill University, Montreal, Canada, 1983

Exploration of the Equatorial LEO Orbit for Communication and Other Applications

E.S. Seumahu, Nanyang Technological University, School of Electrical & Electronic Engineering, Nanyang Avenue, Singapore 639798, Republic of Singapore

Abstract

LEO (Low Earth Orbiting) satellites have been used for several decades for the purposes of scientific research, remote sensing, surveillance, meteorology, satellite imaging and other applications. In the near future, highly inclined LEO and MEO (Medium Earth Orbiting) satellites will also be used extensively for mobile and personal communications.

Tropical countries around the equatorial belt are at a disadvantage from the point of view of highly inclined LEO satellites because of short and infrequent pass times. These countries are also at a disadvantage from the point of view of GEO satellites because meteorological conditions cause severe propagation problems at Ku-bands and there is heavy congestion in the geo-synchronous orbits at C-band.

This paper reports on investigations now being done at Nanyang Technological University (NTU) in Singapore to explore the use of the equatorial LEO orbit to provide improved coverage for geographical areas around the equator. It is predicted that even one satellite in Low Equatorial Orbit (LEQ) can provide 12 passes a day over a point on the equator, with every pass being an overhead pass lasting as long as 27 minutes. Five satellites would be sufficient to provide continuous 24 hour coverage for Singapore. With 6 or more satellites, continuous service could be provided for countries lying between the Tropic of Cancer and the Tropic of Capricorn. The regions to benefit are ASEAN countries (Singapore, Malaysia, Thailand, Philippines, Indonesia..), Southern India, North Africa, Central America, Pacific basin countries and Northern Australia.

Micro and mini satellites in inclined LEO orbit will be used initially to conduct research in the areas of store and forward communications, electronic messaging, remote sensing, remote data collection, enviromental monitoring, meteorology, surveillance, GPS applications, satellite-PCN, and mobile communication. This will also serve as a vehicle for the transfer of technology and satellite expertise to Singapore. Future proposals include the launch of Singapore's own micro- or mini-satellites to pioneer the use of the equatorial orbit

1. Introduction

Artificial satellites launched into circular orbit around the Earth for communications, scientific research, remote sensing, surveillance, meteorology, navigation and other purposes can be classified as GEO (Geostationary Earth Orbit), MEO (Medium Earth Orbit) or LEO (Low Earth Orbit) systems according to their heights and the shape of their orbits.

GEO satellites are satellites launched into a circular orbit above the equator, at a height of about 36,000 km. In this orbit, a satellite circles the Earth once every 24 hours. Since the Earth also revolves with a period of 24 hours, GEO satellites appear to be "stationary" and can therefore be used for telephony, data transmission, TV broadcasting and other applications using dish antennas on the ground which point to a fixed position in the sky. These satellites are big and expensive to build and launch. At such a great distance, the signals reaching the

Earth from the satellite are extremely weak. To transmit a signal from the Earth to the satellite requires a large amount of electrical power.

The use of GEO satellite systems for communications and other purposes is now well established, recognised and appreciated world-wide. Their applications in South-East Asia have been increasing at a great pace, except in the area of Personal Communication Services (PCS).

LEO satellites orbit the Earth at a much lower heights, between 300 km and 1500 km. In this orbit they have a period of about 90 minutes to two hours. They are usually placed in highly- inclined orbits, i.e. their paths pass closer to the North and South poles than to the equator. They appear to be moving continuously in the sky. Many LEO satellites are needed to cover the whole world on a continuous 24 hour basis. The big advantage of LEO satellites is that they are smaller and cheaper to build and launch and, because of their lower heights, they can be accessed very easily, even from mobile or cellular type handsets.

MEO orbits have greater heights than LEO orbits, usually between 5,000 and 15,000 km. The generic term "LEO" is commonly used to refer to both the LEO and the MEO systems. "Little LEO" refers to those systems operating in the frequency bands below 1 GHz, while "Big LEO" refers to those operating above 1 GHz.

GEO technology is now well matured and Singapore has begun recently to exploit this technology for the provision of international telephony, digital and data services, TV relay, direct broadcasting (DBS) and VSAT (Very Small Aperture Terminals). LEO technologies have also been well developed world-wide in non-communication applications such as remote sensing, satellite imaging, meteorology, surveillance, navigation and global positioning. On the other hand, the technology to use the LEO orbits for communication purposes is still in its infancy and will not be in full commercial use until after 1996.

There are two motivations for NTU's interest in LEO satellites:

(a) Singapore would be in a position to take advantage of the predicted growth in LEO applications if the necessary technologies, expertise and infrastucture could be established in the country. So there is a need to acquire these facilities through a program of education, experimentation, hands-on experience and technology transfer.

(b) Tropical countries around the equator are at a disadvantage from the point of view of GEO satellites because meteorological conditions cause severe propagation problems at Ku-bands and there is heavy congestion in the geo-synchronous orbits at C-band. Advanced research projects are being undertaken at NTU to improve satellite telecommunications for equatorial parts of the world. These include satellite beacon experiments, tropical satellite propagation studies and investigation into advanced techniques such as spread spectrum and Code Division Multiple Access (CDMA). Additional research into the use of LEO orbits could lead to economical solutions for equatorial regions.

The countries around the equatorial belt cannot take full advantage of currently used or proposed LEO orbits because of short and infrequent pass times. In these highly inclined orbits, each LEO satellite appears above the horizon for only 10 to 15 minutes per pass on the average. On the equator there may be at most 2 or 3 good passes on any one day. In the tropical regions around the equator where line of sight is often hindered by cloud cover, it is not uncommon for the customer to wait up to 16 days to obtain a clear remote sensing pass from one satellite. To obtain continuous 24-hour real-time coverage world-wide for communications and other purposes, much more than one satellite is needed. For example, 24 satellites are required for global positioning in MEO orbit and between 48 and 840 satellites have been proposed in LEO orbits for mobile and Personal Communication Network (PCN) applications.

Hence there is a motivation for new areas of research and theoretical investigations into the use of LEO satellites in an equatorial orbit.

2. Single Satellite in Low Equatorial (LEQ) Orbit

The basic geometry for one satellite is shown in Figure 1, where

$\mathbf{r}_s$ is the distance from the satellite to the centre of the Earth,
$\mathbf{r}_e$ is the radius of the Earth,
γ is the angular distance between sub-satellite point and the desired edge of coverage,
θ is the elevation angle at the edge of coverage,
h is the height of the satellite,
T is the actual period of the satellite,
Tapp is the apparent period (the satellite is assumed to move from West to East above the equator), and
Tpass is the pass length for points on the equator, i.e. the time taken from satellite-rise to satellite-set.

For a single satellite in LEO orbit, the actual period varies from 90 minutes to 127 minutes and the pass length from 9 minutes to 31 minutes at the equator as the height varies from 270 km to 2000 km. From Singapore, one satellite at a height of 1650 km would always provide overhead passes with each pass lasting for 27 minutes. There are 12 such passes every day. On the other hand, the same satellite in polar orbit would give at most 2 overhead passes a day, each lasting for 25 minutes. Most of the time the polar orbit passes will not be overhead. The average length of these passes is somewhere between 10 and 15 minutes only.

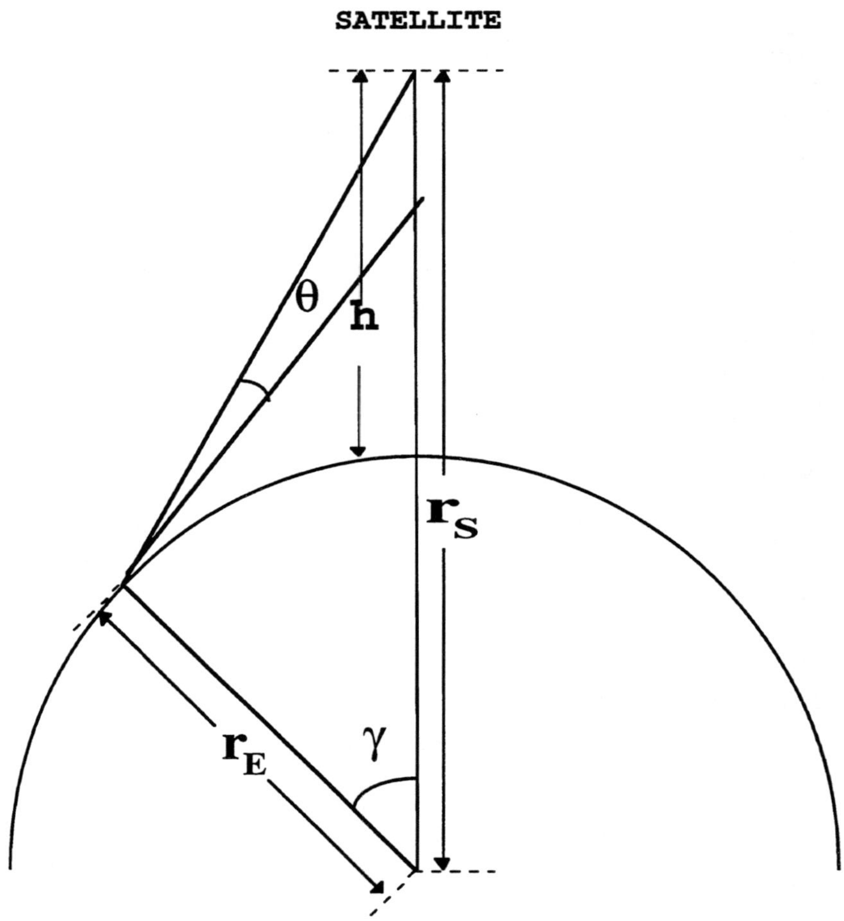

$$r_S = \frac{r_E \cos\theta}{\cos(\gamma+\theta)}$$

$$h = r_S - r_E$$

$$T = \sqrt{\frac{4\pi^2 r_S{}^3}{\mu}}$$

$$T_{APP} = \frac{1440\ T}{1440 - T}$$

$$T_{PASS} = \frac{\gamma}{180} x T_{APP}$$

Figure 1. Geometry for a single satellite in LEQ

3. LEO Satellite Constellation

The geometry for two adjacent satellites in a LEO constellation is shown in Figure 2, where the circles are the points on the surface of the Earth which have zero elevation angles,

> α is half the longitude difference between the two satellites,
> β is the latitude limit, i.e. the northernmost or southernmost latitude where both satellites appear on the horizon at the same time (the geographical areas between the two latitude limits will have continuous cover, and
> Tact is the time when a satellite is "active", i.e. the time between hand-over from one satellite to the next.

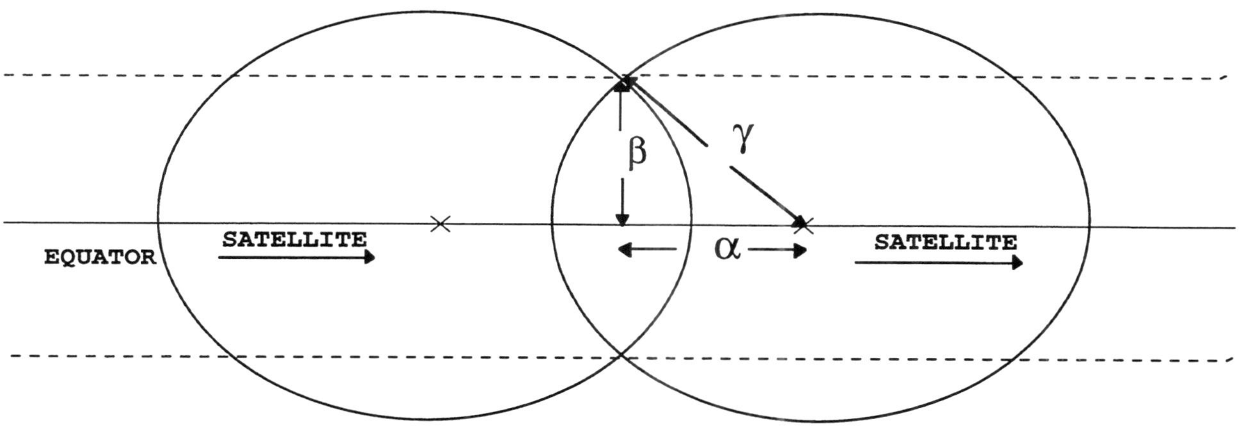

$$\cos\gamma = \cos\alpha\cos\beta \quad ; \quad \alpha = \frac{180^0}{N_S}$$

$$T_{ACT} = \frac{T_{APP}}{N_S} = \frac{\alpha}{180} \times T_{APP}$$

Figure 2. Geometry for two adjacent satellites in LEO

The relationship between orbital height and latitude limit is shown in Figure 3 for different numbers of satellites in the constellation.

The relationship between pass length and latitude limit is shown in Figure 4. It can be seen from these graphs that a constellation of 5 satellites at 1650 km height would give continuous cover for Singapore with regular pass lengths of 27 minutes. A constellation of just 6 satellites at a height of 1650 km would give continuous coverage for all geographical positions between 23.5 degrees North and 23.5 degrees South.

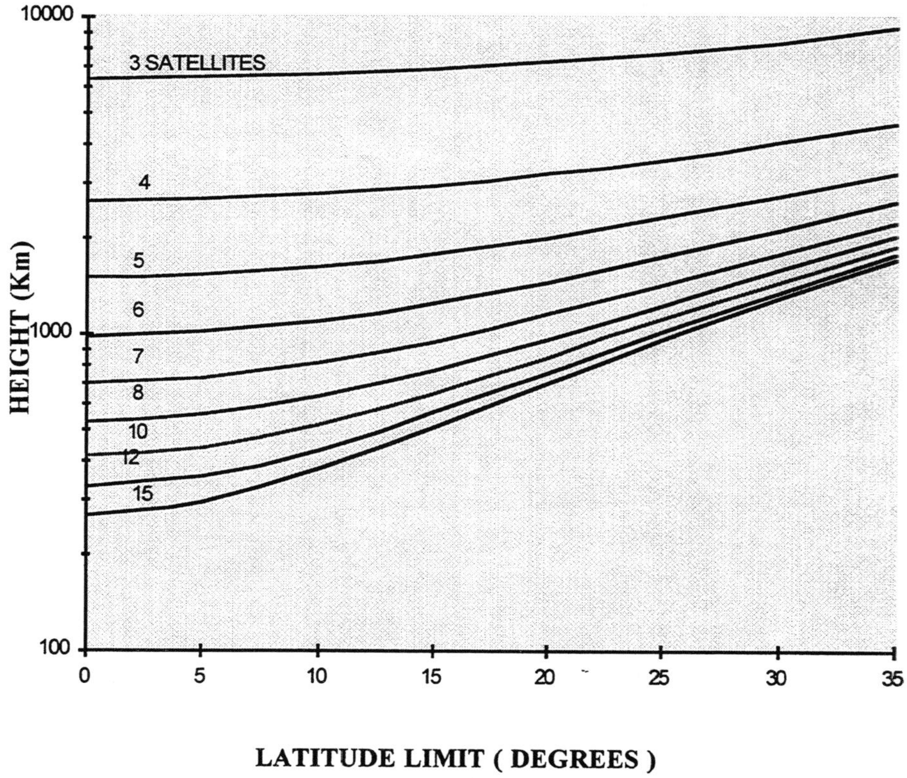

Figure 3. Relationship between orbital height and latitude limit in LEO by satellite numbers

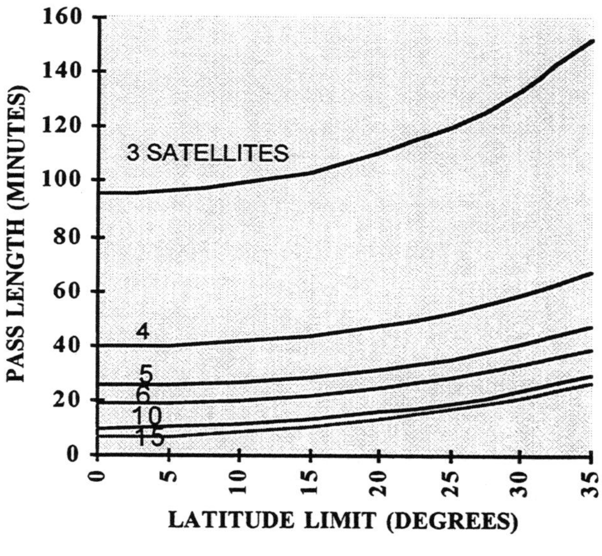

Figure 4. Relationship between length and latitude limit in LEO by satellite numbers

For commercial applications, minimum elevation angles of 30 degrees would be desirable. Figure 5 shows the satellite heights needed to give the required elevations. It can be seen here that, if 5 or 6 satellites are to be used to give coverage between the tropic of Cancer and the tropic of Capricorn, these satellites must be positioned in equatorial MEO (MEO) orbits instead of the LEO orbit. More than 15 satellites would be needed to give the same coverage from the LEO orbit.

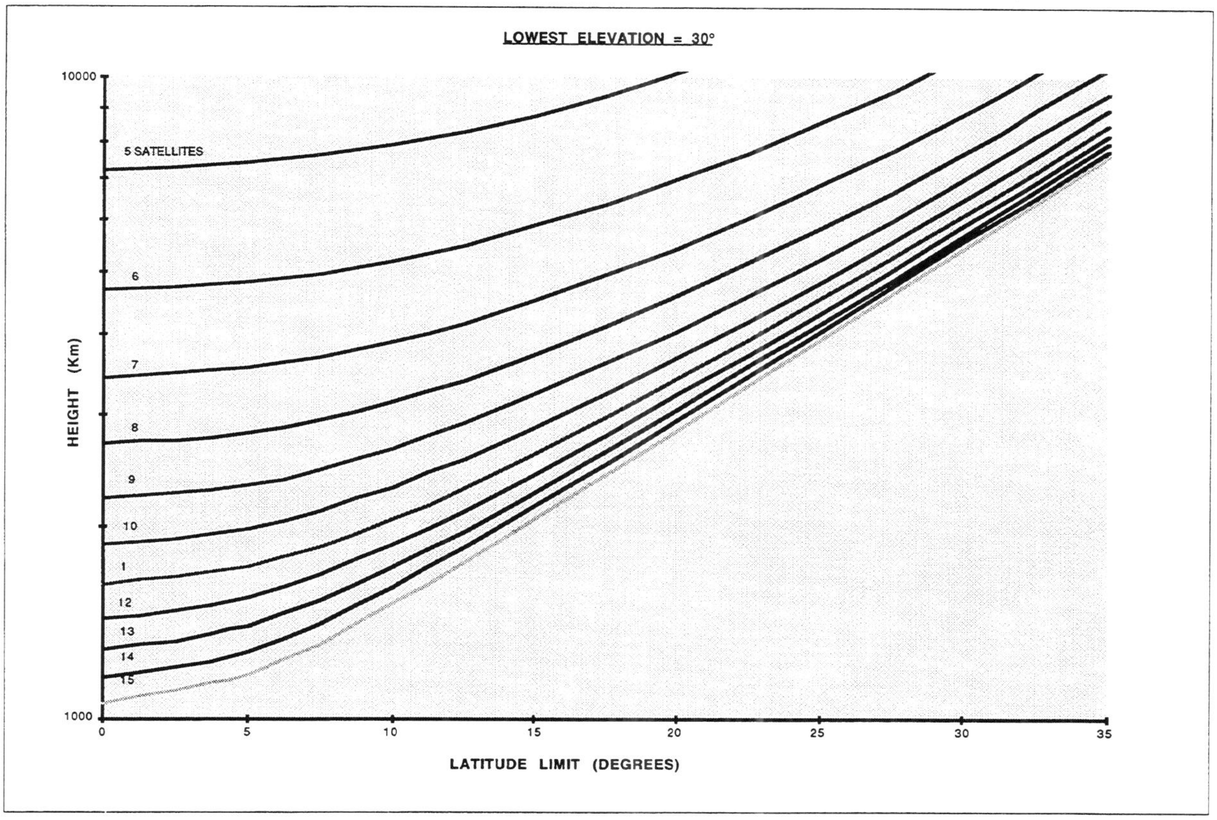

Figure 5. Satellite heights needed to give required elevations

For a fixed coverage between 23.5 degrees North and South, the number of satellites could be reduced if elevation angles less than 30 degrees could be tolerated. Orbital heights for various minimum elevation angles are shown in Figure 6. At 10 degree mimimum elevation, only 10 satellites would be required in the LEO orbit and only 3 or 4 are needed in the MEO orbits.

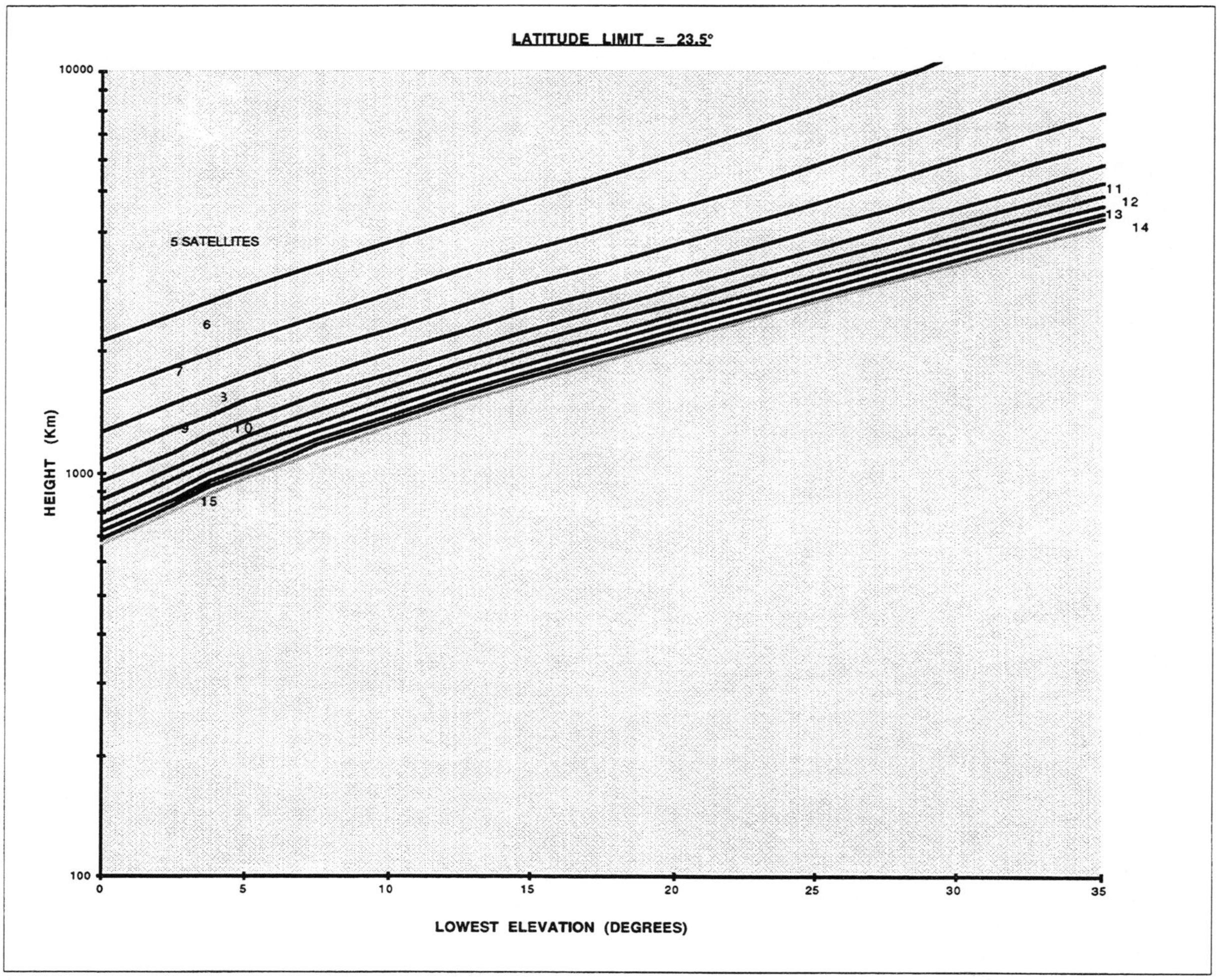

Figure 6. Orbital heights for various minimum elevation angles

Inter-satellite links are often used between adjacent satellites in a constellation.The geometry is shown in Figure 7, where

d is the inter-satellite distance, and
y is the path clearance.

The plots of distances and clearance are given in Figure 8. For instance a 6-satellite system for tropical coverage with zero minimum elevation would have intersatellite link of about 8000 km and a path clearance of about 600 km.

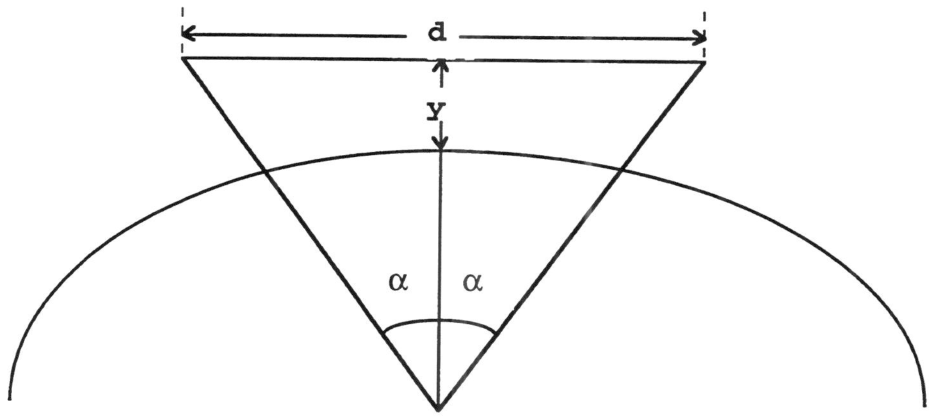

$$d = 2r_S \sin \alpha$$

$$y = r_S \cos \alpha \equiv \frac{r_E}{\cos \beta}\left[1 - \cos \beta\right]$$

Figure 7. Geometry of inter-satellite links in a constellation

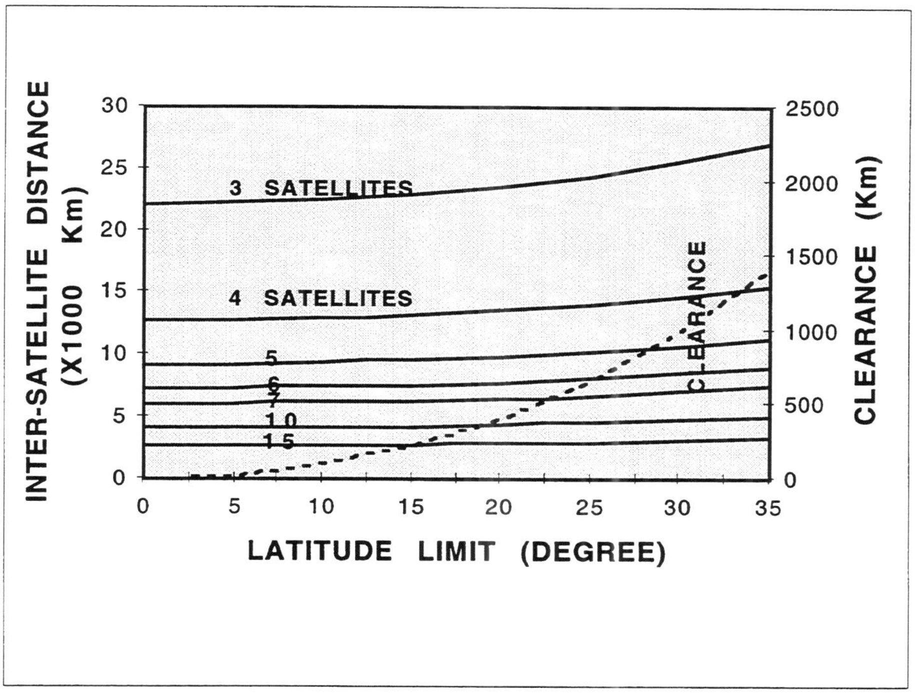

Figure 8. Distances and clearance

4. Current Work at NTU

NTU and the University of Surrey (UoS) have signed a Memorandum of Understanding (MOU) in September 1995 to formalise collaborative R&D programs in satellite research between the two universities. The purpose is to conduct joint research and training in the general fields of satellite engineering and satellite communication and specifically in the area of LEO technology. This will enable NTU to gain the expertise and technology necessary to help Singapore develop its own LEO satellite industry and for the country to be able to play a leading role in the provision of regional and international LEO satellite services.

Following the signing of the MOU, NTU has initiated a LEO research program called the "MERLION" project. The objectives of the Merlion initiative include:

(a) To develop the expertise at NTU in the area of LEO satellite systems, engineering and applications,
(b) To develop ground-segment infrastructure for accessing existing and future LEO systems,
(c) To acquire dedicated space-segment facilities for experimentation and hands-on experience,
(d) To transfer LEO technology to Singapore,
(e) To conduct LEO research and development activies relevant to Singpore,
(f) To train the manpower and provide the intellectual resource for the development of LEO-related industries in Singapore, and
(g) To explore the Low Equatorial Orbit and exploit its potential advantages for communications and other purposes.

To achieve the objectives, the Merlion project initially consists of 2 stages. In the first stage, a control station infrastructure is being established at NTU to enable the tracking, control, telemetring and monitoring of LEO and MEO satellites operating in the VHF and UHF frequencies (Little LEOs). This stage will be completed early in 1996. Funding is wholly by NTU, through an Academic Research Fund. There are many existing satellites which can be monitored, such as those built and launched by UoS, Weber State University in USA, NASA and AMSAT (Amateur Radio Satellite System). The station can also be used to track other satellites that will be launched in the future. It will be indispensable for all later activities of the Merlion project.

The following research and development activities are among those planned for the control station:

- Store-and-forward communication,
- Satellite imaging,
- On-board GPS (Global Positioning System),
- Propagation measurements,
- International satellite messaging (including paging),
- Polarisation measurement and control.

The second stage involves the design by NTU and construction by SSTL (Surrey Satelite Technology Limited) of a communication package, to be flown as part of the UoSAT-12 mini satellite currently being assembled at UoS. This will give NTU staff a hands-on experience of satellite hardware and on-board software. The second stage is being funded by the National Science and Technology Board (NSTB) of Singapore. The Minisat is scheduled for launch in May 1997. NTU's payload package will constitute about 25% of UoSat-12. NTU research and academic staff plan to carry out experiments using their own package on the satellite. In addition, they will also have access to other experiments included on the minisat by UoS. UoSAT-12 is scheduled to be launched in inclined LEO orbit at the end of 1996.

NTU's package onboard UoSat-12 will comprise:

- Transparent (bent-pipe) transponder for L-band uplink and S-band downlink,
- Propagation beacons,
- Processing transponder for L-band uplink and S-band downlink,
- Communication processor for various modulation/coding methods and protocols.

In addition to continuation of the research and development activities started in Stage 1, the second stage will also include:

- Upgrading of control station to work at L-band and S-band frequencies (Big LEOs),
- Channel characterisation,
- Direct international satellite messaging,
- Protocol development for S-PCN and mobile applications,
- (CDMA and spread spectrum communications,
- Real-time mobile communications,
- Satellite linking of vehicles in GPS-based fleet tracking and control.

5. Future Proposals

As a follow-up to Merlion Stage 1 and Stage 2, it is proposed that in the future one micro or mini satellite be launched by Singapore specifically in the Equatorial LEO orbit (the Merlion orbit) for R&D purposes to pioneer the exploitation of this new orbit for communications and other purposes. This could lead to further experimentation using a constellation of 5 satellites or a commercial system involving 6 to 12 satellites. The commercial system can provide continuous services to those equatorial parts of the world lying between 23 degrees North latitude (tropic of Cancer) and 23 degrees South latitude (tropic of Capricorn). This will include all the ASEAN countries and others in Southern Asia, Africa, South America, the Pacific basin and Northern Australia.

Acknowledgements

The author wishes to thank the staff of Surrey Satellite Technology Limited for their the help and encouragement, the National Science and Technology Board for financial resources, and the Telecommunications Authority of Singapore for assistance in licencing and frequency coordination.

Future Development of a Global Space Information and Communications Infrastructure

C.S. Welch and **D.I. Wade**, School of Mechanical, Aeronautical and Production Engineering, Kingston University, United Kingdom

Abstract

Information is a prerequisite for the making of reasoned decisions. In the current era this information can take many forms, for example telephony, video, scientific data, etc. The corollary to the importance of information is that the mechanisms for its provision and distribution are equally critical.

As the world moves increasingly towards the information age, the timely provision of relevant material becomes crucial. However, relevancy depends on the needs of the end-user since the information required may vary, for example, from a simple telephone message, through a weather forecast, to urgent medical data.

To date, many terrestrial and space technologies have been developed to deliver information (e.g. fibre optics and geostationary communications satellites), but the cost of access to such systems has meant that their direct use has been beyond the scope of the bulk of the world's population. Systems to provide telephonic and low rate messaging for the developing world are currently under design. These use substantially different architectures from those employed previously. System providers believe they will offer cost-effective and pertinent services to both the developed and developing countries.

Recently, other forms of information have also become available such as high accuracy positioning data available via GPS and, in the future, GNSS. Furthermore, the volume of information from new Earth-observing and meteorology systems has required the development of data distribution systems which link the space and ground segments very closely.

Characterising these disparate units under the general heading of information systems, it may be seen that, although connectivity exists on the ground side, substantial cross-linking on the space side has not yet occurred. This is partly a result of commercial and politico-legal considerations which have significant effects in controlling information flow. Depending on viewpoint, information/provision may be viewed either as a commodity which may be bought and sold or as a service to improve the lot of the recipient.

This paper considers the above aspects in the context of the general supply of information by space systems, discussing the previously mentioned topics as well as presenting possible future developments of a global space information and communications infrastructure. Implications for the evolutionary impact of this on different nations and cultures are also addressed.

1. Introduction

Information plays a vital role in modern life. Depending on circumstances and context, this information can be a mere convenience or a means of survival. Consequently, the availability of information, or the ability to gain access to it, is becoming increasingly important. Driven by developments in technology the potential availability and use of information has now reached levels that would have been largely unimaginable a few decades ago.

The activities/structures resulting from the need to provide and deliver information throughout the world, the so-called Global Information Infrastructure (GII), offers different parts of the world different opportunities. For inhabitants of the developed world, the GII offers "information on tap" in a variety of forms. For the inhabitants of developing countries, the GII offers the opportunity to substantially improve the quality of life. For commercial concerns, the GII offers a form of "ultimate service industry", potentially unimpaired by geographical or political borders.

As might be expected, and is historically evident, due to the fundamentally global nature of the problem, the space community has, and continues to, address the need for such a provision. The ability to communicate globally using terminals no larger than a briefcase already exists, whilst the imminent launch of a number of satellite constellations offers users the promise of hand-held terminals, providing global voice and paging capability at the least. It is arguable that this will properly initiate the process of providing a truly global information and communications infrastructure. Future developments, however, possess the opportunity to offer an even vaster array of services.

2. Information

2.1 What is Information ?

The term "information" has a number of different meanings depending on context and is sometimes used interchangeably with "data". Although, as far as the mechanical aspects of transmission are concerned, there may be no difference between the two, it is useful to try to define the difference. Numerous "definitions" have been attempted. A useful one is that given by Galliers (Reference 1), who states:

> *"Information is that collection of data, which, when presented in a particular manner and at an appropriate time, improves the knowledge of the person receiving it in such a way, that he/she is better able to undertake a particular activity or make a particular decision. "*

Information, then, may be considered as *"data with meaning"* (Reference 2). What constitutes "meaning" varies, but Nichols (Reference 3) has suggested relevance, availability and timeliness as essential attributes and objectivity, sensitivity, compatibility, conciseness and completeness as desirable. Most of these attributes may not be functions of the putative GII itself, but there is no doubt it will contribute considerably to the enhancement of availability and timeliness. Given that the space segment is primarily a means for distribution rather than processing, it is in these two areas that the space contribution to the GII will have maximum effect.

2.2 Information Formats

Information is available in a variety of formats. These include the printed word, files on computers, television and radio broadcasts, and similar. The particular format used for a given information set depends on a number of factors which include volume, cost, ease of distribution, available technology, required transmission rate, etc. Although, historically, information has often been available in only a single format (e.g. a book, an analogue radio broadcast, etc.), technology has now advanced sufficiently for these formats to start converging around just one. Digital encoding provides a powerful format that enables information to be stored, manipulated, replicated and distributed with comparative ease. Again, it is scope and ease with which information may be distributed that is of interest in the current context.

A useful and obvious example when considering the distribution of digital information over a GII is the Internet/World Wide Web (WWW). Its origins lie in a US Department of Defence project of the 1970's (ARPAnet) designed as an experiment in construction of damage resistant information networks. The key to its success was that it did not require a direct link between source and recipient, but instead broke up each message into a number of addressed packets which were then successively forwarded from computer to computer until they all arrived at their destination and were reassembled. This approach is only possible using digitally encoded information (though, of course, analogue information may be encoded digitally).

Although initially only simple computer files were distributed over the Internet, asynchronous transfer mode (ATM) technology currently enables digitised voice, video and data to be dispatched and switched through networks at enormous rates. It seems logical that many, if not all, forms of information distribution will be carried out by a single integrated network capable of fulfilling all needs (Reference 4). This will mean that information can be distributed with comparative ease and in real time and that users may interact directly with information bases rather than be purely passive consumers. Already in this field new digital techniques are being delivered; digital video compression, video teleconferencing, broadband digital radio, all leading, in the opinions of some commentators, to an eventual convergence of telecommunications, computing, broadcasting and publishing/information services (Reference 5).

Trying to foresee developments even further in the future and establish some form of upper benchmark for systems, the authors propose the possibility of user interactions with information systems over the GII being able to take the form of "virtual reality" (VR). Definitions of VR are many and varied. Here, the working definition used is that

"Virtual reality is a user interface to an information system featuring fully interactive real-time pseudo real-world elements."

In order to support sophisticated VR use over the GII, however, large bandwidth (data rate) communication systems are undoubtedly required and the development of these is likely to prove very rewarding to those organisations active in the area. However, whatever form user interaction takes, in the eyes of some the long term global benefits of the GII are quite clear, and are no less than the revitalising of the economies of the developed countries and the bringing of wealth and prosperity to developing nations (Reference 6).

2.3 Relevance and Benefits of Information

It is useful to look at the information and services likely to be of relevance to the different areas of the world and the benefits that may accrue from it.

The availability of information and the timeliness of its delivery are reasonably easy to define in an objective manner. Very broadly speaking, the more available information is and the faster it can be delivered, the better. Relevancy of information, however, is not a purely objective property but is frequently contextual. Information that is relevant and brings benefits in one situation does not necessarily do so in others. If some medium for its distribution exists, then it is relevance that gives information its currency and value.

Given this, the importance that information has to a nation depends greatly upon the status of that nation in the order of world economics. In the developed world, little day-to-day information is essential to survival, whereas in developing countries the survival of individuals and complete communities can depend on information taken for granted elsewhere.

(a) Developed Countries

Developed countries already have significant access to information. In the future, greater access to information in the home will provide greater convenience and allow increased sophistication in interaction. New information-based facilities are likely to include services which have previously required dealing with a (human) intermediary. Examples of this are the booking of airline tickets or monitoring of progress after a doctor's appointment (Reference 7). Once standards are in place for secure electronic transfer of funds, it is possible that many routine transactions will be handled from home in this manner. Other services which are currently starting to emerge, such as route planning and navigation using global positioning system/global navigation satellite service (GPS/GNSS) information, are likely to become commonplace, indicating and keeping the user informed of roadworks or traffic accidents which may delay their journey.

Information delivery systems can also allow mobility (though usually at the expense of bandwidth). Mobile computers linked to mobile telephones already provide most of the developed world with the technical capability to access the Internet and information bases and use email from almost anywhere inside their borders. In due course, these capabilities will be extended further and fully mobile information terminals will be developed.

(b) Developing Countries

In developing countries, information normally available only at key locations may be critical to future development. A common problem to be overcome is the provision of information to remote centres of population which are not connected to any form of information infrastructure. Education is a particularly important aspect of this.

> *"Education is universally recognised as a form of investment in human beings which yields economic benefits and contributes to a country's future wealth by increasing the productive capacity of its people"* (Reference 8).

However, in all too many developing countries the need to survive can overtake the need for education, resulting in current literacy levels of only a few percent. It is estimated that around the world one billion people have never received an education in reading and writing, including 300 million children (Reference 8), and with the current birth rate leading to a world population of 5-24 year olds of 2.5 billion by the year 2000 (Reference 8), demands for basic education will grow ever more severe.

A strong correlation exists between literacy level and national per capita income as shown in Table 1. This suggests that education should be a priority for developing countries.

	Literacy level	Per capita income ($)
First world	95 %	18 000
Second world	40 - 50 %	400 - 800
Third world	< 20 %	200

Table l. Literacy level vs. per capita income (Reference 9)

Health is also an issue for developing countries. The ratio of citizens to trained medical personnel is high and the latter group must frequently perform their jobs with very low levels of support. Obviously, any way in which they can be supplied with information to assist them will bring significant benefits. A number of experimental tele-education and tele-health programmes have been performed in the past. These include SHARE (Satellite for Rural Health and Education), SITE (Satellite Instructional Television Experiment) and INSAT (Reference 10). It seems certain that the continuation and development of such programmes will be increasingly important in the future.

Other areas of particular importance to developing countries are resource monitoring and disaster warning and control. The former is essential to allow such nations to best utilise their natural assets by helping to identify problems and monitor the implementation and effects of solutions. Disaster warning is more problematical for some countries than for others, depending on the nature of their location, geology and population pattern. However, disaster control, mainly by the provision of effective emergency communications, is likely to be considered important by all countries.

3. Technology and Infrastructure

Current day communication technology relies as much on fibre optic cables as it does on communication satellites. The higher bandwidth available when using fibre optics has driven development of the terrestrial technology required.

High bandwidth is currently only available between fixed points. Mobile communications are now widespread in developed countries. These terrestrial systems require a matrix of terrestrial cells to pass information to the mobile handset, and cannot deliver the bandwidth offered by optical fibre. Both of these technologies require a comprehensive ground infrastructure to be established before they can function effectively.

To establish the ground infrastructures needed for effective communications in developing countries, where vast distances and difficult terrain often separate centres of population, has proved too costly, leaving communities isolated. Comsat technology addressed this issue on one level. Initially this approach was expensive since comparatively large ground stations were needed. As technology has progressed, the size of these has reduced and now smaller terminals allow communication links to be established comparatively easily. Terminals of the size and weight of a briefcase are now readily available, to offer portable, if not truly mobile, services. In parallel, one-way transfer of information has became possible by Direct Broadcast Satellites (DBS) which in turn has engendered the use of Very Small Aperture Terminals (VSAT).

Current portable communication terminals, however, are limited to low bandwidth services. Until recently, space systems have been able to deliver bandwidth, but not truly mobile, communications in real time. Now, as a result of the continuing development of technology in both the ground and space segments, satellites can deliver both real-time broad bandwidth and mobile communications (Reference 11).

In addition to real-time data distnbution, store-and-forward communication satellites have been used, in particular for tele-health. The HealthSat programme (Reference 12), which uses a store-and-forward communications system supported by a University of Surrey micro-satellite (UoSAT), has proved that the provision of tele-health services does not require a costly set-up, and does not necessarily require a continuous link.

3.1 Communication Satellite Networks

Until comparatively recently, it was assumed that only geostationary satellites could offer (near) real-time information distribution services (Reference 4). However, the most recent innovation in satellite communication is the proposed use of satellite constellations in sub-geostationary orbits. The imminent launch of these communication satellite networks will, for the first time, offer global real-time information services that are also fully mobile, wherever the user may be in the world. A number of proposals for different systems currently exist, which may be categorised as follows:

- **Paging and messaging services.** The most fundamental of the systems proposed will offer simple global paging and messaging services to hand-held terminals. Such a system would allow vital information flow without the full expense of stretching to voice circuitry and would, for instance, allow message relay between a fleet of vehicles to optimise a company's delivery operations. One example of such a system is Orbcomm (Reference 13) which, through a constellation of 26 satellites, will provide a global paging, messaging and positioning service.

- **Voice services.** The second array of proposals will offer a full global voice service between mobile phones. In addition, most of the proposals would also allow the transmission of fax and low data rates, between more advanced terminals. It is intended to link most systems into existing ground infrastructure where available to reduce costs to an acceptable level. The Iridium and Globalstar systems are typical examples (Reference 14).

- **Broadband services**. The final category of current proposals contains those systems which will offer broadband services, such as video and Internet access. The availability of such data anywhere on Earth will significantly benefit the provision of telehealth and tele-education, as well as users in developed countries. Teledesic (Reference 15), which will use a low Earth orbit (LEO) constellation of 840 satellites, and Hughes Spaceway (Reference 16), which will utilise 9 geosynchronous comsats, are two of the current proposals, both of which hope to enter service around the year 2000. The potential benefits of such systems are now sufficiently recognised that the bandwidth necessary for their success, allocated by the World Radiocommunications Conference, has been made available (Reference 17).

Table 2 provides further data on a larger number of the currently proposed global mobile communication satellite networks (Reference 14).

Orbital system	Odyssey	ICO	Global-star	Constel-lation	Iridium	Tele-desic	Ellipso	Orb-comm
System Description:								
Orbit class[1]	MEO	ICO	LEO	LEO	LEO	LEO	HEO+ MEO	LEO
Altitute (km)	10354	10335	1389	1018	780	695-705	7846 x520 +8040	775 +730
Number of satellites	12+3[2]	10+2[2]	48+8[2]	48	66+6[2]	840+84[2]		36
Number of orbital planes	3	2	8	4	6	21	2+1	4+2
Inclination (°)	55	45	52	90	86.4	98.16	0+180	45+70
Coverage	Land masses	Global	±70° latitude	Global	Global	Global		Global
Services:								
Service types[3]	Vo, Da, Fx, Pg, Me, Po	Vo, Da, Fx, Pg	Vo, Da, Fx, Pg	Vo, Da, Fx	Vo, Da, Fx, Pg, Me, Po	Vo, Da, Fx, Pg, Vd	Vo, Da, Fx, Pg, Me, Po	Pg, Me, Po
System cost (US$m)	1800	2600	1500	1700	3700	9000	750	
Terminal cost (US$m)	300		750		2500-3000		1000	
Call rates (US$/minute)	0.65	1-2	0.6		3		0.5	
Operation scheduled	2000	2000	1998	1998	1998	2001	1998	

Table 2. Global mobile communication satellite networks (Reference 14)

[1]Orbit class: MEO, Medium Earth Orbit
ICO, Intermediate Circular Orbit
LEO, Low Earth Orbit
HEO, Highly Eccentric Orbit

[2]Number of on-orbit spare satellites

[3]Service types: Vo, Voice
Da, Data
Fx, Fax
Pg, Paging
Me, Messaging
Po, Positioning
Vd, Video

4. Information Systems Connectivity

4.1 Technical Considerations

From the originally fixed disparate terrestrial telecommunication systems developed around the world, adoption of international standards has led to almost complete global connectivity. Terrestrial mobile phone systems are now following a similar trend with the Global System for Mobile Communications (GSM) standard currently in use in Europe, Russia, Australia and the Far East. Existing satellite systems have not thus far been developed to this extent, with many systems being launched for a particular nation or market sector. Intersatellite communications to date have been primarily constrained to systems such as the TDRSS (Tracking and Data Relay Satellite System) and a few experimental systems considering, for instance, the use of lasers for inter-satellite links. One significant reason for this, at least as far as telephony is concerned, is the excessive and disruptive time lag this would introduce into conversations.

The infrastructure to be developed for LEO constellations will provide a much higher connectivity of the space segment, although this may not necessarily only involve the satellite hardware. The physically shorter routing which will occur from the connecting of neighbouring satellites in LEO constellations will lessen the time delay experienced by users of GEO systems and provide near real-time links. Many of the proposed systems will use existing terrestrial infrastructure when available to reduce costs. In particular in developed countries, this may result in signals being routed from the passing satellite directly to a gateway and on to a fibre optic link, from where a terrestrial system is used for the delivery of the information. In more remote areas signals will be fed from gateway to gateway, or from satellite to satellite, to transfer the signal. Whilst this system maximises efficiency for global coverage by current day standards, it may be anticipated that second generation mobile communication satellites will rely on ever greater space segment connectivity (especially as space segment bandwidth availability increases) to link the user and source of information.

Whilst the infrastructure required for global mobile communications is likely to be in place around the turn of the century, socio-economic and political constraints may still prevent the whole world being connected to the GII in its fullest meaning.

Some space information system architectures may feature significant links to terrestrial gateways and it is possible that particular nations may refuse such connections. These, however, are far more easily controlled than architectures which allow direct connection from space into a terminal used by an individual. Such are far more difficult to control and present far greater problems to countries that wish to impose controls on certain forms of information flow. Some of the reasons for wishing to restrict connectivity/information flow are outlined below.

4.2 Political Considerations

Issues of information flow have been widely debated over the last few years. The spread of easy access to the Internet and the sudden proliferation of information sources on it has focussed many minds on the availability of information and whether or not some form of control is required, or even possible. A good example of this is the activities that have surrounded the USA's Communications Decency Act.

The amount and nature of information available over the GII may be considered inappropriate for citizens by some governments This might not necessarily be as a form of repressive censorship - it has been suggested that it is impossible for a government to meaningfully govern if it is not able to have first sight of some information ahead of its citizens to enable it to form policy in a rational, rather than reactive manner (H. Wolff, personal communication).

Nevertheless, if governments choose to put in place mechanisms (of any nature, legal, technical, etc.) to act as significant barriers between their citizens and the rest of the world, the crosslinking of information systems will be impeded. The overall effects of this depend on the level and type of restriction. However, if global and/or mobile access to the GII is available directly through the space segment without first passing through a national information infrastructure, then the ability of governments to control information flow is potentially much reduced.

4.3 Economic Considerations

In addition to political constraints, economic considerations will also affect the nature of connectivity in different areas of the world. In some parts of the developed world, legislation exists that makes the operator of the physical hardware that constitutes the information system responsible for the content, and hence legality, of the information carried on it. Since this would otherwise require great vigilance on the part of the service providers, it encourages them to extend their activities into the provision of information as well as the delivery of it. This, in turn, encourages the development of large proprietary information systems with minimum cross-system connectivity. This may well discourage the entry of new players into the field and prove anti-competitive. It is certainly viewed with hostility by some (Reference 18). Furthermore, trying to ensure that national regulatory frameworks across the planet allow the right combination of incentives and controls and take account of local variation in culture will be a vital, but diffficult task (Reference 19).

As already outlined, access to a space GII provides huge potential benefits for developing nations. These are achievable, however, only if the countries in question are capable of obtaining the elements to make up the ground segment. For some nations this may prove more problematical than for others. Given the benefits that could accrue, it would be unfortunate if the up-front investment required prevented such countries from taking advantage of space access to the GII. There are a number of mechanisms (commercial, intergovernmental, etc.) that could help in this case by the discounting/subsidising of community terminals for developing nations.

4.4 Social/Cultural Considerations

Although the GII may make global access to information a reality, there is no a priori reason to assume that all of Earth's inhabitants will necessarily wish to utilise it. Some have suggested that the desire to implement a GII is merely another form of cultural imperialism by the countries of the West (Reference 20). If this is a view voiced in the developed world, it seems almost certain that it will be shared elsewhere and that some groups/countries may well prefer not to participate in a truly global GII, but either to develop smaller counterparts or go without completely.

5. Future Developments

Future developments in global communications can be seen from two viewpoints, that of the user on the ground and that of the infrastructure that exists to service the user.

5.1 The User

In the past, it has been suggested that the aim of global communications should be to provide access to information from any location in the world. However, at the moment if this provision is to be to a mobile terminal, then broadband is precluded. None of the systems currently proposed plan to supply global mobile broadband services. However, once broadband information services are available via the space segment of the GII there is little doubt that some users will want to take as much advantage of these as possible and will desire to do this while mobile.

Given this, and the earlier discussion of VR as a benchmark for information systems performance, the grail to be sought from the user perspective is then full VR access via the GII through a mobile terminal from any location in the world.

If the goal of mobile VR is achieved, the appearance of a GII terminal could be radically different from that of current mobile telephones and personal mobile computers - possibly a pair of "multimedia sunglasses" and two virtual reality gloves. Whether this would find favour with users, however, remains to be seen.

5.2 The Infrastructure

Providing mobile broadband information services from space will undoubtedly present technical challenges. However, one driving factor in their development will remain the power needed by a mobile terminal to communicate with a satellite in GEO. This is at least 103 times greater than that required to do likewise with a satellite in LEO. Taking into account the disadvantages due to signal path length using geostationary satellites, the appeal of LEO constellations is compelling. Combining this with improvements in technology that are currently planned, such as spread-spectrum techniques (intended for use on Globalstar) combined with soft definable cells (intended for use on Teledesic), makes the potential for mobile broadband information services seem unassailable.

For mobile, space-delivered VR applications, even the small time lag on a LEO constellation may prove to be a problem. This may be solvable by distributing the information and processing throughout the GII so that information that requires near real-time processing is dealt with near at hand, either in the VR terminal or elsewhere, maybe in one of the satellites, while information held remotely travels the full distance. This requires extensive cross-linking between the components of the space segment.

This sort of architecture is a form of space-borne computer network in which each of the nodes would possess a certain degree of processing power itself, and maybe substantial memory as well. Each satellite/node is a computer in its own right, but is also part of a greater whole. Under such circumstances the difference between the individual computer and the network connecting it to others starts to become blurred; to borrow a slogan, *"the network is the computer"*.

So far the largest number of satellites proposed for a LEO constellation is 840 for Teledesic. However, whether there is any practical upward limit to the number is not known. Certainly, work has been done at the Jet Propulsion Laboratory on very small spacecraft with masses of the order of 10 kg or less (References 21, 22). These were designs for interplanetary spacecraft, but demonstrate that substantial miniaturisation may be practical in the future. It is possible that, in the next century, LEO constellations with thousands of satellites will be the norm. If this is the case, then the satellites that make up the space segment of the GII will no longer be mere "telephone exchanges" but rather may be an "intelligent sky" with which users can interact directly.

Having discussed the future of LEO constellations, what will be the future of geostationary comsats? It is the opinion of the authors that these will not disappear from use. They will still provide a highly effective and cost-efficient way of transferring information that does not require real-time delivery. Other niche markets will continue to exist. As always, the driving force will be the user and economic forces will dictate what types of services will prosper. Digital data packets will be able to take a variety of routes from source to destination, depending on the priority associated with the timeliness of their delivery. The key point is that all of the delivery systems will form components of a larger space GII (similar to the present terrestrial telephone system) and that this space GII will also be tightly and multiply linked to its terrestrial equivalent to form one genuine global information infrastructure.

This infrastructure might well include such space information systems as meteorology and earth resources satellites and global positioning systems. If an infrastructure exists to distribute high bandwidth information globally, then the requirement for dedicated systems to support such missions might be obviated and a more cost-effective solution obtained. It is possible that even telemetry, tracking and command (TT&C) functions could be dealt with in this manner.

6. Impact and Implications

The new services and convenience offered to the developed world by global mobile access to the GII will probably ensure its ready acceptance in these countries. Meanwhile, the populations of the developing countries will, for the first time, have access to a sophisticated information infrastructure. Space systems of the sort described will bring broadband access to everyone in the world who wants it, at the same price.

6.1 Impacts on Nations/Cultures

(a) Developed nations

Any GII will permit people to remain in near-constant contact with work, family and information systems should they wish to. As a result, it is likely that work patterns will continue to move in the direction of telecommuting. However, the fully mobile aspect of the services offered by a space-based GII will mean that this is accentuated even further. Since all locations will be equally information-accessible, then decentralisation of many functions may occur. This will allow previously small centres to grow and flourish. Many commentators predict that the new "information age" will stimulate the economic revitalisation of the economies of the developed countries as much as the earlier industrial age did.

Two particular societal concerns may affect developed countries. Firstly, the access and use of a GII may either enhance or reduce the tendency currently visible for society to split into two main parts, a techologically well-educated upper half enjoying a pleasant lifestyle and a poorly educated remainder living at a much lower level. It is to be hoped that the developed nations will take steps to utilise the GII to enhance the lives and help fulfil the aspirations of all their citizens. This is less a matter of technology than it is of policy.

Secondly, as a result of the advances in healthcare and the reduction in birthrate in the developed world, the percentage of the population over the age of 65 is increasing all the time. By the early twenty-first century, this age group will account for almost one third of the population. At the least, it seems likely that specialised services for this group could create new markets for information providers. Certainly, the GII could enhance the lives of elderly citizens otherwise confined through frailty or ill health. For those who wished to continue working, information-based employment would continue to be practical after more physically demanding jobs became impossible. Given the problems predicted in the cost of caring for the elderly in the future, this might become a very important factor.

In the longer term, the effects of the GII on society are difficult to judge. Information systems have major effects on societies. It is only necessary to look at the effect of the telegraph, telephone, radio and television to see this. There is no reason to assume that the development of the GII will have any lesser effect than the aforementioned.

(b) Developing Nations

It is the developing nations that stand to gain most significantly from the space segment of the GII, since it provides them with a ready system for communications to which they otherwise would not be able to have access for economic reasons.

As previously mentioned, the key benefits that this will afford will be in the fields of health and education. Equally important, at least in the everyday sense, for the inhabitants of developing countries will be easy access to communications of all sorts. The importance of mere telephony, let alone mobile VR, in such circumstances cannot be over-emphasized The correlation between the number of telephones and GDP per head, respectively, is very strong. While it is not clearly demonstrable that having more telephones causes a country to prosper, the evidence from countries such as Thailand, Malaysia, Argentina and China, which have invested heavily in this, seems compelling.

Through education and an increase in trade, developing countries should gradually witness an improvement in the standard of living. As the standard continues to increase, access to a wider range of information will be available, providing new services upon demand. This may result in an extremely rapid development, at least for certain sectors of the population, which it is hoped, will have a beneficial effect for the nation.

There remains the question of the effect that "global culture" will have on previously isolated communities that suddenly gain access to the GII. As well as the undoubted benefits, the volume and quality of the information may prove overwhelming. Even a few years ago, gaining information from the Intemet was described as "like trying to sip from a firehose" (G. Spafford, original source unknown). Additionally, one of the quoted strengths of the Internet is the scope for individual expression. As a result, a proportion of the information available is of a low quality, and a certain degree of education is needed to filter the good from the bad. If the Internet was to be used as a model for the GII this could have unfortunate cultural effects and implications for unsophisticated "first generation" users.

Given these factors, access to the GII for developing countries can be viewed as a Pandora's Box containing many things that are problematical but, in the end, bringing great benefits.

7. Conclusions

It has been shown that there are significant benefits to humanity from a global information infrastructure acting as a single system to provide a variety of different types of information and communication.

It is anticipated that the future development of the GII will feature space GII systems very heavily. The space information infrastructure will consist primarily of a combination of low earth orbit constellations containing very large numbers of small satellites and smaller numbers of larger geostationary satellites. These will all be "intelligent" and heavily cross-linked to form, effectively, an extended space-borne computer network which will be tightly integrated with a terrestrial counterpart. The system will be capable of delivering broadband information to mobile terminals at any point on the globe. However, there are a number of political, economic and cultural considerations that may impact its effectiveness in certain areas.

The benefits to the developed world will be mainly in terms of convenience and access to an increased number and quality of services, though some societal and economic benefits are also likely.

The benefits to the developing world are likely to be much more important, providing access to routine communications, enabling significant improvements in health and education and stimulating substantial economic growth.

In the longer term it may be hoped the provision of global information services via a space information infrastructure will tend to bring together and improve the lot of the nations, cultures and individuals that constitute the human race. In the words of Arthur C. Clarke (Reference 23):

> *"When all men, wherever they may be, have equal access to the same vast communications network, they will inevitably become citizens of the world ."*

References

1. Galliers, R.: *Information Analysis: Selected Readings*. Addison Wesley, Wokingham, United Kingdom, 1987
2. Lindsay, J.: *Information Systems -Fundamentals and Issues*. Available on the World Wide Web at http://infosys.kingston.ac.uk/isschool/staff/papers/lindsay/book/contents
3. Nichols, G.E.: "On the Nature of Management Information." *Management Accounting*, Chartered Institute of Management Accountants, London, United Kingdom, April 1969
4. Holebrook, S.: "Long-Term Developments in the Global Telecommunications Infrastructure." *JBIS*, Vol. 44, pp. 243-250. British Interplanetary Society, London, United Kingdom, 1991
5. Chong, R.B.: "Trends in Communications and Other Musings Our Future." *Federal Communications Law Journal*, available on the World Wide Web at http://www.law.indiana.edu/fclj/v47/no2/chong.html
6. Gilder, G.: "The Bandwidth Tidal Wave." *Forbes ASAP*, Forbes, Inc., New York, USA, 5 December 1994
7. Flower, J.: "Downhome Technopia." *New Scientist*, No. 2013, pp.33-36. IPC Magazines Ltd., London, United Kingdom, 20 January 1996
8. Rao, U.R.: *Space and Education*. Presented at the Plenary Session, 46th Congress of the International Astronautical Federation, Oslo, Norway, October 2-6, 1995
9. Rao, U.R.: *Space and Rural Development*. Paper presented at the Plenary Session, 45th Congress of the International Astronautical Federation, Jerusalem, Israel, October 9-14, 1994
10. Rycroft, M. (Editor): *The Cambridge Encyclopedia of Space*. CUP, Cambridge, United Kingdom, 1990
11. Goldstein, I.: *The Evolution and Promise of Satellite Communications*. Paper presented at the Communicating Via Satellite Symposium, George Washington University, Washington DC, 16 October 1995
12. University of Surrey: *UOSAT and SSTL Microsatellite Missions*. Available on World Wide Web at http://www/ee/surrey.ac.uk/EE/CSER/UOSAT/missions.html
13. Orbital Communications Corporation: *System Description*. Orbital Communications Corporation, Dullis, USA, 5 April 1995
14. Wisloff, T.E.: *A Tabulated Overview of Big LEOS*. Available on World Wide Web at http://www.idt.unit.no/torwi/synopsis.html
15. NASA MSL: *Teledesic..* Available on World Wide Web at http:/Leonardo.jpl.nasa.gov/msl/QuickLooks/teledesic.html
16. International Space University: *Global Access Tele-Health and Education System, Final Report*. International Space University, Strasbourg, France, 1994
17. Foley, T.: "WRC Backs Broadband Satellite Plan." *Communications Week International Online*, available on World Wide Web at http://techweb.cmp.com/cwi, 24 November 1995
18. Johnson, N.: "Jefferson on the Internet." *Federal Communications Law Journal*, available on World Wide Web at http://www.law.indiana.edu/fclj/v47/no2/johnson.html
19. Blumenfeld, S.D.: "Developing the Global Information Infrastructure." *Federal Communications Law Joumal*, available on the World Wide Web at http://www.law.indiana.edu/folj/v47/no2/blumenfe.html
20. Frezza, W.: "Satellite Data: On Top of the World." *Network Computing*, CMP Media, Inc., Manhaset, USA, 15 November 1995
21. Collins, D. and Kukkonen, C.: *Miniature, Low-Cost, Highly Autonomous Spacecraft - A Focus for the New Millenium*. Paper IAF-95-U.2.06, presented at the 46th International Astronautical Congress, Oslo, Norway, October 2-6 1995
22. Jones, R.M.: "Microspacecraft Missions and Systems." *JBIS*, Vol. 42, pp. 448-454. British Interplanetary Society, London, United Kingdom, 1989
23. Clarke, A.C.: *Profiles of the Future*. Victor Gollancz, London, United Kingdom, 1962

Report on Panel Discussion 3: Telecommunications

A. Berinstain, D. Gendre, 1995/96 Master of Space Studies (MSS) students, International Space University

Session Chair: **H. Chasia,** International Telecommunication Union (ITU), Geneva, Switzerland

The issues covered in the discussion included regulation, future strategies and techniques for communication satellites, and applications of telecommunication services to humanitarian services.

The questions put to the panel led to the following statements:

- **W.L. Pritchard.** The Low Earth Orbit (LEO) and Geosynchronous Earth Orbit (GEO) communication satellites have different - complementary - roles, and the market is wide enough to accommodate both.

- **T. Howell.** Countries should be prepared to face lobbying at ITU.

  The day of the big international organization is almost certainly finished, and private systems are taking over.

- **E. Seumahu.** There is room for a dedicated equatorial LEO communication system. The current demand is low, but the predicted growth is higher than anywhere else in the space communications sector.

  Singapore's program is still at the level of a technological demonstrator. Singapore will look for international cooperation as soon as the service proves viable and more frequency bandwidth is needed.

- **W.L. Pritchard.** ITU put aside some frequency bands for all LEO systems, not just for Teledesic. The LEO systems are universal, and not limited to equatorial LEO and European systems. There is room for competition.

  The solar power satellites, if they are ever launched, will not be a problem to accommodate within the radiofrequency spectrum, because they only require one frequency and no bandwidth.

- **R. Jakhu.** Attention should be paid to the contracts, because a majority of countries do not have the right to grant licenses. Western laws and the competitive approach which is prevalent in Western countries cannot be applied to developing countries.

- **C. Welch.** Virtual Reality in itself is not important. It is just a tool. Equal access to information is important.

- **D. Herlemont.** The telemedicine design project at the 1996 ISU summer session in Vienna could pay attention to future options, study the possibility of dedicated satellites for emergencies, and evaluate the political, legal and financial aspects.

- **H. Chasia.** The liberalization of telecommunications is on the way, and ITU has an increasing number of private members (who bring money). The ITU, which was set up as an international organization, must find ways to involve these private firms in funding and other activities.

  ITU wishes to ensure that advances in telecommunications are of real benefit to developing countries. ITU gives them advice and tries to help them cope with the new demands.

- **W.L. Pritchard.** In areas such as mobile services, broadcasting and quickly set-up systems for small countries, optical fibers cannot replace satellites. Different needs and uses exist for optical fibers and satellites.

- **T. Howell.** Optical fibers and satellites will remain complementary, and satellites will be especially good for fast set-up.

- **H. Chasia.** Following recent reforms within the ITU, the Union plans to streamline procedures which have until now proved complicated for countries wanting to launch satellites.

Session 4
Space of Service: Further Applications

Session Chair:
G. Soffen, Director of University Programs, NASA Goddard Space Flight Center, USA

Proposal to Apply Results of NASA-Sponsored Research for Space Operations to Improve the Quality and Quantity of Goods and Services

R.S. Hornstein, COST LESS Team, NASA, Washington, DC 20546, USA
J.K. Willoughby, The Kendall Group, Littleton, CO 80122, USA

Abstract

NASA-sponsored research has produced a basic framework for characterizing space mission operations and for structuring systems that enable cost-effective operations planning and execution. The researchers never wavered from their focus on simplicity and generality as the means of realizing demonstrable operational effectiveness and life cycle economies. As a result, the framework can be transferred beyond the space program communities to domains that were not originally the targets of the effort.

The story line of this paper is a proposal to transform the space operations results into innovative solutions for preserving Earth and improving life. The authors describe the space operations framework and suggest that its methodology, associated technology, and decision-support strategies can be applied in managing environmental factors, human capabilities, and natural resources that are the determinants of the quality of life on Earth. A concept for reusability of space mission program elements is also presented. Called "Fundamental Reusables for Enterprise Deployment" or "FRED", this was formulated in response to a challenge from the NASA Administrator to confront the status quo (of developing mission-peculiar, project-unique, and organization-specific systems) and recommend a target architecture for providing cross-cutting and cost-cutting Agency services for space mission operations. The authors propose to explore the relevance of a FRED-like concept for sustainable exploitation of a nation's resources. The paper is authored by members of NASA's COST LESS Team for Mission Operations, an international, inter-disciplinary, and non-parochial team of government, industry, and academia, charged with reducing the cost of success for space mission operations.

1. Introduction

With the explicit encouragement of the NASA Administrator, a volunteer team, drawn from several NASA organizations, aerospace contractors, US military space organizations, academic institutions, and international space operations experts, has been working on re-inventing the processes used to conduct space mission operations. This team, called the COST LESS Team for Mission Operations, has recognized that managing the operations of space missions is not dissimilar to managing the operations of any other enterprise. The basic tasks deal with allocating scarce, expensive, and shared resources to competing objectives, satisfying conditions on their utilization, and time-phasing specific activities that impact these resources. Those nations wishing to use space missions to upgrade the quality and quantity of their goods and services have two similar management goals, one embedded in the other: (1) they must manage the operations of their space missions, and (2) they must manage the use and preservation of their natural, social and cultural resources. A framework for the effective management of space mission resources might serve equally well as a framework for the management of the broader resources of a nation.

The COST LESS team began with a working premise that the complexity apparent in how NASA has managed space mission operations in the past has resulted from failures to recognize similarities among functions. Sometimes these inabilities to recognize similarities have resulted from true technical myopia or a lack of experience; at other times they may have resulted from built-in incentives to view each new circumstance as unique in order to justify a budget, an organizational expansion, or local control of a project. The COST LESS team believes that whatever the causes may have been, unnecessary complexity is the consequence. Today's fiscal realities call for a simplified framework for the operational management of space missions.

Reducing complexity should not imply a reduction in either mission capability or operational effectiveness. Simplifications in mission operations should result from defining and implementing fundamental building blocks, from minimizing the number of building blocks necessary, and from re-engineering processes so that they use and reuse the building blocks for synthesizing the systems needed to perform space operations. The building blocks can be software components, hardware devices and subsystems, and even procedures used by operations personnel. The minimal set of building blocks is sometimes referred to as a "canonical set", borrowing a concept from classical physics that emphasizes its fundamental, reusable, and generic nature.

In the following sections, the COST LESS team presents its early work on identifying and evolving the building blocks necessary to support space mission operations. Then, a possible user/provider structure is suggested which has designed-in incentives for encouraging the use of the building blocks, for upgrading their implementations, for keeping the costs minimized by means of competitive market forces, and for maximizing the reuse of the canonical set. Finally, an analogous approach for managing a nation's natural, social and cultural resources is suggested. Based on simplification and reuse of fundamental building blocks, several analogies are provided which may be beneficial to any government or agency that is rethinking how it might use space missions to improve the quality and quantity of available goods and services.

2. Discovering Two Categories of Reusable Components

In 1992, Charles T. Force, NASA's Associate Administer for Space Communications, commissioned a "Blue Team" to review how space operations might be changed to effect the Administrator's challenge for conducting space programs in a faster, better, and cheaper manner. That Blue Team, led by Rhoda Shaller Hornstein, an author of this paper, was the predecessor of the COST LESS team. The Blue Team's findings, recommendations, and strategic initiatives were published in 1993. One of the more important findings was the realization that all activities or processes used to perform space mission operations could be grouped into two categories. The first category is *Information Handling* which includes all activities that transfer, store, retrieve, transform, and analyze data. The second category is *Resource Management and Control* which includes all activities that plan and command the mission. Diagrams of the highest levels of this simple taxonomy are shown in Figures 1 and 2. The importance of this straight-forward (non-complex) abstraction lies in the fact that none of the typical organizational decompositions common in NASA's current operational practices is imposed at the top levels. The typical dichotomy of spacecraft and ground systems is not imposed, nor is the usual split of science (payload) vs. platform (spacecraft) systems. Lead-time and mission-phase distinctions, such as long-range planning vs. real-time planning, launch operations vs. on-orbit operations, primary vs. extended mission operations, cruise phases vs. encounter phases for interplanetary missions, are also not imposed. Not even the customary distinctions between crewed and robotic space flight are present in the top levels of the functional breakdown. This is not to say that specialized requirements that depend on the characteristics or stage of a mission are unnecessary, but rather to point out that imposing specializations too high in a functional breakdown is a major cause of unnecessary complexity and duplication. The imposition of specialization too high in a decomposition of tasks or functions creates activities that add no value.

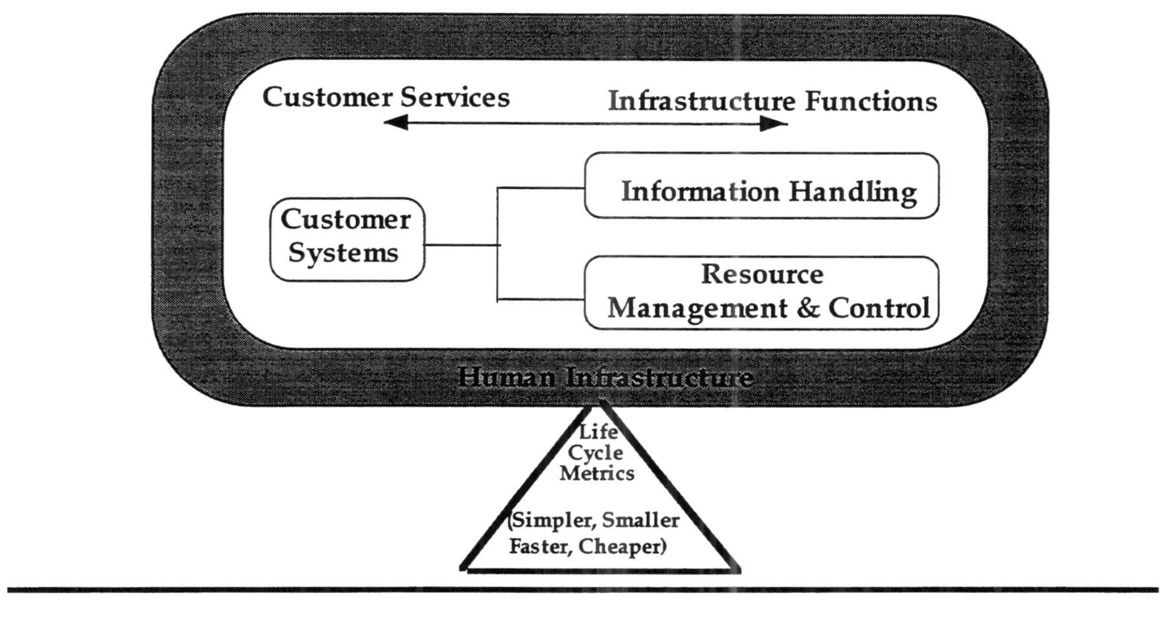

Figure 1. Highest level decomposition of mission operations

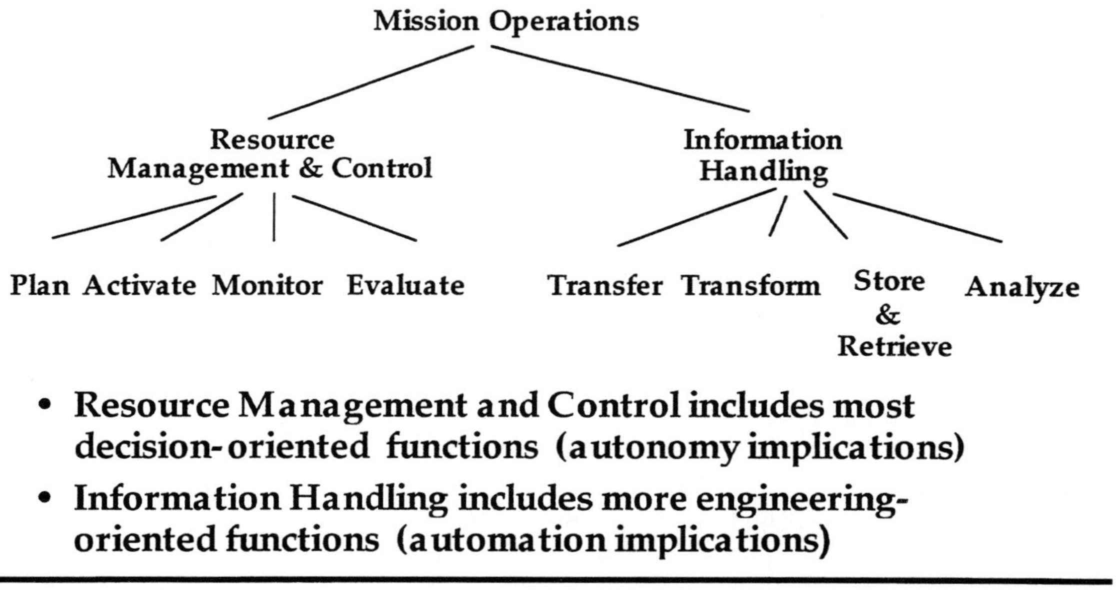

Figure 2. Second level decomposition of mission operations

A comparison among the mission operations activities at the second level of decomposition reveals the prominence of decision-making in the *Resource Management and Control* category. Engineers know how to perform the mechanics of manipulating data and converting them to information. Once it has been determined what data are to be moved, stored, etc., the systems to perform these handling functions can, to a large extent, be mechanized. The need for direct human involvement in routine *Information Handling* tasks is almost non-existent using current technology. If operational difficulties occur, humans are used to help diagnose the problems and to assign alternate resources that re-establish routine operations. However, these diagnostic functions fall more naturally into the *Resource Management and Control* category. In that category, we find planning, scheduling, status monitoring, trouble shooting, command, and control. All of these functions require some degree of experiential knowledge, technical judgment and other forms of intelligence that current practice is just beginning to support with elementary artificial intelligence systems. The natural separation of decision-oriented tasks from the more engineering-oriented or mechanizable tasks is useful in recognizing the nature of the fundamental functions that will be needed in the canonical set.

More has been done in developing useful decompositions in the *Information Handling* category than in the *Resource Management and Control* category. A familiar example is the seven-layer data management model often used in communications and data base management system designs. Recently, efforts have been initiated to develop standards that apply to those command and control functions which fall into the *Resource Management and Control* category. In addition, members of the COST LESS team have developed a set of functional components that can be used in the area of planning and scheduling, and have been evaluating the premise that command and control functions are similar to planning and scheduling if the time constants involved are regarded as changeable parameters. These efforts should lead to the recognition of reusable components in the *Resource Management and Control* category.

It is a goal of the COST LESS team to continue developing the functional taxonomy of mission operations functions, keeping in mind two important objectives. The first objective is to keep the elements in the decomposition relatively free from peculiarities that originate from functional or organizational "stovepipes," that is, specializations, organizational boundaries, or historical biases that interfere with the recognition of common functional requirements. The taxonomy should contain specializations only at the lowest possible levels in its hierarchy.

The second objective is to achieve a functional decomposition that is useful as a blueprint for defining and implementing reusable components that meet the criteria defined earlier. This means that the components must be sufficiently basic that they can be used as building blocks for constructing any operational system for any space mission. The vision is that *future space mission operations systems should be constructed via the synthesis of reusable components rather than by functional decompositions of requirements followed by specialized development of one-of-a-kind systems.*

Space utilization objectives aimed directly at improving the quality of life, such as determining the most productive use of natural resources, the dissemination of science information to schools, or the provision of early and accurate weather information, can be accomplished by the synthesis of a combination of *Information Handling* and *Resource Management and Control* functional components. If NASA could develop a mechanism for identifying, developing, disseminating and supporting the reusable components, and if this mechanism relied on incentives to ensure that the components were used, reused, updated, and eventually made available from non-governmental sources, then such a mechanism should be relevant to enterprises external to NASA.

3. A Supply/Demand Framework for the Reusable Components

NASA is currently organized as five Strategic Enterprises, each responsible for executing its part of the Agency's mission. These Enterprises, depicted in Figure 3, have the responsibility for setting priorities and for defining and conducting space missions. It can be seen that no Enterprise-specific activities appear in the top levels of the hierarchy of functions. It is the COST LESS team's perspective that the Strategic Enterprises could implement their space missions using reusable components from a common source. In fact, it is our explicit assertion that any existing duplications within the Enterprises could be phased out and replaced by a concept of development by synthesis of reusable components. These components are, of course, the same ones taken from the yet-to-be-completed hierarchy that has a top-level structure shown in Figure 1. Begging, for the moment, the question of how the reusable components would be provided, it is obvious that the NASA Strategic Enterprises represent the users or customers for those components. That is, NASA's Strategic Enterprises represent the demand side of a supply/demand framework. In the planning of the supply side, awareness of the characteristics of this demand side is essential. Who are the users and what do they really need? These are the critical questions that must be answered as the functional taxonomy is developed and the reusable components are defined.

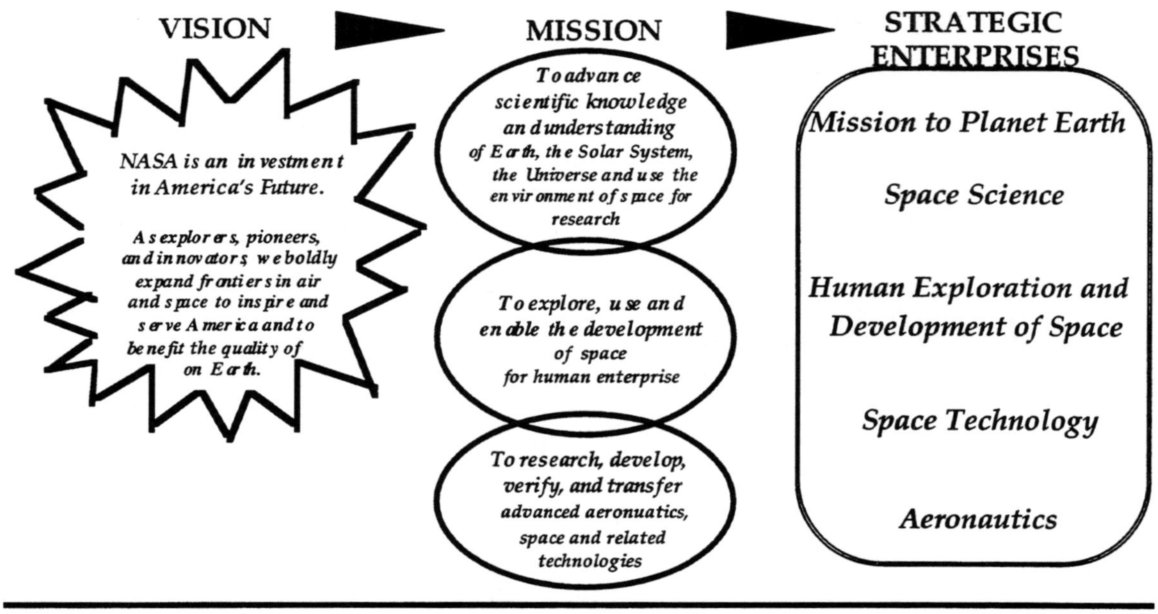

Figure 3. NASA's strategic framework represents the demand

In assessing the requirements for the supply side of a supply/demand framework, the COST LESS team has assumed that the NASA Strategic Enterprises exist to produce scientific knowledge that is related to, or derived from spaceflight. Furthermore, the team has assumed that the reason for the continued existence of the Strategic Enterprises is the on-going production of the most useful and effective science possible for the funding available. While the Enterprises have no consensus definition of "useful" or "effective," if science is their goal, they should embrace mechanisms that support tasks for the lowest cost and with the fewest impediments to performing meaningful scientific activities. Stated in modern free enterprise terms, the Enterprises will be eager to reduce non-science costs and to outsource the functions that do not directly add science value.

4. Supplying a "Storehouse" of Reusable Components

The COST LESS team has spent some effort conceptualizing how the supply side of the supply/demand framework might be constructed for NASA. The framework was initially given the meaningless working title of "FRED," but after a replacement title would not gain popularity, FRED was afforded the dignity of becoming a meaningful acronym: "**F**undamental **R**eusables for **E**nterprise **D**eployment."

FRED can be understood, in-part, by making an analogy to a store. Like a store, FRED sells items (in this case, reusable components) to customers, and must keep its shelves stocked with products (components) that its customers want. As with any store, the consequences of having incomplete or undesirable merchandise are customers who take their business elsewhere. Therefore, to keep FRED responsive to the demands from NASA's Strategic Enterprises, a competitive atmosphere must be present. To assure this competitive "supply" environment for NASA, the Enterprises would be encouraged to seek satisfaction of their "demand" in the open market place. When no equivalent components exist in the market place, FRED becomes the sole supplier and has the potential for making a healthy profit. Conversely, if this profit potential is recognized by commercial developers who believe that the demand for certain reusable components is high and will be sustained, then they will choose to become competitors of FRED. When private enterprise provides an adequate supply of any high-quality reusable component, FRED ceases to offer that component as a product.

Additionally, FRED offers services to help customers make the best use of the components it sells. These services are also subject to competitive forces, since third-party service companies can provide useful support functions such as training, product selection advice, integration, and maintenance. Some companies can offer services that relieve the customers from buying equipment. This is analogous to the local phone company offering a voice-mail service that precludes the need for customers to buy and install a personal answering machine. When FRED's customer service becomes non-competitive, FRED gets out of that aspect of the services business.

To carry the analogy a step further, FRED must publicize the availability of its products and services. Catalogs and instruction manuals must be available to assist the Enterprises who are developing systems to select the appropriate components or building blocks necessary to fly their missions. In some cases, installers and integrators can be dispatched to the users' locations to assist in the rapid synthesis of available capabilities.

Finally, FRED must anticipate demand and provide the reusable components that are not available in the open market. This is a research and development function. Spaceflight pulls technology out of the laboratory faster than the open market can see sufficient sales forecasts to justify private development. Yet a vigorous commercial sector will have amazing visionary skills. FRED must be endowed with truly long-range vision in order to be a bridge between the needs of the Enterprises and commercially viable suppliers. In doing so, FRED also provides the private sector with technological proofs of concept.

For a storehouse concept like FRED to work best, it is important that the customer Enterprises become smart buyers. For example, the safety requirements that permeate space system operations may be similar to those found in public transportation businesses, such as airline operations. "Cheaper" must not rule in these industries to the extent that "Better" is sacrificed. The reliability, safety, operability, and maintainability of space missions cannot be ignored. Yet these same quality metrics exist in many other industries currently supplied by open markets. NASA's Strategic Enterprises may need to undergo some culture transformations, however, to establish the habits and procedures necessary to become smart buyers instead of smart builders of operational space systems.

There is another aspect of the supply/demand framework that does not fit the store analogy. Most stores would try to provide a product line as long as buyers were available. FRED, however, must have an incentive to stop providing the reusable components when the private sector can provide them more competitively. Incentives to stop doing something are challenging to devise. The COST LESS team envisioned a periodic publication of FRED's specifications for the components so that any provider could bid on the opportunity to supply them. If sufficient providers emerge, the reusable component would be dropped from FRED's inventory; the Enterprises would then use the open market for their supply. Personnel working for FRED would have an incentive to discover commercially available sources. When their discoveries in the market place led to a discontinuation in a FRED product, they would be reassigned to more cutting edge research and to the development of new reusable components. This would attract the most creative talent to the FRED structure. Soon, FRED would be the home for the best non-science-specific innovators, e.g., the advanced technologists and the methodologists. This would parallel the situation within the Strategic Enterprises, where the best science innovators would be attracted. Those who did not wish to keep pace with the rapidly changing profile of reusable components provided by FRED might choose to migrate to the commercial sector where product innovation would be occurring at a more calculated pace and where cost, reliability, and customer service are discriminators that will cause a product to stay on the market for a longer time. The commercial providers are the sustained supply source who replace FRED as the initial supplier of reusable components.

5. Reusable Components and FRED for Managing Other Resources

The concept of FRED is structured to create a win-win situation for a central organization and for commercial organizations that have a symbiotic relationship to the central organization. In conceptualizing the supply/demand framework called FRED, the COST LESS team was thinking of NASA as the central organization, NASA's Strategic Enterprises as the users of reusable components, FRED as a quasi-commercial initial supplier of the components, and commercial entities as the subsequent and continuing suppliers of the components. Space mission operations was the application paradigm. The COST LESS team believes, however, that the concept has broader applicability. Suppose, for example, that a country is seeking to improve the quality and quantity of goods and services to be made available to its population. In order to do that, it

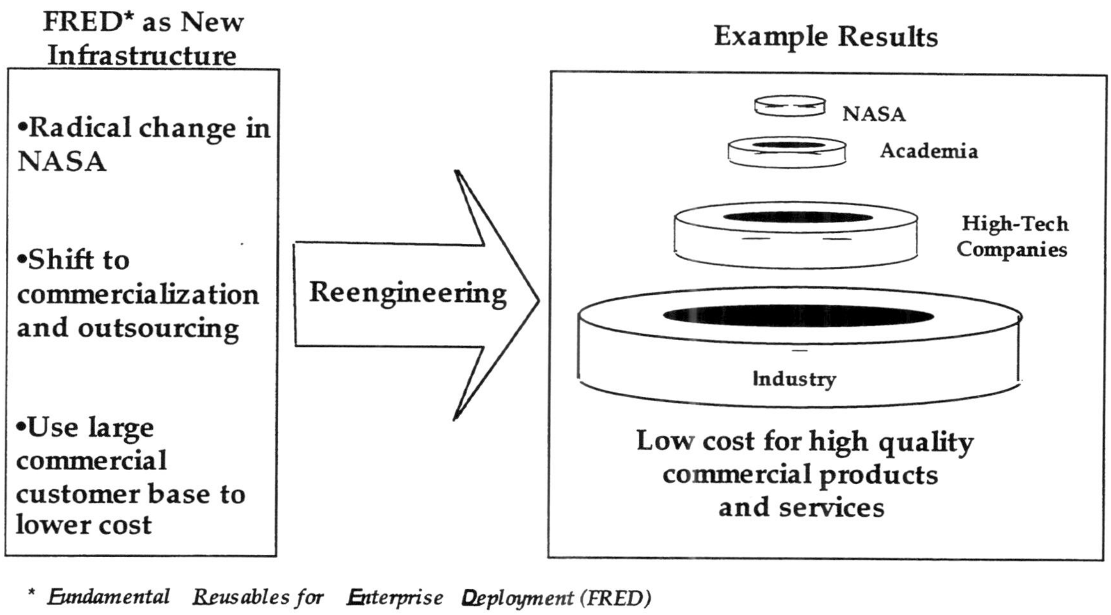

Figure 4. Radical new approach

wants to exploit space technology, and hopes to stimulate its economy by finding a productive balance of government and commercial initiatives. The challenge is to get market forces working on the aspects of the problem that can be commercialized, so that public moneys can be focused on those aspects of the operations that will not occur if left solely to market-place incentives.

Public funds should be spent in such a way that unnecessary duplications among organizations are avoided. That important requirement forms the genesis of the concept for finding the truly reusable components and for supplying them through a lean distribution channel such as FRED. Without a conscious effort to abstract the processes needed to manage operations, the resulting consequences are numerous duplicative implementations of systems and organizational structures that try to justify their existence through claims of uniqueness. The

COST LESS team believes that incremental and evolutionary developmental processes may not lead naturally or inevitably to the recognition of generality. The gradual evolution that usually accompanies the development of a new frontier tends to be myopic and often sees peculiarities before it sees commonality. Only after experience provides enough repetition does wisdom about duplication begin to emerge. By that time it may take critical shortages of resources to force innovative, discontinuous changes instead of minor incremental adjustments.

New space programs should have an advantage in this regard. Perhaps they can learn early the experiential lessons of the pioneering programs that have now become unnecessarily complex and duplicative. The older programs must re-engineer themselves in ways that make use of the minimal set of fundamental reusable components rather than an enormous inventory of special-purpose components.

The World Bank has recently announced the continuing development of a new model to assess the "wealth" of a nation. Presumably the goal is to do an assessment of loan-worthiness based on the metrics of a nation's potential success rather than on traditional metrics such as per capita income. Factors such as produced capital, natural capital, human resources, and the creation of social institutions are examples of the new metrics. This paper suggests that expansions of these types of metrics to include reusability might be very useful to assess a nation's wealth and its loan worthiness. Although the COST LESS team has focused on space mission operations, the metrics for the successful management of scarce, shared, and expensive resources for conducting space mission operations are not fundamentally different from those needed to manage a nation's resources in a manner that provides better goods and services to its population. This paper implies metrics for assessing how well any major operational endeavor is managed. These metrics are based on the broadly applicable concepts of simplification, avoidance of duplication, reuse of common components, and efficient distribution. Among them are:

- recognition and use of fundamental and reusable building blocks to simplify operational activities,

- the presence of a viable distribution system that encourages the reuse of common components to support operations across multiple sectors of an institution, agency, nation, etc.,

- the creation of incentives to enable the distribution system to stay at the leading edge of technologies and techniques that would otherwise not emerge because the economic rewards were not yet obvious to commercial providers,

- the presence of additional incentives for the centrally-oriented distribution system to drop components that, in fact, can be provided via commercial providers, i.e., an economically-based technology transfer program that is intentional, not coincidental.

Without professing expertise in areas beyond the limits of its experience or training, the COST LESS team hopes that the concepts underlying its work for NASA may find utility for those intending to use space as a means to improve the quality and quantity of goods and services available to humankind.

6. Summary

The COST LESS team is developing a simplified framework for reducing the cost of performing space mission operations. It is based on the notions of identifying and using fundamental building blocks to synthesize mission operations systems, rather than on building unique systems to solve similar problems. The definition of the fundamental reusable objects is an on-going activity for the group. The emphasis is on simplicity derived from quality abstractions and generalizations. The team is striving to find the minimum set of building blocks that will remain stable and reusable in at least their design and functional interfaces, even though the technology for implementing these components continues to change.

In order for the components to be used and reused, an effective framework for supply and demand must also be in place. This framework, called FRED, is also being refined. Organizational entities, operational procedures, financial accounting standards, and personnel motivators, along with the exploitation of competitive influences, and the relationships with commercial and academic product providers are important considerations in the conceptualization of this framework.

The on-going work of the COST LESS team seems relevant to broader goals, such as those of this international symposium for Space of Service to Humanity. The use of space missions to increase the quality and quantity of goods and services is a practical goal if duplication and complexity can be minimized and the reuse of common components can be maximized. Other agencies like NASA are seeking to learn to conduct space mission operations at a fraction of the costs expended in the pioneering years of such endeavors. Looking to the future, any organization with a vision for the practical exploitation of space, must find faster, better, and cheaper processes for providing goods and services.

Matching Space-Related Intellectual Property Rights to Space Industry Needs

B.L. Smith, Société de Services de Propriété Industrielle, Alcatel Alsthom Intellectual Property Department, 14 rue de la Baume, 75008 Paris, France

Abstract

Intellectual Property Rights (IPR) are an insurance for high-risk industrial investments in technology. Obtention and use of IPR in space fields is problematic because of uncertainties about applicable legal framework, especially nationality and territoriality, both for obtaining and for enforcing IPR for space activities This is not conducive to private investment, in particular for large sums typical of space endeavors.

Space industry is thus hampered because of limits on R&D funding and the spectre of third party infringement proceedings. Legal certainty is prerequisite for any full scale exploitation of extra-terrestrial industrial opportunities. Until now, only ad hoc solutions have been advanced: the US Space Bill, the Inter-Governmental Agreement (IGA) for the International Space Station, US case law (Hughes Aircraft Corporation vs. X), and doctrine based on maritime law. Existing patent law has no firm basis for territorial extensions to extra-terrestrial activities on orbit and beyond.

Legal and practical uncertainties abound in enforcement of IPR : 1) applicable law, because of its territorial nature ; and 2) how to discover and prove infringement to recover damages.

Industry needs legal certainty arising from applicable, internationally harmonized space-specific legislation. Patchwork territoriality thwarts legal certainty.

Legal certainty is necessary to attract private investments and to assess infringement liability. A workable enforcement mechanism is also required, free from territorial considerations. We need an international legislation, with an international enforcement body, such as an international board of arbitration or court. This is a gigantic undertaking, but more likely than harmonizing disparate national laws. Nationalist tendencies may subsist but uniform legislation must undoubtedly be proposed rapidly on a global level.

1. Introduction

Space activities have their historical roots in national, captive markets governed by specific public procurement practices. Almost all current main industry players have grown out of defence-related industries, national champions nurtured primarily by defence contracts of the cold war era. This history has contributed to traditions which continue today, in spite of leaner budgets.

Important for considering the use of intellectual property rights (IPR) in space related industries is the fact that a significant majority of the industry has been (and remains) essentially not for profit, for example infrastructure, scientific endeavors, and most non-recurring development efforts. This means that applicable pricing is most often approximately at cost, or slightly above - or even below cost, in many instances.

Historically, launch services and contracting have been subject to similar considerations, with little consensus on the actual costs of launch. With the emergence of commercial launch services, accusations of market-skewing government subventions still continue to be heard among the major players on the commercial launch market.

A further consideration is the fairly widespread use of the principle of "juste retour" as applied in procurement practices by treaty organizations and multinational agencies such as ESA. Even in the case of procurement by national agencies, similarly arbitrary contract awards have frequently been used to "distribute the wealth" among two or more national players. This is again a tradition from cold-war defence funding of the "military-industrial complex".

During the cold war period, competition in space activities was much stronger between nations than between individual firms. Because the major players were in general traditional national champions, the non-recurring development costs were typically heavily subsidized. Therefore, the real costs of major realizations in the space sector are still difficult to determine. This results in distortions to fair applications of calls for tender and an open bidding process.

The space industry has recently evolved into a business sector which is adopting a more orthodox behavior. Commercial contracts are awarded on tender, ranging from launch services to satellite procurement. Deregulation of telecommunications is being undertaken around the world, allowing private entities to offer telecommunications services. Some services are being offered via privately owned satellites. Telecommunications satellites are becoming a market commodity, serving private, national or regional entites, many of whom do not have a self-sufficient space industry capability. This has led to the dawn of commercial competition between firms for the supply of each of the component systems of a satellite, up to the entire satellite or even an entire turnkey satellite telecommunications system.

Due to the fairly limited number of programs and players, competitors on one program will be partners on another, customers or main contractors on another, subcontractors on still another, in every imaginable combination. The customers of the space-related industries are also somewhat atypical. Because of the large investments required, and also because of regulatory hurdles for orbital slots, frequency allocations and the like, customers of the space industry are still mostly government-affiliated agencies, or multi-governmental treaty organizations. Procurement policy continues to rely on political will, where free market competitive mechanisms do not apply.

From an industry analyst's point of view, some of the contracting policies of the major agencies and treaty organizations further undermine the competitive mechanisms of a free market economy.

2. Prospective Developments

Some trends in the development of space activities are easy to predict. Such trends will impact the needs for applicable IPR as will become apparent in the following paragraphs.

We will see increasing international cooperation in space activities, to the point where national borders no longer separate the different players. There will also be increasing global competition for supplying hardware and services, further eroding the geopolitical frontiers as applicable to different components of space business.

Former government activities will be increasingly spun off into the commercial sector, as can already be seen in different fields such as launch services, weather forecasting, remote sensing, telecommunications, localization (GPS), and eventually, perhaps, space-based manufacturing. At the same time, privatization and commercialization of activities of treaty organizations are gathering speed. Privatization of entities or of specific projects is foreseen, for example, for Inmarsat, Intelsat, Comsat, Eutelsat, ...

An explosion of new applications will also fuel the market expansion, for example Direct TV Broadcast by Satellite (DBS), mobile communications, GPS, direct radio broadcast, multimedia, etc. Last but not least, future industrial activities in outer space are actively in the planning stages : the International Space Station (Alpha, formerly Freedom), lunar mining operations, solar energy orbital platforms, Mars exploration and exploitation, etc....

All of these factors will lead to increased awareness of the need for IPR in outer space. More and more investment will be required to realise all of these plans, and much of the financing willl be required to come from the private sector. Such private investment will be motivated by hopes of financial gains, which can only be ensured in a favorable competitive environment. This is where IPR becomes a competitive "weapon".

2.1 Industry Expectations from IPR

Industry and practitioners alike cite the alleged "protection" of R&D investment. This may mean simply the recovery of the R&D investment, or, on the other hand, may be extended to include the future fruits of such an investment. Visionaries and optimists may go so far as to speak of ensuring not only the autonomy but also the very survival of the enterprise as something which can and should be protected by the acquisition of IPR. A more justified expectation is that IPR should confer a competitive advantage, either to win contracts or to exclude competitors from using a cost-saving innovation or an improved technical solution.

Further commercial and marketing advantages are also often cited by industrial firms and the inventors themselves. For example, the overall image of a firm is enhanced by the IPR generated within it. Granted patents and even pending patents constitute a proof of competence in the field which, upon closer examination, may be shown to be only casually related to the filed patent or to the invention itself.

Perhaps the most realistic expectation which we could have is the usefulness of IPR in a defensive mode - against third party attacks on the basis of third party rights. Having a broad patent portfolio can be useful when coming to a friendly agreement after being approached by a hostile third party, if it can be argued that the third party is also a potential infringer, or that he could benefit from accepting a cross-license instead of continuing with litigation.

2.2 Problems for IPR in Outer Space

One major problem is the difficulty of obtaining the enforcement of IPR in space applications. The territorial nature of IPR legislation leads to territoriality of jurisdiction for infringement proceedings, with the practical result that enforcement will almost always be pursued in foreign courts under foreign legislation. An exception is possible if the jurisdiction for litigation proceedings happens to be that of the patent owner, but this is quite unlikely (unless the owner happens to be American).

A second major difficulty is that of discovery and proof of alleged infringement. On orbit, infringement is understandably difficult to detect. On the other hand, at contractor's sites or at launch sites, secrecy reigns (a throwback to the cold war) and infringement can go undetected on the ground as well.

Finally, it must be mentioned that existing patent law is intended to apply only within a national territory. However, it is now seen that unilateral extensions of sovereignty may open possibilities of infringement of national laws on orbit, thus leading to increased exposure to infringement liability risks.

Examples of such extensions include the US Space Bill, or common law interpretations of "use" in the territory to include downlinks to the territory, or common law interpretations of "control" to include uplinks originating from the territory. Completed with an attitude of refusal of the doctrine of temporary presence (cf. Hughes Aircraft Corporation vs. X), the United States has become the world's most likely jurisdiction for infringement proceedings in outer space cases, merely because of its efforts in unilateral territorial extensions.

Meanwhile, on the ground, agencies and treaty organizations have collected extensive IPR and options including sublicense rights, originally consented by industry to non-commercial entities. In this regard, it should first be noted that quite often the R&D has been funded by an agency or other organization, under contractual clauses which attempt to protect or recover the agency investment, but offer little or no protection for the accompanying portion of industry investment. As for the recovery of R&D costs, under contractual clauses in widespread use, often a successful patented technology will also result in the patent proprietor paying royalties to the funding agency for the use of the patent, considerably slowing the rate at which the industry investment may be recovered. Agency and similar third party R&D funding contracts also commonly reserve licensing rights with accompanying extensive sublicensing rights. This generally means that the agency will be able to use patented inventions for other purposes than those foreseen by the original contract, and also that the agency can grant sublicences to the originating firm's competitors, effectively defeating the hoped-for competitive advantages.

Furthermore, this IPR under licence or option is increasingly regarded as a source of revenues for the agency or treaty organization, to the detriment of the originating company. If and when non-commercial entities become commercial or privatised, they have a tendency to adopt a more aggressive stance for the exploitation of such IPR.

Finally, as treaty organization signatories evolve from state chartered entities into the private sector, massive legal loopholes are opened for defeating patent protection, as numerous sublicensing possibilities have already been set forth within the original founding treaty. Another dilution of potential competitive advantage comes from the fact that quite often tender bids do not necessarily reflect true cost, and maybe therefore do not reflect cost savings achieved through patented technology. In addition, standardization of patented technology may lead to the loss of monopoly through compulsory licensing.

So how about the contribution of an IPR enhanced company image to commercial and marketing success? The effect should be undeniably positive, even if the magnitude of the effect is subject to speculation. However, it may argued that a similar effect may be obtained much more cheaply. A widely filed, single patent can easily cost the owner up to several hundred thousand French Francs, whereas open publication in a refereed scientific or technical journal will be more widely read and appreciated by clients and competitors alike, and cost no more than a few thousand French Francs.

Can a broad patent portfolio be effectively used to protect a firm from third party aggressions? This can hardly be guaranteed, once again because of the lack of legal certainty and the resulting uncertainly of predictions of the successful outcome of infringement proceedings. Because of the huge sums which may be earned in a successful infringement suit, attempting an attack becomes attractive enough to raise venture capital to pay the legal fees of the attacking party, even if the chances of success are slim. In the US at least, lawyers working on a purely speculative basis on potentially lucrative lawsuits are far from uncommon.

Yet another subtlety of territorial-based patent law comes up when one seeks to define which acts constitute infringement. Does transfer of ownership on orbit constitute infringement? What about the use on orbit of the satellite having an infringing component or subsystem aboard?

Finally, further problems are easily imagined concerning activities which may be carried out aboard the International Space Station. What constitutes a disclosure (novelty bar) and how can this be avoided in practice? Which state may make national security claims? What provisions must be made for the use of third party inventions when such a use may not be reasonably avoided? Is some type of compulsory licensing scheme necessary?

2.3 Which is the Applicable Law Today ?

There are three possible answers to this question in existing legislation. The first solution is that of multilateral consensus treaties, such as those ratified under the auspices of the United Nations. This is perhaps quite satisfying for the signatories, but until now has not given any practical consequences regarding IPR problems in outer space.

A second type of agreement is exemplified by the Inter Govenmental Agreement for the International Space Station. This agreement is the bare minimum necessary for building and operating the space station. It is a rather awkward agreement to put into practice, and does not even attempt to address the broader questions raised above. It does have the advantage of allowing work to proceed on the station, without waiting for a more elegant and complete legislation. It is possible that difficulties will arise in application.

In the absence of these first two solutions, any state may declare a unilateral extension of territory for the application of its own IPR legislation. This is exactly what has been enacted by the US Congress under the name of the US Space Bill. Germany has also introduced similar legislation. Private entities or firms may also attack on the basis of case law and doctrine, both of which have allowed extensions of territory in the past to resolve disputes concerning vessels outside territorial waters, or aircraft entering claimed airspace, whether intentionally or not.

2.4 What Improvements are Required?

First of all, and most importantly, we need legal certainty arising from appropriate, applicable legislation. It seems imperative that such legislation be conceived with international harmonization for space use. Patchwork territoriality leads inevitably to inconsistent application and spawns legal uncertainty.

We are also of the opinion that doctrine and case law are not sufficient, in themselves, to obtain the necessary legal certainty. We therefore need some space-specific international legislation. Legal certainty is a necessity in order to attract private investors to finance space industry efforts in the commercial sector.

This legal certainty is also necessary to enable industry to assess the risks of infringement liability properly. This may be a key element when bidding on commercial supply contracts, particularly when the buyer is a service provider, and when infringement suits may call for royalties based on operating revenues ("use" of an infringing technology) instead of only the value of the hardware provided under the supply contract.

In addition to legislation, we also must establish a workable enforcement policy and mechanism. Many present problems must be overcome for this to become a reality. First, the present territoriality patchwork means that enforcement actions are always to be taken in foreign courts under foreign laws, with only one exception per plaintiff when, by pure chance, the infringement is in his own country. Secondly, the detection of infringement is presently next to impossible with the confidentiality which reigns in the industry.

We must all, as world citizens, be particularly vigilant concerning the unilateral extensions of territorial or national sovereignty, whether they be legislative or common law arising from decisions in national courts. In this respect, we note that state law (e.g. California law, Texas law, etc.) may also apply in addition to federal law, if a dispute embraces non-patented know-how or other proprietary information such as trade secrets.

In fact, the legal void may lead to dangerous interpretations before national courts. It is equally dangerous when a single national legislation tries to establish, alone, a legal framework which intends to embrace all of its space activities as well as those of foreign nationals whose activities fulfill unilaterally dictated criteria.

3. Modest Recommendations

Whereas the space industry is a truly global endeavor at present, the playing field is not level for all of the players. This is related to several of the problems stated above, plus favouritism of some governments and contracting organizations towards their own national champions.

The best we could hope for in a free market economy of space-related endeavors would be to reap the (until now) illusory expectations of industry and private investors as exposed above. This would surely be limited in practice. However, industry would like to be able to use IPR to protect its R&D investments to gain competitive advantages such as marketing and commercial advantages, and to be able to continue industrial pursuits without fear of third party aggressions on IPR issues.

As exposed above, a major and necessary step towards achieving such a situation is to *establish legal certainty* for IPR issues in space-related activities. Obtention of IPR must be clarified in view of disparate territorial principles.

An even more difficult problem is that of the exercise of IPR which obviously requires special "fair" rules for space-related activities. Here, we would like to draw a distinction between two components of the problems to be solved. Questions of legislation, such as how to achieve uniformity in the obtention of IPR, would appear to involve a component which is almost exclusively logical. On the other hand, questions such as "fairness" in the application and enforcement of IPR in a global context appear to rely on ethical issues, which would call components from another sector, not relying purely on logic.

The obvious question to be resolved is that of the universality vs. territoriality or nationality question. However, in view of the general interest of space-related activities to mankind as a whole, as reflected in the terms of the Outer Space Treaty, we should consider the possibility of developing a fair scheme which strikes a balance between the private interests which generate innovation at the expense of financial and human resources, and which hope to capitalize on those investments, and the wider interests of the planet, perhaps through some sort of compulsory licensing scheme.

We point out that the space-related industry is not the only one to face such problems of general interest to all mankind; perhaps we can collaborate with, or take inspiration from, IPR practitionners in other fields where similar debates are raging. The biotechnology field comes immediately to mind, with its research on the human genome, attempts to patent gene fragments (by a government agency !), resulting diplomatic incidents, etc.

It would appear that some issues are just too broad to let any country decide alone what should happen, even within its own borders. And we notice yet again that the first move towards appropriation of as yet unclaimed territory comes from the same part of the world !

It seems to us that there is no quick fix. The best solution might be to *establish space and its accesses (launch sites, vehicles) as a single territory with a single, uniform law*. This will always be easier than harmonizing the internal national laws of the player countries. However, this may be a utopic proposition, because it would then also be necessary to *establish a single, universal enforcement body* such as an international court of law or an international arbitration authority. It may take decades to overcome nationalist tendencies.

In order to try to level the playing field between space industry competitors, on the one hand, and to integrate the newly privatized entities into the commercial space sector, on the other, the rules of terrestrial IPR and related contracts should be revised to take into account the specificities of the space activities. Everybody should be able to benefit from the knowledge and wealth generated by space activities in an equitable fashion according to the terms of the OSA (outer space agreements). Industry interests must balance those of the general population. But space industry, which now requires more and more private investment, will also require a reasonable return on that investment.

As is already the long standing tradition for terrestrial activities, it becomes increasingly urgent to ensure that the space industry benefits from a workable IPR legislative framework which can guarantee a minimum level of security for investments and for future activity.

Integrating Wireless Power Transportation and Solar Power Systems Studies and Society

G. Pignolet, CNES, 2 Place Maurice Quentin, 75001 Paris, France

Abstract

Microwave energy beam technologies have to move from the laboratories into the social and economic arena. A suitable site has been found for a relevant case study in the mountain village of Grand Bassin, on Reunion Island.

Several challenges have to be met in this case of limited scope, typical of the challenges that must be considered for larger scale projects. Satisfactory solutions must be found to the problems of the actual energy needs of a population, the social acceptability of a new technology, and its proper integration into the environment. The interactions with society during a case study lead to an evolutionary understanding of solar power systems (SPS) technologies; if implemented, the project will undoubtedly be a major step on the path of sustainable and clean energy from space.

1. Background

It may be a hundred years between the emergence of a concept and the time when it becomes commonplace (References 1, 2). It may well be a hundred years after the 1960's when Peter Glaser first presented his idea of SPS (Solar Polar Satellites - or Systems), before the time when the major centers of the world receive their energy directly from space. It is almost a hundred since Nikolai Tesla predicted WPT (Wireless Power Transmission - or better "Transportation") when today we are considering, as a case study, the first possible operational WPT link to power a small village in the mountains of Reunion Island (References 3, 4).

In 1979 and the early 1980's, the Department of Energy in the USA and the Space Studies Institute of Princeton University conducted extended design studies to demonstrate the feasibility of the SPS. A new incentive came from the Ministry of International Trade and Industry (MITI) of Japan and from many Japanese organisations, and more especially from the Institute for Space and Astronautical Science (ISAS) with its SPS-2000 strawman concept. The first functional model was built at ISAS in 1994 to demonstrate the end-to-end operation of a SPS system (Figure 1). A second, improved SPS functional model has been built in 1995 .

In 1991, a major world symposium was organised by the French Société des Electriciens et des Electroniciens (Reference 5) and many experts pointed out the need for an operational implementation of WPT as an essential step in the development of SPS systems. Proposals were made in Alaska and in Reunion Island. After preliminary studies, a site suitable for a relevant case study was found at Grand Bassin (Figure 2). A case study refers to a real situation, but without any commitment to effective implementation. But a subsequent implementation would be made easier by a successful case study.

Figure 1. The SPS-2000 #1 functional model demonstrated at ISAS and at the Stella Matutina Museum in Reunion Island.
It features all the elements of a SPS system, from top to bottom : photons, photovoltaic cells, microwave generator and projector, microwave beam, rectennas, power grid, end users.

The most important issues in the Grand Bassin case study are the environment and social acceptability. It should be clear that WPT and SPS systems are future solutions to major environment questions.

After the WPT 1995 conference in Kobe, Japan (Reference 6), a world task team was formed to discuss further the case of Grand Bassin, with participants from Japan, Russia, Ukraine and USA. Because the focus is on the environment and society, some important aspects of the SPS concepts have been made more obvious: WPT and SPS are environmentally friendly, they are clean, they are safe for people and for society. A fact worth noticing is the great synergy between these different aspects which derive from the physical design itself.

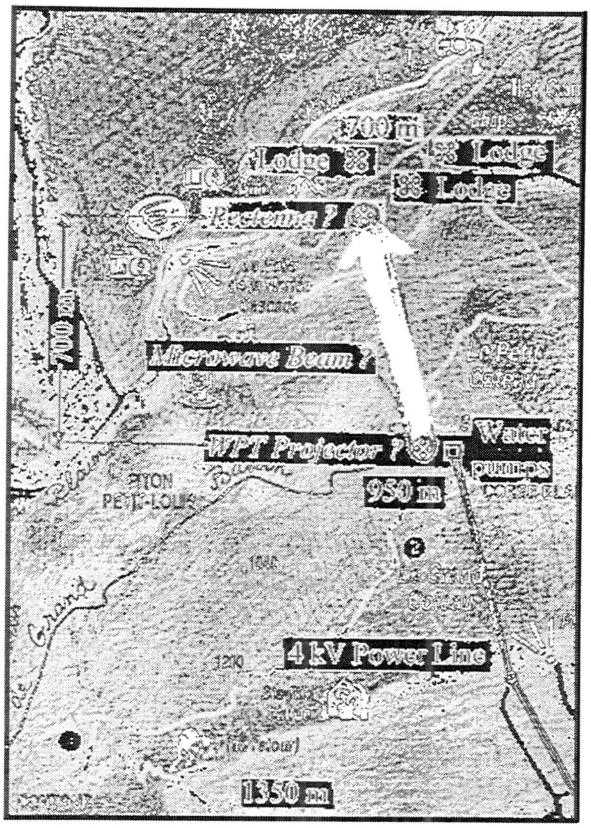

Figure 2. The Grand Bassin case study in Reunion Island
From bottom to top : a 4 kV power line would feed a 20-m diameter multi-magnetron projector, a microwave beam would carry the energy 700 m down across a steep canyon, a 10-m diameter cyclotron wave converter and rectenna combination array would turn the microwave field back to DC energy, a power conditioning system would eventually provide 220-Volt AC current to three tourist lodges.

2. SPS Technology

The preferred frequency for WPT and SPS is 2.45 GHz. Beams of this frequency are little affected by adverse weather conditions. Further, this frequency is already tagged as an energy frequency with more than 50 million microwave ovens operating on this frequency around the world. With a wavelength of 122 mm (together with a half wave of 61 mm and a quarter wave of 30 mm), these are practical basic dimensions for the single units that will be multiplied by thousands and even millions in actual array systems.

The rectenna arrays will look like nets, and allow shapes that can closely follow landscape forms. They may be painted.

Sunlight will go through the WPT/SPS receiving arrays. Therefore, plants can grow under the receiving arrays, for human food, for cattle, or even, in the case of offshore rectennas, for fish farming. Therefore, WPT/SPS technologies can be very profitable for society, because they allow for a dual use of land space, for energy and for food (Reference 7).

Another important point about microwave power systems is that it is believed that there are no biological effects, besides heating. There should be no genetic damage in the use of microwaves. Most studies for WPT/SPS systems at 2.45 GHz consider using densities well below the overheating danger level (5 mW/cm2 workplace safety standard, 25 mW/cm2 WPT/SPS projects, to be compared with an average 100 mW/cm2 clear day sunlight energy density). WPT/SPS systems should be safe for humans and for all living beings.

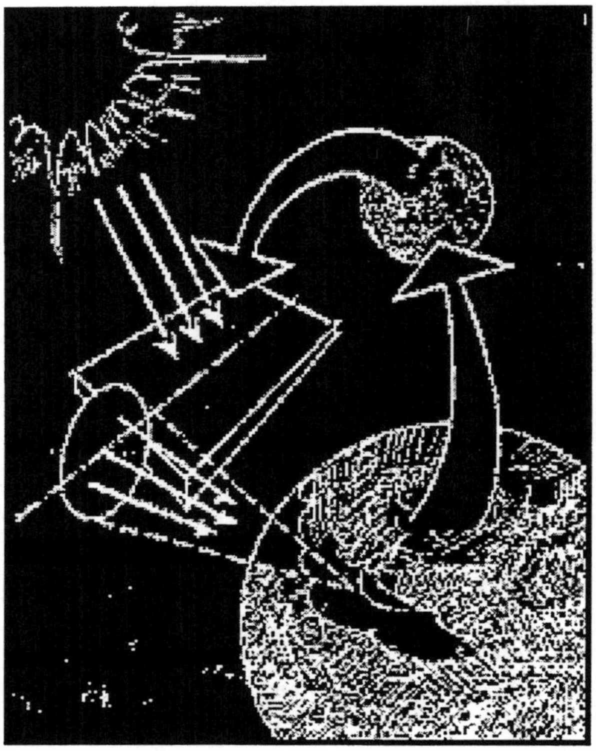

Figure 3. Energy for the 21st century
A counter clockwise look at the staged evolution of the system : construction of a Moon base for the production of SPS materials, construction of the SPS in space, collection of sustainable solar energy and projection to the Earth for the benefit of the Earth's megacities. The Grand Bassin case study may be an important step in this long process where environment and public service will be key issues.

3. A Key to Major Space Development

The advantages of WPT/SPS systems have become clear during the Grand Bassin study while discussing the different questions with the local authorities, the regional department of the French Ministry of the Environment and potential users of the systems.

Although operational WPT may be close, actual SPS may still be far away, cheap access to Low Earth Orbit being another requirement before full scale lunar development and then SPS development (Figure 3) can take place. But, on the other hand, SPS may be the most reasonable motivation to proceed with such developments (Reference 8).

Counting down from the future with Peter Glaser, we may envision in-orbit reflectors and Low Earth Orbit demonstrators, High Altitude Long Endurance Platform (HALE) systems, and point-to-point energy delivery systems (Reference 9), all of them using microwave energy beams. The Grand Bassin case study is an important step in the process.

Addressing the social and environmental issues as the most important ones has already brought important results since the SPS'91 (Reference 5). Other future major activities in space have been discussed by Pignolet (Reference 10).

References

1. Gaudin, T.: *2100, Un Récit.* Payot, Paris 1990
2. Von Muldau, H.: *Notes for the 5th IAF Seminar on the Future of Space Education*, PFIAT, 1995
3. *Proceedings of the Medium Term Case Studies Workshop -Wireless Power Transportation,* SPS-IdR'94, Sofiom,1995
4. Pignolet, G.: "SPS-IdR Case Studies in Réunion Island", *Solar Energy Review,* Vol 56 No 1, 1996
5. *Proceedings of SPS-91 Power from Space.* Société des Electriciens et des Electroniciens, Paris, France, 1991
6. Pignolet, G., Kaya, N., Lan Sun Luk, J.D., Naruo, Y. and Vanke, V.: "Demonstrating SPS technologies on Earth : SPS-IdR studies in Reunion Island towards point-to-point operational WPT. WPT'95, Kobe", *Solar Power Review,* Endicott, New York, USA,
7 Centre National d'Etudes Spatiales: *L'espace comment ça marche ?, à quoi ça sert ?* CNES Publications Department, Toulouse, France, 1995
8. Pocé, O.: "Are France and Europe in a good Lunar mood?", *Les Colonies de l'Espace,* Nazelles-Négron, France, 1991
9. Glaser, P.: *The Window of Opportunity for Wireless Power Transmission.* CNES DRE/OM/PB, Paris, France, 1994
10. Pignolet, G.: *Les Enfants de l'Espace.* CNES DRE/OM/PB, Paris, France, 1994

Impact of Space Medicine Experimentation on Terrestrial Medical/Clinical Applications - The Austrian Experience

F. Gerstendbrand, M. Mossaheb, A. Saleem, J. Schneider and **C. Stepan,** Austrian Society for Aerospace Medicine, Hohe Warte 19, A-1190 Vienna, Austria

Abstract

In 1988, the Austrian Federal Government initiated a bilateral programme with Russia with the aim of flying an Austrian Cosmonaut to the Space Station MIR to be accompanied by approximately 170 kg of Austrian developed and manufactured scientific/medical experiments. This cooperative effort, which reached its climax with the launch ofthe Austrian Cosmonaut Franz Viehbock with the crew of Soyuz TM-13 to the Space Station MIR on 2 October 1991, came to be known as the AUSTROMIR project. Due to the success of the mission, the quality of the experimental results, and the fruitful collaboration between Austrian and Russian researchers/scientists the Austrian Federal Ministry for Science Research (and Art) supported the founding of the Austrian Society for Aerospace Medicine (ASM) & Life Sciences. Since its inception and according to its founding mandate, ASM has striven to become the focal point for all Austrian activities in space medicine and space life sciences. Recently, under the management of ASM from the Austrian side, Austria has flown improved versions of all the major experiments from the AUSTROMIR project within the framework of the Russian National Space Program. This collaborative effort, referred to more commonly as the Russian Long-term Flight (RLF) Project, was centered around the Russian Physician Cosmonaut Valery V. Polyakov who broke the space endurance record with 437 days and 17 hours of spaceflight. The main goal of his mission was to investigate the effects of long-term spaceflight - especially weightlessness - on the human body. As a result of the AUSTROMIR Project, the RLF Project and their respective continuation phases, Austrian researchers/scientists have accumulated (in total) over 2 years of experimental data on long-term spaceflight. This means that Austria, in this specialized field, has attained a greater knowledge and experience than many of the larger space faring nations and agencies with respect to long duration spaceflight. This paper, by giving an overview of ASM, the AUSTROMIR and RLF projects, will focus on two main areas. First, how Austria is implementing the results and reaping the benefits from its space based experiments within terrestrial medical/clinical applications for the benefit of humanity. This will be highlighted by connecting the results from the Austrian space based experiments to terrestrial applications in the areas of radiation therapy, radiation protection, rehabilitation, and the early detection of cognitive disorders, etc. Secondly, the paper will demonstrate how Austria as a small nation with limited resources can actively, effectively and successfully participate in microgravity activities.

1. Introduction

In space the human organism is exposed to a physical environment which can neither be adequately simulated nor reproduced on Earth. Gravity affects the function of cells and tissues by applying pressure and stresses. Therefore, when it is removed the body's chemistry is altered, muscles are weakened, the loss of calcium in bone tissue is accelerated, fewer red blood cells are manufactured, inner ear and vision mechanics are altered, and the body's ability to regulate fluid volume changes is affected (Reference 1).

The driving goal of space medical experimentation is to better understand these processes and to develop appropriate countermeasures so that humans may live longer in space. Space experimentation may, for example, include the study of various physiological and neurological functions independently of gravitational influences (such as the functions of individual muscles, tremor development, motion control, etc.), the assessment of the reaction of humans to extreme situations and high levels of stress (cerebral performance, toxicological and immunological effects due to isolation and weightlessness, etc.) as well as the determination of biological effects of cosmic radiation. This also leads to an improved understanding of the functioning of the human organism as a whole and the knowledge gained through this understanding may be used to improve terrestrial medical/clinical applications for the general benefit of all mankind.

By giving an brief overview of the Austrian activities in the field of space medicine, within the framework of bilateral cooperative projects with Russia, this paper will demonstrate the transfer of space experimentation to terrestrial medical/clinical applications. The paper will also highlight how a small space faring nation, such as Austria, may successfully contribute and participate within microgravity activities.

2. AUSTROMIR

Due largely to the close cooperation between Austrian and former Soviet Union scientists/researchers in various space orientated activities over a period of more than 10 years, Austria was invited in mid-1987 to participate within the framework of activities connected with the Space Station MIR program. Austria's objectives were to participate in two main ways, namely:

- to fly an Austrian Cosmonaut to the Space Station MIR, and
- to collaborate on the design and development of experimentation to be flown on the Space Station MIR.

This Austrian participation, within the scope of the two activities mentioned above, came to be known as the AUSTROMIR project. In detail, the activities that were covered under the contract between Austria and the former Soviet Union, included:

(a) Austrian Responsibilities

- Selection of Austrian Cosmonaut candidates, supported by the Institute for Bio-Medical Problems (IMBP) in Moscow.

- Development and manufacture of 14 experiments and instruments (9 medical, 5 material and physical sciences), designed in cooperation with Russian partners, weighing a total of approximately 170 kg .

(b) Russian Responsibilities

- Training of two (one backup) Austrian Cosmonaut candidates by the IMBP and the Yuri Gagarin Cosmonauts Training Center (CTC).
- Transport of the payload and Austrian Cosmonaut to the Space Station MIR.
- Guaranteeing of a minimum total of 42 hours of experimentation by the Austrian Cosmonaut and other crew members of the Space Station MIR during his 8 day flight.
- Return transport of the scientific results with a total weight of 3.3 kg (Reference 2).

The entire development from the conception to the delivery of flight hardware and software was realized in a period of just 3 years, followed by a period of dissemination and publication of the results. As the experiments were co-developed by the Austrian and Russian scientists/researchers, it was agreed that the experimental data would be jointly evaluated for the purpose of publication. This spirit of true cooperation with mutual respect and understanding was undoubtedly one of the key factors for the success of the AUSTROMIR project and subsequent further cooperation between the parties involved.

The project reached its climax with the launch of Cosmonaut Franz Viehbock on 2 October 1991 with the crew of Soyuz TM-13. After six days of intensive experimentation Cosmonaut Viehbock returned to Earth on the morning of 10 October 1991.

As originally intended the project did not end with the return of the Austrian Cosmonaut. Instead, under the supervision of the Principal Investigators from the AUSTROMIR project, the experimentation continued within the framework of the AUSTROMIR-E and MED-F follow-on phases confined to medical experimentation. In the AUSTROMIR E phase nearly all the AUSTROMIR medical experiments were in continuous use by the Russian Cosmonauts between the period from November 1991 to March 1992. In the MED- F phase a few of the AUSTROMIR medical experiments were selected for use by the Russian authorities between the period from March 1992 to August 1992.

3. Austrian Society for Aerospace Medicine & Life Sciences

Following the successful completion of the AUSTROMIR project, the Principal Investigators approached the Austrian Federal Ministry for Science, Research and Art seeking support to continue activities in the space medicine domain - to this end the Austrian Society for Aerospace Medicine (ASM) was established in 1991. ASM was founded to pursue the following aims (Reference 3):

(a) To plan, coordinate and implement scientific research in the field of aerospace medicine and space life sciences on a national and international level.

(b) To foster inter-disciplinary research approaches.

(c) To adapt instrumentation and methods employed in space medical research to terrestrial applications for medical diagnosis, therapy, prevention and rehabilitation.

(d) To issue publications and provide information to the general public concerning aerospace medicine and space life sciences.

(e) To collaborate with partner institutions at a national and international level.

Essentially, ASM was founded to provide a basis for multi-disciplinary approaches to spaceflight biomedical research. ASM was to act as a platform where diagnostic, prognostic and elective, operational and preventative countermeasures could be conceived and developed. Recently, ASM has flown modified and improved versions of all medical and life science experiments from the AUSTROMIR project within the scope of the 14 month "Russian Long-term Flight Project (RLF)."

4. The Russian Long-term Flight Project

Most recently, as part of the Russian National Space Programme, ASM has at the invitation of its Russian partners IMBP contributed 9 medical experiments within the Russian Long-term Flight (RLF) project - using the Austrian terminology. As the longest Russian spaceflight by a Russian Cosmonaut (437 days and 17 hours of spaceflight) the RLF project was considered by Russian space authorities as an important milestone in a future manned flight to Mars. The scientific/medical part of the complete RLF project was managed, controlled and coordinated by the IMBP on the Russian side. On the Austrian side this was managed by ASM.

The RLF project started on January 8, 1994 when the Russian Physician Cosmonaut Valery V. Polyakov lifted off from Baikonur with the crew of Soyuz TM-18 to investigate the effects of long-term spaceflight, especially weightlessness, on the human organism. Dr. Polyakov returned from the Space Station MIR on March 22, 1995. During his stay on the Space Station MIR, Dr. Polyakov and rotating crews made regular use of the ASM medical experiments. In point of fact, the ASM medical experiments were the only medical experiments that were used over the entire period of Dr. Polyakov's stay on the Space Station MIR.

In theory, using the crew medical examinations taken before, during and after each flight, as well as comparative medical data obtained from control subjects on Earth, the following results are expected:

- to access the mechanisms and extent of change in the human organism caused by the long-term exposure to the space environment;
- to evaluate the process of adaptation to the space environment and re-adaptation to the Earth's environment after landing;
- to develop improved countermeasures against adverse effects of long-term spaceflight;
- to gain valuable insights for a planned future mission to Mars.

The analysis and evaluation of the data collected are currently being performed in close cooperation with IMBP and will be ready for publication in late 1996.

Table 1 demonstrates the accumulated experience gained from the AUSTROMIR and RLF projects in total approximately 12 years of experimental data.

Project Crew	No. of Cosmonauts	Approx. Mission Duration/Months	Total/Man Months
AUSTROMIR	1	0.2	0.2
AUSTROMIR-E	2	4	8
AUSTROMIR MED-F	2	6	12
RLF	1	14.5	14.5
Crew 15	2	6	12
Crew 16	2	4.5	9
Crew 17	2	6	12
Crew 19	2	3	6
Crew 20	25	5	10
Crew 21	2	6	12
Total	**18**	**51.2**	**95.7**

Table 1. Comparative analysis of AUSTROMIR '91 and RLF data as well as data from parallel terrestrial investigations is currently on-going
(Note: Total number of months of dosimetric measurements aboard MIR: 35.2)

5. Terrestrial Applications

5.1 Introduction

Based on the findings and experiences gained in the course of space experimentation, a number of studies and pilot projects have been, and will be, implemented by ASM to adapt the methodologies and instrumentation developed for space use to terrestrial applications and clinical practice.

Depending on the type of experiment, this transfer process concentrates on a number of steps, for example:

- studies to optimize experimental and test setups;
- acquisition of on-ground reference data and standard reference values;
- adaptation of the instrumentation to the needs of clinical practice, i.e. more robust design without limitations in weight, size or packaging;
- cost-effective manufacturing by use of non-space components, etc.;
- adaptation and simplification of methodologies and handling procedures in order to be able to accommodate large numbers of patients (e.g. development of patient data banks, optimization of data storage, etc.);
- software modifications to obtain real-time on-screen results for evaluation by responsible medical personnel.

Benefits from the application of the results of space medical experimentation can, amongst others, be expected in the following areas: improvement of the effectiveness of radiation therapy and radiation protection, muscular training, neurological diagnostics and rehabilitation, further basic research to improve the understanding of disease development and to develop effective preventative measures, and development of portable non-invasive diagnostic tools, etc....

In the following paragraphs, five experiments from the AUSTROMIR and RLF projects have been selected to demonstrate these developments.

5.2 Experiment ADLET - Improvement of Radiation Protection Measures and Radiation Therapy

(a) Aims

Measurement of ionizing cosmic radiation with miniaturized passive detectors.

(b) Method and Equipment

The absorbed dose and average Linear Energy Transfer (LET) is measured with both standard and specifically developed Thermo-Luminescent Dosimeters. In a simple, specifically developed evaluation process, simultaneous information of the magnitude of the absorbed dose and the average LET of absorbed radiation, i.e. the equivalent dose, can be obtained. These newly developed methods are then compared with standard methods.

(c) Examples of Terrestrial Applications

- Evaluation of the quality of radiation protection measures via determination of the absorbed dose and quality of radiation in mixed fields.

- Improvement of the effectiveness of radiation therapy using protons or heavy ions via determination of depth-dose-curves and variations in the quality of radiation with depth.

- Measurements of dose distributions and average LET on humans and electronic components in high flying aeroplanes and spacecraft with crews as a basis for the development of improved protection measures.

5.3 Experiment COGIMIR - Clinical Diagnostics and Rehabilitation in Neurology

(a) Aims

Detailed scientific investigation of cerebral performance during spaceflight and due to cerebral disorders.

(b) Method and Equipment

The experiment consists of various simple computer-controlled tests which present visual stimuli on screen. The test person has to react as quickly and precisely as possible to these stimuli by pressing appropriate keys on a reaction box. The reaction is recorded and later evaluated. The tests assess both simple and complex forms of reaction speed, performance of the frontal lobe and, in particular, visual-spatial processing.

(c) Examples of Terrestrial/Clinical Applications

- Improvement of diagnostic measures and methods of classification of cerebral disorders (e.g. cranial-cerebral traumata, strokes) via examination of visual-spatial functions, measurement of psycho-motor reaction speeds, measurement of frontal executive functions, etc....

- Neurological rehabilitation of patients with cerebral disorders (e.g. reading disorders due to a cerebral lesion in the area of the primary visual centres or vigilance disorders) via application of individualized training programs.

- Application of improved diagnostic measures in the resolution of focal epilepsy.

- Occupational medicine, e.g. evaluation of the risks associated with reductions in concentration due to shift-work.

5.4 Experiment MICROVIB - Development of Portable Non-Invasive Diagnostic Instruments

- **Aims**

Investigation of the origins of micro-vibration, the physiology of "holding-tremor" development and the interaction between muscle fatigue and disturbances of fine-motor performance.

- **Method and Equipment**

Highly sensitive sensors and a specially developed microprocessor unit (KYMO) are employed to record tremor and micro-vibrations of the arm and the ankle joint. The sensors are located in pockets of a waist-belt and a foot-pedal, respectively, minimizing movement impediments and also enabling work in free-floating weightlessness. Pre-defined load profiles have to be executed by the Cosmonauts.

(d) Examples of Terrestrial/Clinical Applications

- Development of a clinical instrument to improve neurological, muscular and psychiatric diagnostics via integration of various non-invasive functional tests coupled with real time analysis of data (e.g. measurement of all types of physiological tremor and micro-vibrations, postural sway and fatigue, reaction times and speeds, muscular endurance and fatigue, etc.)

5.5 Experiment MIRGEN - Basic Research into Stress Related Diseases

(a) Aims

Diagnosis of the effect of various physical and psychological stress factors (e.g. weightlessness, isolation, etc.) on the immune system (genotoxic and immunotoxic effects).

(b) Method and Equipment

The white blood count of blood samples was investigated, in particular with respect to the amount of killer cells and the ratio of the immunglobulins IgA, IgG and IgM and, additionally, for damage and repair of DNA (Deoxyribonucleicacid).

(c) Examples of Terrestrial/Clinical Applications

- Further investigation of the effects of stress factors on the immunological system and genetic integrity.
- Occupational medicine: investigation of the correlation between occupational stress and the frequency and duration of sick leave.
- Intensive care: investigation of the causes of deterioration of the state of patients under intensive care due to hypokinesia-related reductions in the strength of the immune system.
- Oncology: study of new aspects of tumour development in relation to stress and development of new measures to prevent carcinoma development.

5.6 Experiment MOTOMIR - Muscle Training and Rehabilitation Instrument

(a) Aims

Development of appropriate countermeasures against muscular atrophy due to weightlessness.

(b) Method and Equipment

A specifically developed Dynamo-Ergometer was employed, which enables alternative exercise of the arms or legs. The actuators (pedals or handles) are driven by electro-motors at pre-set speed patterns. Depending on the selected program, Cosmonauts have to resist or accelerate the movement of the actuators. The applied forces are measured, enabling the analysis of various parameters of the strength of the muscular system.

(c) Examples of Terrestrial/Clinical Applications

- Sports medicine: development of individualized training programs for muscle build-up specific to a sport.
- Rehabilitation: improvement of rehabilitation success and optimization of recovery interval by application of specific training measures after accidents or surgery.
- Geriatrics: development of exercises to maintain agility into old age and to prevent the development of osteoporosis.

6. **Austrian Experience** (Reference 4)

Austria, although a member state of ESA, decided not to participate within the framework of the ESA Microgravity Program. (The reasons for this are many and complex, and are not considered within the scope of this paper.) However, at the same time the Austrian Federal Ministry for Science and Research has stimulated, guided and encouraged bilateral cooperation with Russia. As tradition dictates, these activities fit well with Austrian national interests of maintaining close ties with countries of Eastern Europe.

Some of the main reasons for this approach are summarized below:

- Austria is a small nation, with limited resources, and so optimization of existing resources is always a strong project criterion and consideration. For example, to build an independent space in-orbit capability for experimentation clearly takes major commitment and funding, even within the framework of multi-national partnerships.

- By initiating a bilateral approach it was and is possible for Austrian scientists to access the Space Station MIR laboratory. This was and is successful because "cooperation" occurs in the true sense of the word, i.e. with mutual benefit for both parties involved. This avoids the unnecessary "re-invention of the wheel" which is all too common in multi-national space programs.

- Austria was and is able to control directly its project funding so as to optimize the scientific/research aspects, with the overall goal of advancing the "state of the art" in this particular field. Even with the best of intentions, this is not always possible in such a direct way through multi-national partnerships.

- Austria has a long and strong history in the medical sciences with many leading experts residing within the country. To tap into this reservoir of expertise obviously has major benefits, i.e. the maximization of available resources.

- The Austrian research medical infrastructure was and is well suited to the types of projects implemented in cooperation with the Russians. In effect, although at the height of the both AUSTROMIR and RLF projects more than 100 project personnel were involved, the majority of these personnel were drawn from research institutes, clinics and centers. This meant that they maintained their regular employment while also working on the AUSTROMIR and RLF projects. When the projects finished these personnel continued with their normal activities, thus avoiding the need to build up a large permanent project team requiring major resources.

- As a side benefit, as some of these project personnel were students, this form of exposure relatively early in their careers enabled them to appreciate the inter-cultural aspects and associated difficulties of working within an international environment.

- From the Austrian experience, which undoubtedly applies worldwide, scientists want to maximize science and minimize bureaucracy. Unfortunately, bureaucracy appears to be well connected to the "western" approach to international projects. Working in a bilateral fashion with the Russians minimized the paperwork and management overheads.

7. Summary

The policy decision, to collaborate bilaterally with the Russians in the field of space medicine, taken by the Austrian Federal Ministry for Science, Research and Art has resulted in the following benefits for Austria:

- Austria has an established partnership with the Russians which may be utilized for future projects. Indeed, both sides have overcome the "cultural" and "ways of working" barriers to establish a high degree of mutual respect and understanding. For example, IMBP as the partner responsible for the medical aspects within the framework of the International Space Station, has invited ASM to contribute to this development. Details are currently under discussion.

- Austria is utilizing and reaping the benefits from the results of its space research in terrestrial medical/clinical applications, in:

  - radiation therapy;
  - radiation protection;
  - development of non-evasive diagnostic instruments;
  - rehabilitation;
  - early detection of disorders, etc....

  In a global sense, as spaceflight grows longer the knowledge gained from space experimentation will be used to answer deeper, more difficult questions about the nature and fate of human life on Earth (Reference 5).

- Austria, as well as acquiring knowledge in the field of space medicine, is also in parallel building up a reservoir of expertise in terms of the individual Austrian scientists/researchers that have worked on the AUSTROMIR and RLF projects. This expertise is further supplemented by a large number of relatively young engineers who have gained valuable experience in the development of complex flight instrumentation. An important aspect of this process is that continuity of activities in this field is maintained.

- In terms of long term medical spaceflight experience, Austria now stands second in line to Russia - a large feat for a small nation.

References

1. Foley, T.M.: "Learning from living in space." *Aerospace America*, AIAA, Washington, DC, April 1995
2. BMWF:*AUSTROMIR Handbuch*. BMWF, Vienna, August 1991
3. Austrian Society for Aerospace Medicine: *Health from Space Research: Austrian Accomplishments*. Springer, Vienna, 1992
4. Gerstenbrand, F., Mossaheb, M., Saleem, A. and Schneider, J.: *The Austrian Society for Aerospace Medicine and the Russian Long-term Flight Project*. Paper IAA-95-IAA.3.5.07, presented at the 46th International Astronautical Congress, Oslo, Norway, October 2-6, 1995
5. Holloway, H.: "Life without boundaries." *Aerospace America*, AIAA, Washington, DC, April 1995

Utilization of Satellite Communications, Remote Sensing, and Space Imagery Data for Wind Power Resource Mapping in Developing Countries

S. Ximenes, Lockheed Martin Communications Systems, 1 Federal Street, Camden, NJ 08102, USA (samuel.ximenes@den.mmc.com)

Abstract

Obtaining energy from wind power is recognized as a means of controlling energy-related environmental impacts at competitive costs. As a prerequisite to the widespread use of wind energy systems, wind resource assessments are performed to identify a country's wind resource potential. Data derived from ground-based wind mapping methodologies can be augmented with space technology in areas of communications, imaging, and space-based Doppler lidar for wind measurements. Technology advances are required to achieve surface wind measurements, and integration of various sets of data. Wind prospecting is an innovative application.

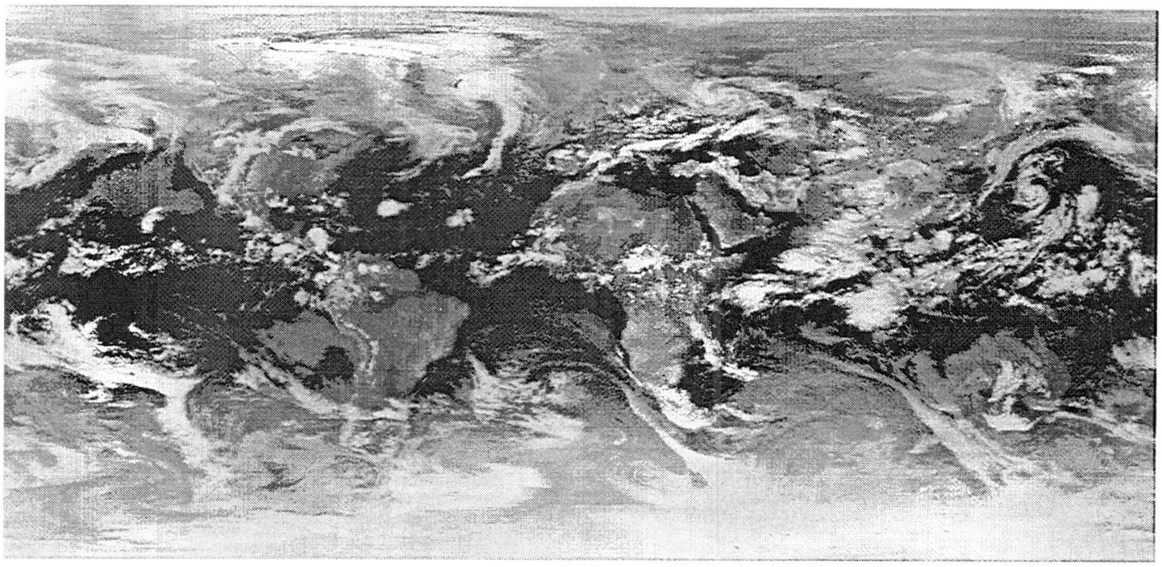

Figure 1. Mosaic of global cloud coverage indicating prevailing winds

1. Wind Energy

Satellite technology has enabled images of clouds above the Earth's surface (Figure 1) to be obtained every 30 minutes. The changes of the clouds' positions and their evolution indicate the winds that blow.

Within 15 years, it is estimated that the total energy consumption in the developing world will surpass that of the United States, Western Europe, Australia, Iceland, Japan, New Zealand and Turkey combined (Reference 1). To meet the developing world's needs for electricity, wind energy is recognized as a means to control energy-related environmental impacts at competitive costs. A single stand-alone wind power system is often sufficient to raise the standard of living for many of the world's poorest rural inhabitants. Larger systems in combination with other renewable energy resources such as photovoltaic arrays can meet the needs of an entire village.

It is now realized that wind power generation has begun to compete with conventional energy generation. Among the factors making utility-scale wind generation competitive are energy costs of $0.05/kilowatt-hour or less, wind's environmental benefits, new reliable technology, and a better understanding of identifying a nation's wind resource potential. In the US (Figure 2), the wind resource potential is large enough to produce more than 4.4. trillion kWh of electricity each year, or half as much electricity as the entire nation used in 1990 (Reference 1).

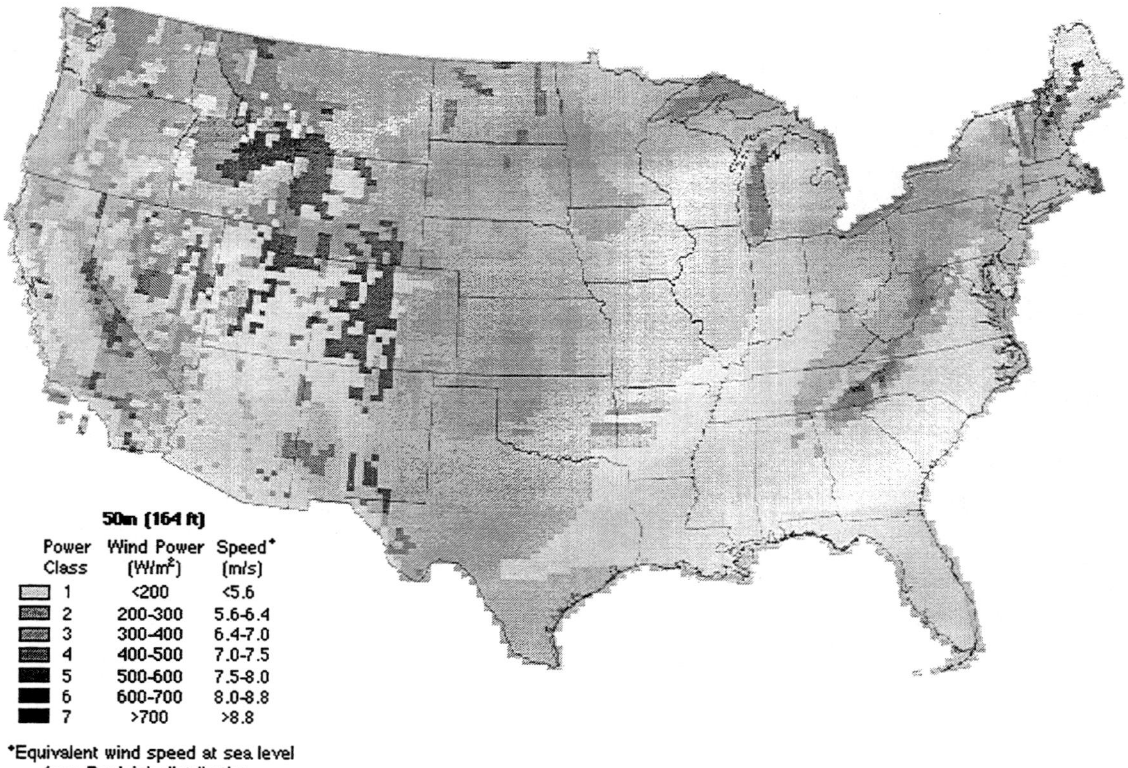

Figure 2. Wind energy resource atlas of the United States

2. Wind Prospecting

The development of a wind resource map is a prerequisite to the widespread use of wind energy systems. Wind resource assessments incorporate various sources of summarized surface wind data to produce such wind resource maps. These maps identify wind power classifications for utility-scale applications or rural power applications for the developing country (Figure 3).

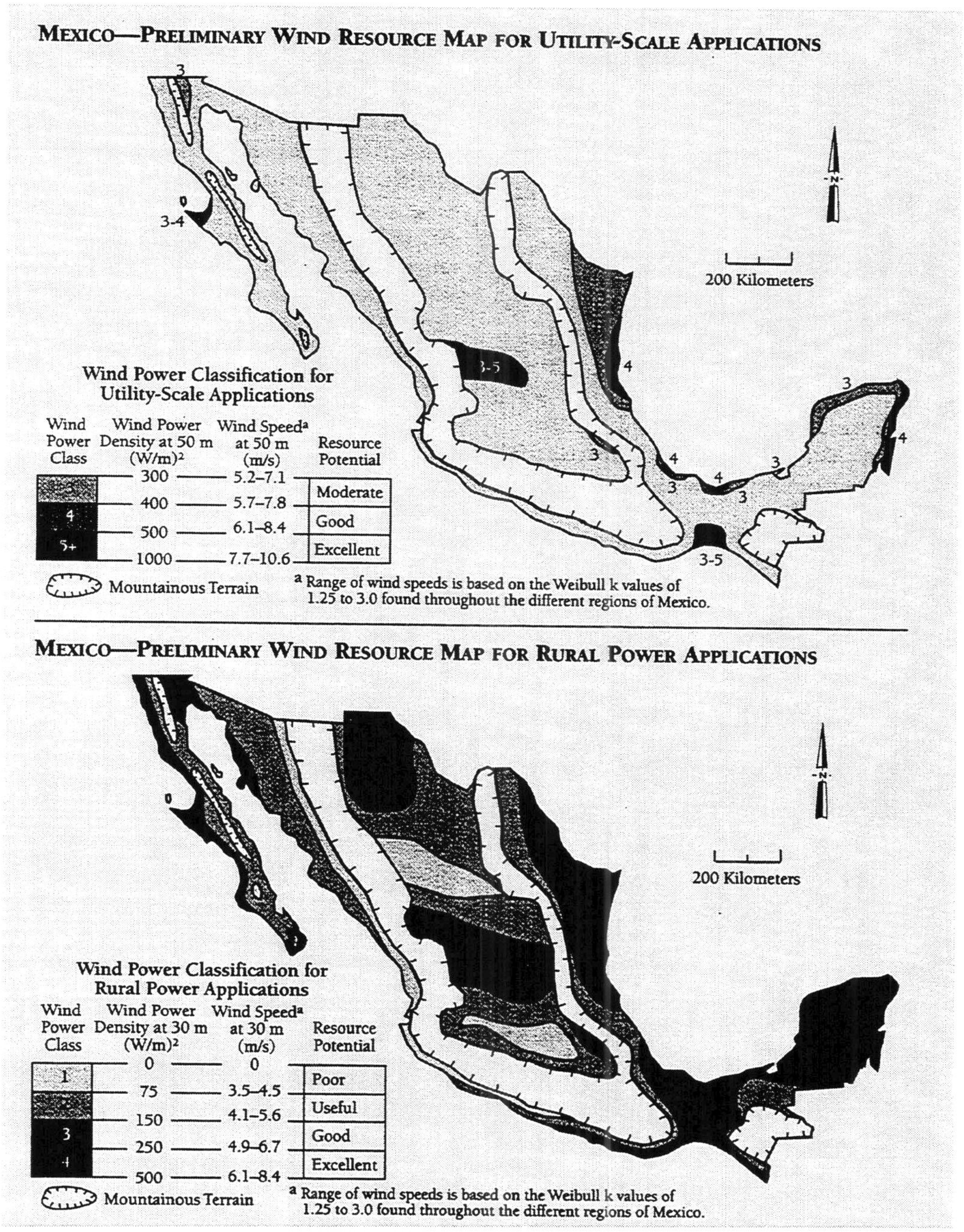

Figure 3. Preliminary wind resource map for Mexico

3. Wind Resource Mapping Methodology

Conducting a wind resource assessment involves a variety of methodologies, instrumentation, and techniques for collecting and interpreting data on wind speed, wind power density, direction, and other variable attributes such as seasonal changes. Typical sources of data and methods employed can include:

- Historical summarized climatological data from the country's national meteorological, aviation, or educational institutions.

- Meteorological data sets such as the Surface Climate Database from the US National Data Center, which are primarily for the period 1973 to 1993. Hourly surface weather observations have been collected, for example using the Global Telecommunications System, and stored.

- Statistical summaries of wind characteristics extracted from hourly observations of meteorological station anemometers (at heights of 10 m). For the wind resource assessment of Mexico, surface wind data were obtained from 186 locations, of which 86 were located at airports (Figure 4). Future plans call for installing as many as 600 new meteorological stations (Reference 1).

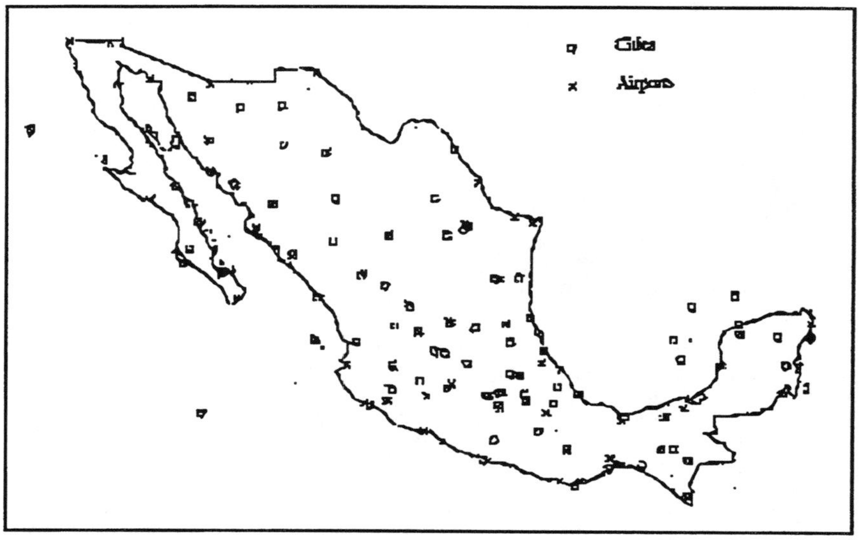

Figure 4. Locations in Mexico for surface wind data measurements (Reference 2)

- Extrapolated marine wind data, largely based on observations by ships, are used to estimate wind resources for offshore areas, as well as coastal and inland sites well exposed to the prevailing direction of the ocean winds.

- Measurements of upper-air-wind data obtained using balloons, useful for estimating wind resources just above the surface, and for estimating vertical profiles of wind speed.

To produce the resource map, the primary statistical summaries of surface wind data used are:

- station and sea-level pressure by month,
- mean wind speed by direction,
- frequency distribution of the wind speed,
- average wind speed and power on an annual and monthly basis,
- interannual variability of the wind speed and power,
- diurnal variability of wind speed and power,
- number of observations on an annual, monthly, and hourly basis.

Wind resources for rural power applications (mean wind speed >3.5 m/s or mean power density >60 W/m^2) are analyzed in more detail. Quality checks on the mapped data plots are performed by visual inspection to reveal trends and/or peculiarities from the observed data. Degradation of the anemometer, increased vegetation or man-made obstructions which make the site less exposed to the prevailing wind are examples of the changes which can occur. In the case of the Mexican wind resource map (Figure 3), representative periods of record for wind measurements at various sites were found to be between 3 and 7 years long (Reference 2); the complete 20 year record was not required. After each station's representative annual wind speed and power densities are determined they are plotted on a map. To confirm and adjust the surface wind resource pattern of the map, upper-air-data measurements for a profile of the vertical wind speed are used.

These maps may be general enough to identify areas of good wind resources, but may not show changes induced by local terrain features (e.g. coastal, hilly, and mountainous terrain) which can cause the wind to vary considerably over short distances.

Figure 5. Wind turbine technology used in wind power plants

4. Utilization of Space-Based Measurements

Developing countries interested in establishing grid-connected wind energy systems (Figure 5) as well as off-grid village power and isolated systems often lack reliable wind data, especially in remote undeveloped areas. Data obtained from space-based systems can be used as an adjunct to ground measurements to offset the lack of reliable historical data; space systems can also aid data collection from remote areas. Incorporating space technology in areas of communications, remote sensing and satellite imagery leads to effective methods for validating ground-based surface wind measurements. Wind prospecting is believed to be an innovative service for advancing space technology toward commercial applications.

5. Space-Based Measurements for Global Surface Wind Profiles

5.1 Space-Based Lidar

Obtaining global wind measurements from space is possible with current Doppler lidar technology (lidar is an acronym for Light Detection and Ranging). It is a remote sensing technique that uses laser light in much the same way that sonar uses sound, or radar uses radio waves. A satellite deployed in polar (sun-synchronous) orbit would transmit pulses of light from a laser to Earth. The pulses of light backscattered by the Earth's atmosphere or surface and received by a telescope are spectrally analyzed to recover Doppler shifts caused by the movement of airborne targets such as clouds or aerosols. With lidars it is possible to measure winds by targeting cloud particles and naturally occurring aerosols which move at approximately the speed of the wind (Reference 3). Current space-based lidar wind measurement concepts advance the benefits of measuring the global tropospheric wind field for understanding and predicting possible climate change and increasing the skill of numerical weather prediction (Reference 4).

Obtaining global surface wind measurements is theoretically possible using Doppler lidars in space. However, the minimum requirements for a space-based Doppler lidar to measure surface winds are beyond the ability of current laser types with regard to the size needed and the costs to develop. Preliminary requirements for space-based surface wind measurements (Reference 5) include:

- horizontal spatial resolution of wind speed and direction measurements of no more than 25 km,
- wind speed accuracy of 1 m/s or better,
- reliable wind measurements for a layer centered approximately 50 m above the surface (the hub height of a wind turbine),
- monthly averages of wind speed and direction, plus data on diurnal variation of the wind.

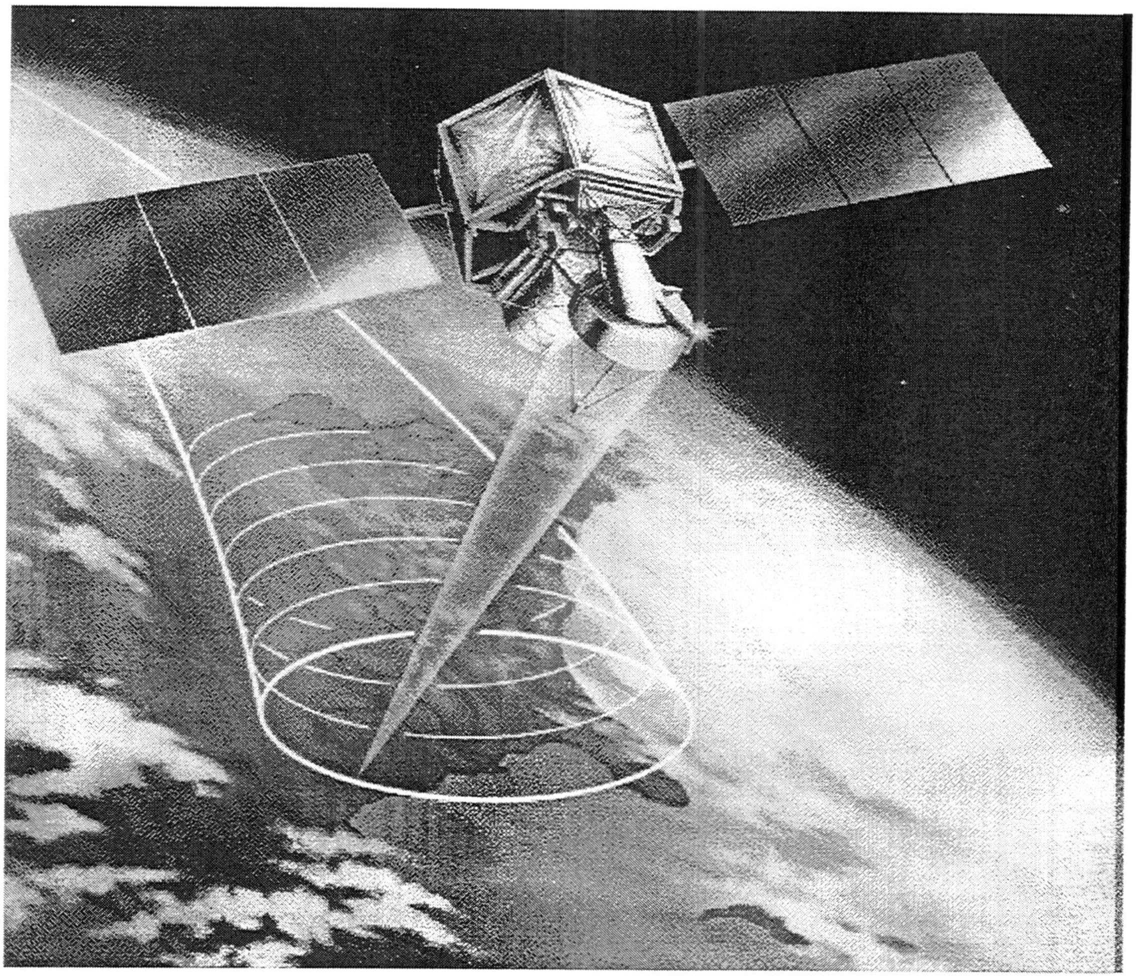

Figure 6. Space-based Doppler lidar (Reference 3)

5.2 Satellite Imagery

The coming generation of commercial satellite images should offer 1 m resolution, with the ability to get pictures to customers in hours instead of weeks. Wind indicators from these remote sensing images can be linked to topographic features. Gaps through mountains are known to accelerate the wind passing through them.

Visual interpretation of eolian features from high resolution images include a variety of wind indicators, i.e., vegetation, geomorphological, and cold climate features (Figures 7 and 8). Other medium-scale imagery such as Synthetic Aperture Radar (SAR) and Side Looking Airborne Radar (SLAR) can vividly illustrate the shape and elevation of landforms.

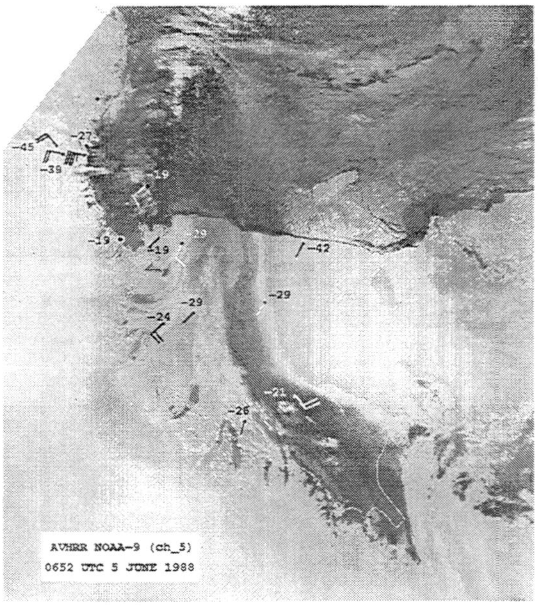

Figures 7, 8. Antarctic wind regimes

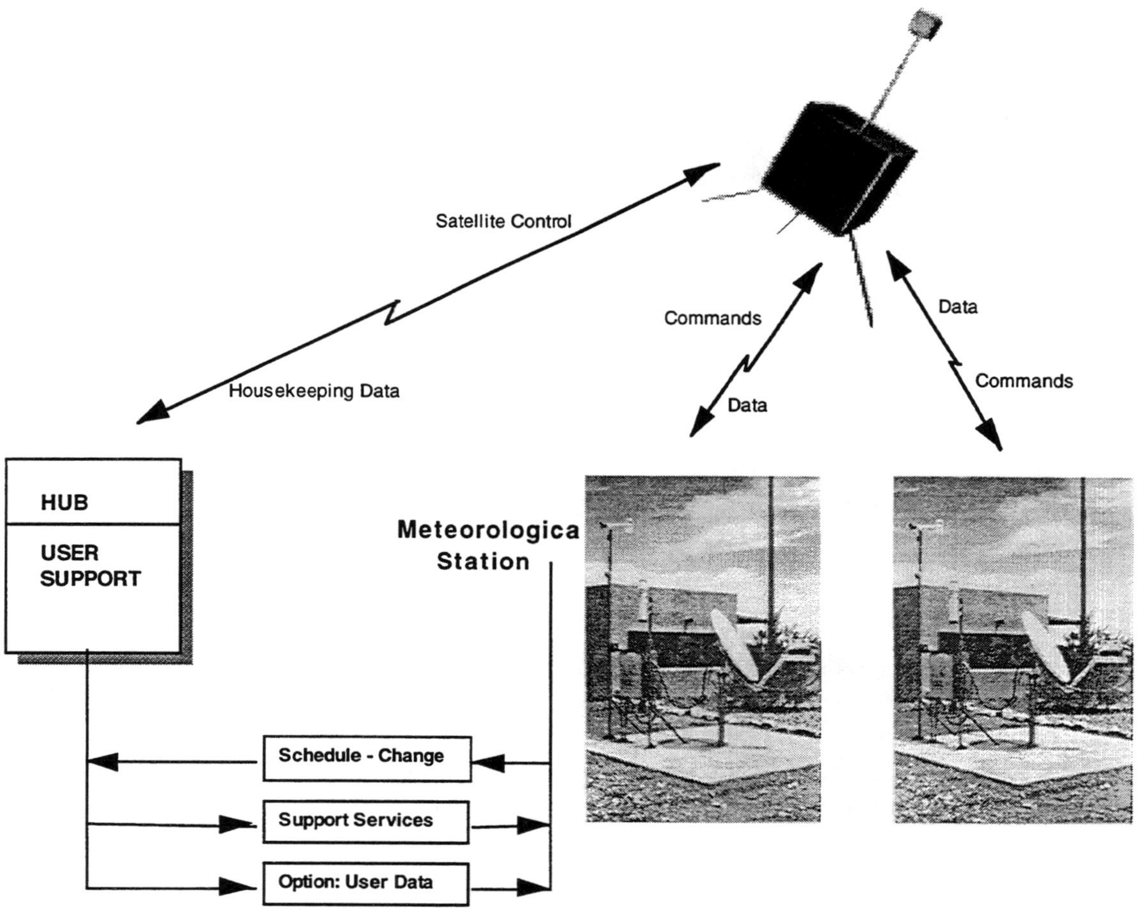

Figure 9. Meteorological station communications link

5.3 Space Communications

Space communications assets could be utilized to interface with meteorological stations established for the wind prospecting assessment. For example, if all sites proposed for the resource assessment of Mexico (Figure 4) had anemometers with transponders, these stations could transmit data to a communications satellite and be received at a ground operations center. A simplified scenario is illustrated in Figure 9.

6. Conclusion

Large amounts of Earth observation data on global scales are now obtained using orbiting satellites. Space-based remote sensing, space imaging and space communications are all contributing to the explosive growth in information technology. Enabling technologies in the above-mentioned areas are at varying degrees of development, marching toward the eventuality where end users in industrial as well as developing countries will have at their fingertips an integrated complement of data for management of the environment. Global wind prospecting is an innovative application with economic benefits for driving advances in space technology for Earth observation.

References

1. US Department of Energy: *Wind Energy Program Overview..* DOE/GO-10095-071, US Department of Energy, March 1995
2. Schwartz, M.N. and Elliott, D.L.: *Mexico Wind Resource Assessment Project.* National Renewable Energy Laboratory Report No. NREL/TP-441-7809, December 1995
3. Baker, W.E. et al.: "Lidar-Measured Winds from Space: A Key Component for Weather and Climate Prediction", *Bulletin of the American Meteorological Society,* Vol. 76, No. 6., June 1995
4. Cordes, J.J.: *Economic Benefits and Costs of Developing and Deploying a Space-Based Wind Lidar, Final Report, D-9502.* NOAA Contract No. 43AANW400223, National Weather Service, NOAA, US Department of Commerce, March 1995
5. Personal Correspondence, National Renewable Energy Laboratory, January 1996

Prelude to a New Space Renaissance

T. Ohue, External Relations Dept., National Space Development Agency of Japan, World Trade Center Building, 2-4-1 Hamamatsu-cho, Minato-ku,Tokyo 105-60, Japan
M. Mori, Rocket System Corporation, Hamamatsu-cho Central Building, 1-29-6, Hamamatsu-cho, Minato-ku, Tokyo 105, Japan
H. Nagayama & T. Hanyu, New Science Dept., Mitsubishi Research Institute, Inc., 3-6 Otemachi 2-Chome, Chiyoda-ku, Tokyo 100, Japan

Abstract

The technological advances of the space age owe much to the political and military pressures of the Cold War between the US and Soviet Union. The recent collapse of the Soviet Union, and the ending of the Cold War, however, have made it increasingly difficult for countries to justify continued investments solely on the basis of political prestige or scientific advances; in today's environment, space activities must also be perceived as a wise economic investment.

The National Space Development Agency of Japan (NASDA) believes that the future success of the public sector space efforts will depend upon objectives which go beyond technological advancement for its own sake; rather, future efforts should provide tangible social and economic benefits that can be readily identified and appreciated by the society which provides the funding for them. The key to realizing such goals is a flexible, adaptable and inexpensive space infrastructure (the foundation upon which all other activities are built), an infrastructure which is viewed as a means of stimulating the utilization of space, rather than an end in its own right.

Based on these understandings, NASDA has been conducting studies for several years on an innovative space infrastructure. This paper analyzes the impact which an innovative space infrastructure could have, not only on the Communication/Broadcast Satellite markets which exist today, but also upon the potential new markets which would emerge. The paper assesses potential market character and size using an original quantitative prediction model. Furthermore, this paper examines the development of future scenarios for exploitation of lunar resources and solar energy, as well as attempts to verify the actual effectiveness of the proposed fully reusable space infrastructure.

1. Introduction

The technological advances of the space age owe much to the political and military pressures of the Cold War between the US and the former Soviet Union. The collapse of the Soviet Union and the end of the Cold War has changed the political, social and economic environment. Concurrently, growing constraints on public spending in many countries have made it increasingly difficult for any one country to justify a continued investment in space solely on the basis of political prestige or human exploration. Space programs are expected to yield measurable public and economic returns.

Despite the recent successful development of such advanced technology programs as H-II, Japan possesses limited experience in manufacturing commercial satellites that can compete on the international market. This is partially because the enforcement of the "Super 301" provision has limited Japan's domestic development to satellites oriented towards research and development.

However, the smooth and steady expansion of space activities in Japan will ultimately require benefits in excess of the intellectual property gained from the technological development. We must propose a program plan which achieves a national consensus of support; this plan must embody the principle that achievements will complement and extend existing public services, and will extend our economic potential.

As an effort to consider NASDA's future objectives, the agency initiated a review of the traditional methods to develop and operate space vehicles and systems. A major conclusion of this review was that today's activities are highly specialized, with designs specialized for individual missions. In contrast, it was noted that a low cost, reusable infrastructure would better meet the objectives of lowering mission costs, enhancing profit potential, and creating the necessary conditions for a sustainable 'space economy'.

Based on these understandings, NASDA has recommended that the future space systems be fully reusable ones, and analyzed their impacts not only on the current markets, but also on the potential emerging markets. The impacts were evaluated in terms of cost/pricing impact, marketability, and required technology.

In developing this report, we express our gratitude for valuable data, discussions and contributions by Professor Tanabe and Associate Professor Nakasuka, Engineering Department of Tokyo University; Mitsubishi Research Institute; CSP Japan, Inc.; Mitsubishi Corporation ; and three world-renowned consultancies, CSP Associates, Inc. (US); GRC Inc. (US), and; Euroconsult (France). Their assistance and efforts enhanced the accuracy of the analysis and evaluation.

2. The Advent of a Techno-Economic Paradigm

Global space activities have emerged from an era in which technological competitiveness drove competition between two superpowers. The dissolution of the Soviet Union in 1991 and the establishment of a more cooperative relationship between US and Russia have ended the 'space race', Such drastic geopolitical changes require spacefaring nations to seek a new framework in which to pursue space activities.

Some space analysts point to a "new techno-economic paradigm". This concept suggests a closer link between activities in space and activities in the economic and public service spheres on Earth. In particular, the significance of the global space activities centers on the following five aspects.

2.1 Alternative to Existing Public Services

Many studies note that the impact of population growth and economic development may deplete natural resources such as clean water, energy, and food sources.

In order to achieve sustainable development, our global society will require new and more efficient means to meet requirements for food, energy, transportation and communications. A primary application of space technology should be to support these needs. In meeting these primary objectives, the development of space and technology industries will also provide domestic economic returns via a revenue and tax base.

2.2 Acquisition of New Insight

Expanding the realm of activities in space could enrich human lives by providing insights on the universe, Earth and life. In this way, space development could foster a more sophisticated and mature society.

2.3 Expansion of Human Activities

Efficient and economical resources provided via space activities could expand the scope of human activities and habitation, which is currently limited by terrestrial resources.

2.4 Participation in International Cooperation

The voluntary participation of each spacefaring nation in international cooperative space affairs provides an additional opportunity to develop greater interdependence and understanding, thereby enhancing peaceful relations and evolving a greater sense of our 'global community'.

2.5 Stimulating the Imagination and Curiosity of the Next Generation to Support the Advancement of Human Society

Advancement of mankind is brought about by an incessant need to challenge the unknown. Challenging the unknown universe is one of the most important themes for the next generation.

In order to expand and upgrade future space activities, training and securing of talented personnel are essential. Inheriting the dream and challenge of space by the next generation through space development will produce young promising people talented not only in science and technology but also in other fields, who will contribute to the maintenance of active economies and societies.

3. Problems of Present Space Activities

In order to achieve a new objective for space activities within the existing geopolitical climate, the following points will have to be considered.

3.1 Establishment of a Long-Term Vision for Commercial Gains

During the Cold War, the objective of the civil space activities mainly lay in technology development. This priority made it difficult to measure the economic and social returns to the public. In the post-Cold War climate, nations are competing for economic and global market strength as much as technological dominance. In formulating a long-term vision for space, the commercial gains as well as the public returns must be considered. Thus, while it is easy in principle to adapt economic advancement as a motivation for public sector space activities, it is important to note that the competitive aspect must also be addressed - i.e. what are the appropriate limits of government support? For example: how far into the market should government space activities extend; can (or should) governments try to restrict the benefit of their investments to local companies, or should they try to promote international consortia?

3.2 Generic Issues Affecting Space Activities

- In the absence of the Cold War conflict, national governments can no longer justify large space investments for national prestige and security alone. The disappearance of two superpowers conflicts and declining space budgets have called heightened attention to the 'economics of space investments.'

- One factor limiting access to space is the expendable launch infrastructure. Although the existing launch vehicles aptly deploy satellites into their required orbits, neither the satellite nor the launch vehicle can be serviced or reused. As long as this regressive approach continues, space activities will not become a viable means by which to serve public needs or seek alternative resources.

- Current space programs are fragile, due to their large scale, long lead-times, and substantial investment requirements. The complexity of large-scale programs introduces such uncertainties as operational timelines and cost overruns. In today's world, the problems encountered in one project can frequently spill over into others, even across international borders.

- Conventional expendable space systems are creating orbital debris, which is quickly becoming an 'environmental hazard' threatening the safety of manned space flight, and potentially degrading the near Earth space environment for astronomical observation.

3.3 Obstacles to Future Japanese Space Activities

Since the enforcement of the 'Super 301' provision in the late 1980s, NASDA has focused its satellite projects on scientific satellites (including Earth observations) and on experimental applications satellites.

Although the global demand for satellites remains robust, Japanese companies are ill-positioned to compete in competitive international bids. Japan's small domestic market and 'late-comer 'entry on the international market suggests that it will be difficult, if not impossible, for Japanese companies to achieve the necessary economies of scale to compete with dominant US and European satellite providers.

3.4 Competition with Converted Intercontinental Ballistic Missiles (ICBMS)

China and USSR/Russia have been converting existing stockpiles of ICBMs into space launch vehicles. The US too, has a large supply of decommissioned ICBMs, although to date US policy has restricted the pace of their conversion, and limited their use to supporting US government launch requirements. In addition, both China and Russia are offering expendable launch vehicle services at prices which are far below those of the Western spacefaring economies. This sudden surge of relatively low-cost launch vehicles destabilizes the international launch market, making the demand for US and European launch vehicles difficult to determine.

4. Outline of an Innovative Space Infrastructure Architecture

Future space activities should contribute to mitigating the existing problems and competing with the public services.

4.1 Concept of Innovative Space Infrastructure

The proposed infrastructure is based on a "fully reusable launch vehicle (RLV)" that would travel between Earth and space "like an airplane". The infrastructure would also comprise an Orbital Servicing Vehicle (OSV), a Fuel Station (FS) which would provide satellite maintenance in space, and an Orbital Transfer Vehicle (OTV) which would transfer payloads between low-Earth orbit and geosynchronous orbit. All systems are conceived as fully reusable systems. The infrastructure's primary functions would include crew and cargo transportation to low-Earth orbit as well as on-orbit satellite service and repair. (See Figure 1).

4.2 Approach to Innovative Space Infrastructure

Shifting suddenly from the existing expendable launch vehicle (ELV) system to an innovative infrastructure would be difficult, both from a technological and budgetary perspective. Hence, a more appropriate strategy would entail an evolutionary, incremental approach in which the existing ELV infrastructure is gradually replaced with reusable systems (see Figure 2).

The first phase of the innovative infrastructure would entail modifying existing ELVs into partially reusable vehicles. More specifically, the H-II Orbiting

Plane (HOPE) would be configured as a reusable 'upper stage' for ELV launchers. Through this process, the necessary experience in reusable technologies and in-orbit operations would be acquired. Concurrently, research and development efforts would focus on a fully reusable systems. In addition, mission plans for inserting satellites into low Earth and medium Earth orbits as well as logistical support and retrieval transportation to and from ISSA shall be developed in the initial phase.

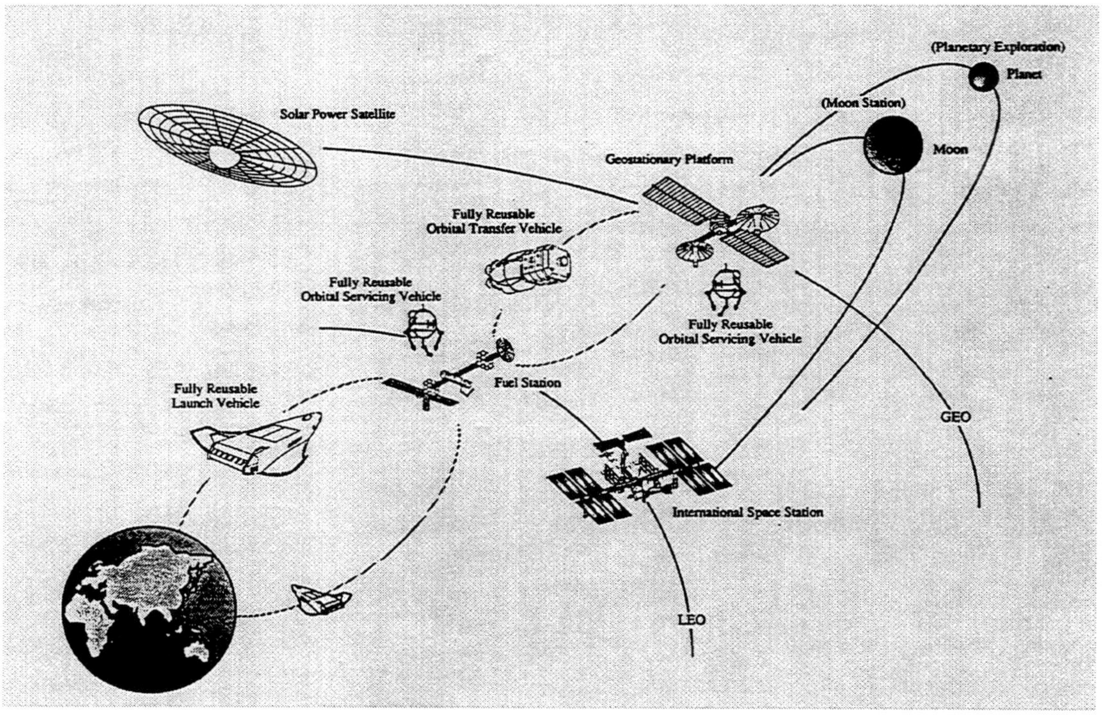

Figure 1. Innovative space infrastructure

Phase two of the implementation scenario would involve a gradual transfer to a fully reusable system. An OTV and FS would be developed and deployed, enabling payload transport between various orbits. At the same time, satellite designs would shift to accommodate Orbital Replacement Modules (ORMs) meeting standard interface requirements which would facilitate in-orbit servicing and change-out.

4.3 Reduction of Space Activities Costs

In conceiving the innovative infrastructure, reductions in development and operational costs were estimated. The requisite technologies to achieve such costs, however, were not considered. The following cost estimates assume an architecture comprised of one RLV, seven FSs and six OTVs (see Table 1 and Figure 3).

	Development & Production(B$)	Operation	Note
Fully Reusable Launch Vehicle	25 (1 Vehicle)	Refer to Figure 3	Payload : 10 ton Turn Around Time : 1week Refer to "Sänger" Model
Fully Reusable OTV	7.92 (Total 6 Vehicles)		Payload : 1 ton (Average) 100 times reusable Fixed by TRANSCOST
Fully Reusable Fuel Station	12.6 (Total 7 Vehicles)	371(M$/year)	Refer to Estimation of "European Autonomous Station"

Table 1. Cost model for innovative space infrastructure

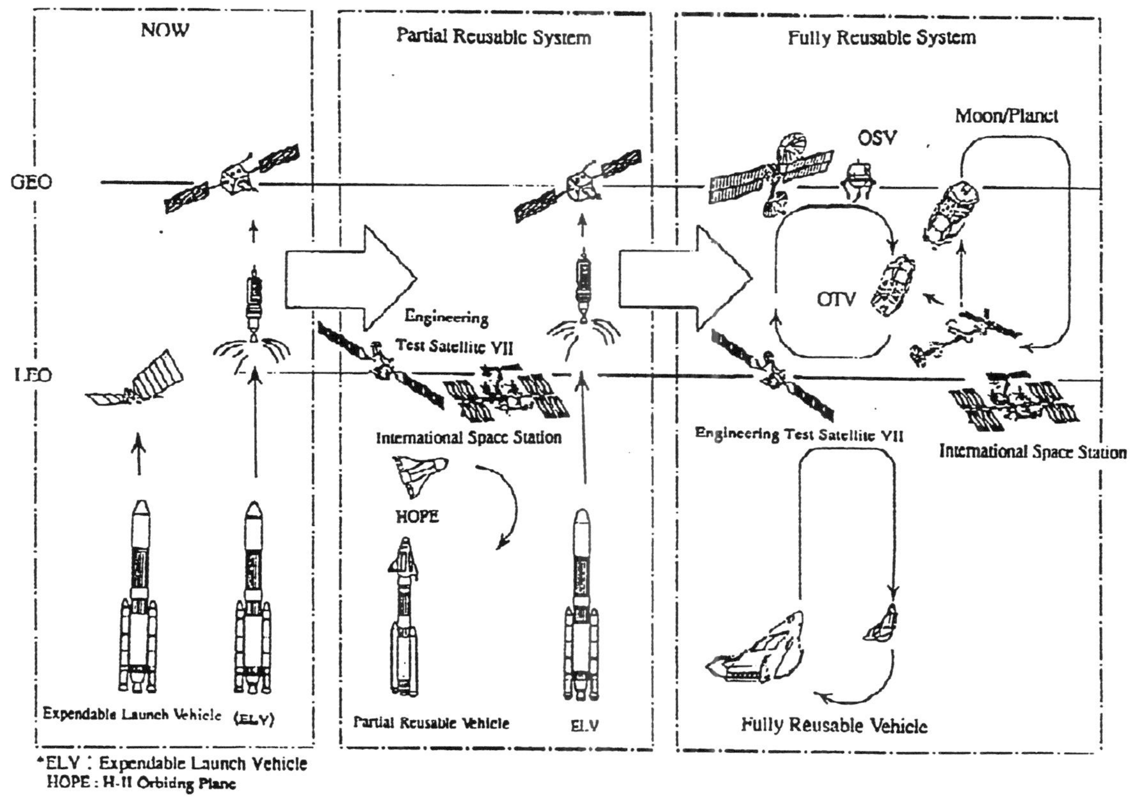

Figure 2. Approach to innovative space infrastructure

According to the estimates presented in Table 1, the innovative infrastructure would achieve transportation costs of $3,000/kg (~$1,360/lb.). This cost structure represents a 75% reduction in the least- expensive LEO launch prices, assuming a 50-vehicle annual launch rate (see Figures 4 and 5).

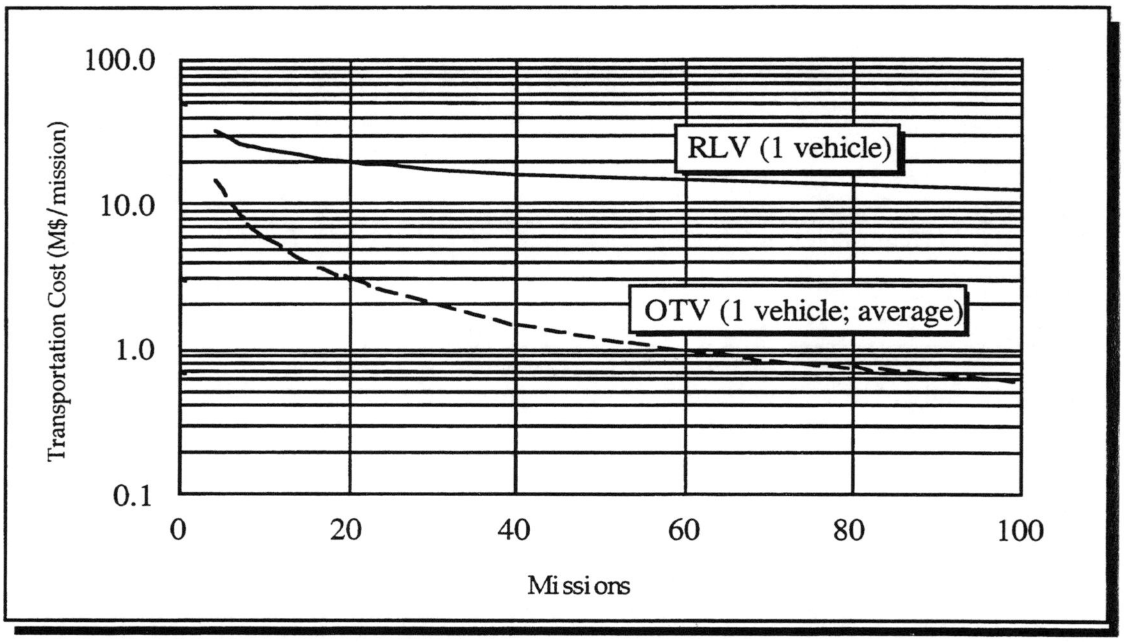

Figure 3. Operational costs of RLV and OTV

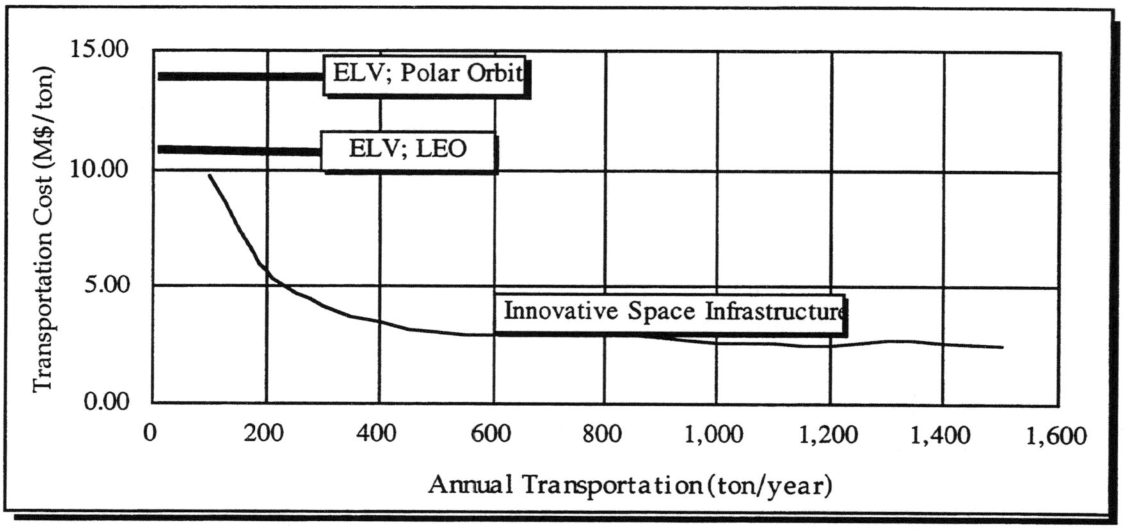

Figure 4. Function between annual transportation and costs

The annual operation costs of existing ELVs are estimated at $6.5 billion; average lift capabilities for expendable vehicles are approximately 300 metric tons. As envisioned, the innovative infrastructure would reduce current launch operations costs by as much as 84% ($1.2 billion) while offering the same lift capacity. When the annual transportation demand rises to more than 1,000 tons, the operation costs could be reduced to one-tenth ELV costs, or about $2,500/kg.

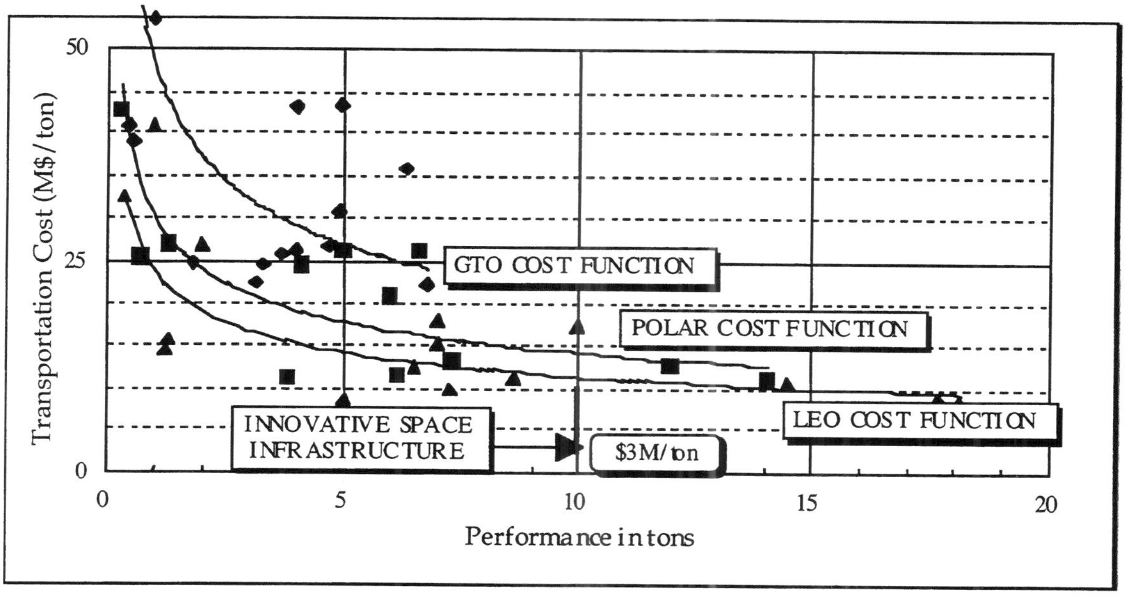

Figure 5. Comparison of transportation costs

With substantial reductions in launch costs, spacecraft designers would be encouraged to consider incurring the mass penalties associated with such cost-efficient manufacturing processes as satellite modularization, standardization and multifunctional design (hereinafter satellite platform; see Table 2).

Cost Reduction Factor	Cost Reduction Rate of Unit Weight	Weight Increase of Satellite
1. Cheap material and redesign	- 40%	+ 20%
2. Standardization	- 25%	+ 10%
Total	- 55%	+ 32%

Table 2. Cost model for satellite platform
(scenario:90% cost reductions in infrastructure costs; immediate delivery cycle)

The cost model in Table 2 shows that the current cost of launching a 3.5 ton communications satellite, approximately $175 M ($86.5 M procurement, $86.5 M launch), which could be reduced to $63 M ($52 M procurement, $12 M launch) when using the innovative launch infrastructure. In this scenario, a communications satellite venture could maintain a nominal 20% Internal Rate of Return (IRR) break-even with a 15% reduction in transponder costs because of a shorter turnaround time (from 3 years to 1 year).

4.4 Evolution of New Space Markets

An innovative space infrastructure would reduce the cost and increase the access to space. These factors will encourage growth in existing space markets as well as the development of new space markets. Growth opportunities are likely to occur in the following space application areas:

- **Communications/broadcasting**
  Data transmission, mobile communications, remote health care and education
- **Space transportation**
  Logistics, crew/cargo to the orbiting facility (e.g. ISSA), space tourism*
- **Earth observation**
  Meteorological observation, resource exploitation, terrestrial observation, ocean observation, crustal disturbances monitoring, survey and locating
- **Space utilization**
  New material development, life science/engineering*
- **Resource exploitation**
  Photovoltaic power generation, lunar resources exploitation (Helium 3), energy transmission*
- **Earth environment protection**
  Nuclear waste disposal, farmland development through afforestation, maintenance of natural environmental balance*

* Underlined items are new space activities.

5. Impact of Space-Related Markets

Microeconomic analysis indicates that the innovative space infrastructure would have an impact on the evolution of existing space markets and on the viability of new commercial opportunities.

5.1 Communications

Satellite transponders can handle rates of as high as 30 Mbps and some communication ventures in Japan provide their data transmission services at the rate of 256 kbps to 1.5 Mbps. While transponder services are unaffected by transmission distance and terrestrial disasters, their operational costs amount to about $36,000 per month, including the ground station.

Terrestrial alternatives offer similar types of services; however, fees are calculated according to the distance served (e.g. $10,000 per month for 100 km distances to $70,000 per month for distances of 3,000 km or more).

If transponder prices are reduced by 50%, commercial satellite communications ventures would achieve the 20% IRR break-even position. (See Figure 6).

Under the existing satellite development and launch scenario, transponders are more cost-effective than terrestrial alternatives at distances greater than 1,000km (e.g. Tokyo to Hokkaido or Kyushu and beyond). In applying the economics of an innovative infrastructure, transponder transmissions would become cost-effective at distances of 250 km and beyond.

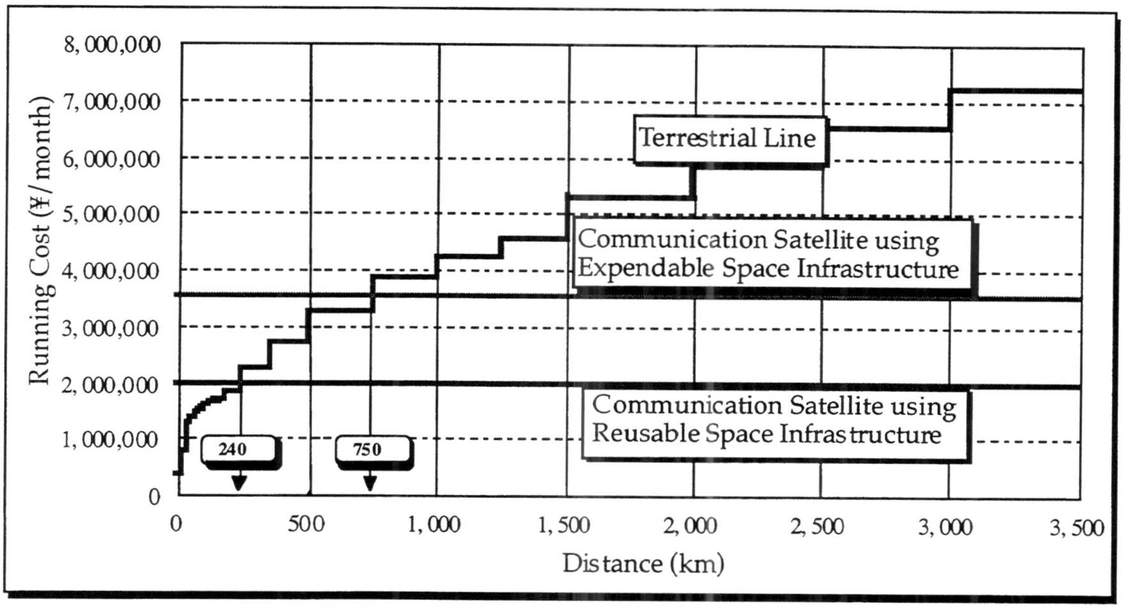

Figure 6. Cost comparison between transponder and terrestrial lines

At present, satellite transmissions constitute only about 0.02 % of ground digital communications traffic. If one assumes, however, that satellite communications handle 80% of the traffic for distances of 250 km or more, then the space-based communications market would increase to 700 times its current size. This expansion would require an estimated 100 satellites carrying a total of 4,000 36MHz transponders (see Figures 7 and 8).

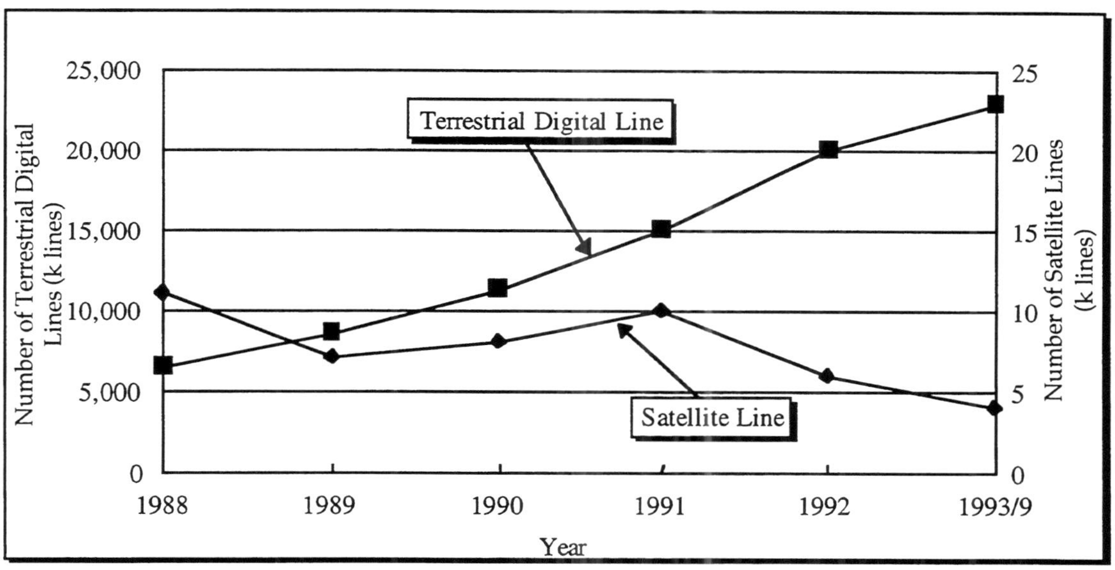

Figure 7. Subscriptions in satellites and ground communications

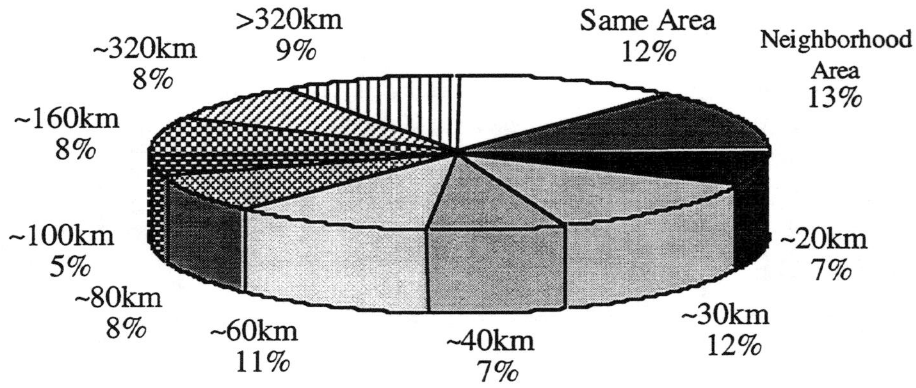

Figure 8. Communication distance and average traffic

5.2 Earth Observation

The following analysis illustrates the impact of an innovative space infrastructure on the estimated IRR of a hypothetical commercial remote sensing venture operating for a 12-year period.

The hypothetical operation includes three Earth observation satellites, developed at 1 ton with a 5-year design life, to obtain and market images of the Earth's environment, natural disasters and global climate. Image prices were estimated at $5000 per image (comparable to the cost of a Landsat image and about one-twentieth the cost of an aerial photograph mosaic of similar size). In the first year of operation, the business is projected to sell approximately 5,000 images. Demand growth will be constant after two years.

An innovative space infrastructure would result in a 90% reduction in operation costs for the remote sensing venture. This marked reduction in operational costs would reduce the initial investment requirements for a commercial remote sensing business by as much as 50% and raise net profits by some 130% (see Table 3).

	Fully Reusable Space Infrastructure		Expendable Space Infrastructure
Satellite Life Time	5 years		5 years
Business Period	12 years		12 years (+ procurement period 3 years)
IRR	38%		20%
Initial Cost	$39.4M		$77.9M
Gross Sales		$1,164.1M	
Net Profit	$417.9M		$322.9M

Table 3. Comparison of investment efficiency

The price necessary to maintain a threshold IRR of 20% falls to $3,300 per image; in turn, the lower price stimulates the demand for images and makes the venture more profitable (see Figure 9). In addition, high resolution sensors could provide observation data comparable to that acquired from aerial photographs. According to market estimates, the high resolution imagery market is projected to surpass the current aerial photography market (about $1 B per year in Japan or Europe).

5.3 Resource Exploitation

The candidates for new space markets are photovoltaic generation; lunar base construction; nuclear waste disposal; and farmland development. This section will focus on photovoltaic power generation, a potential means by which to generate clean energy.

The key element of the proposed Space Power System (SPS) is a quantum box driven semiconductor laser. This laser would be directly pumped by solar energy, and its conversion efficiency would be more than 40%. The laser beam would be transmitted to the rectenna and converted to electricity by a solar cell array. The outline of this SPS and Rectenna systems and their efficiencies are shown in Tables 4 and 5.

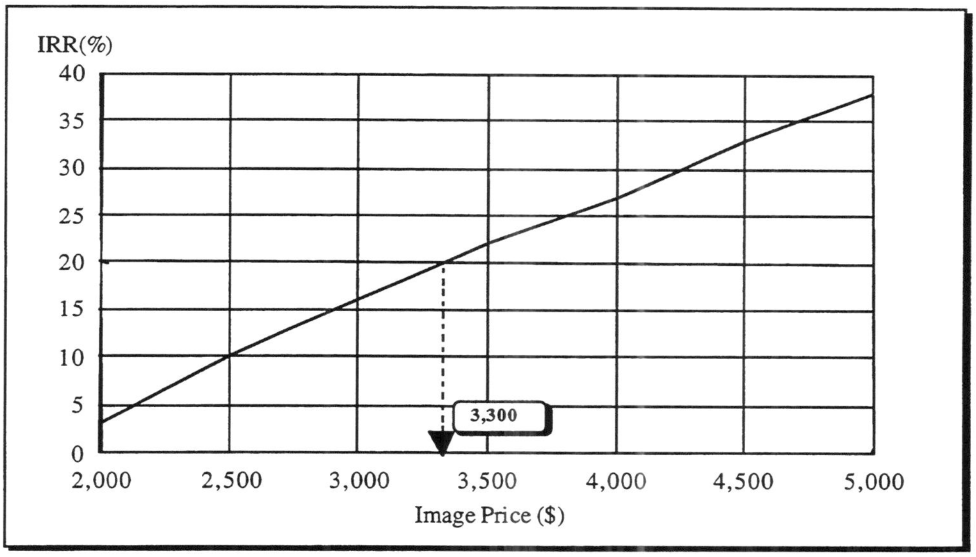

Figure 9. Function between image pricing and IRR

The SPS would weigh about 2,500 tons and consist of several modular components. RLVs would boost the components into LEO and OTVs would transfer these components to GEO where OSVs would assemble the SPS. Once located in the proper orbit, the SPS would transmit converted solar energy via a laser to the ground facilities.

While the SPS structure would be permanent, the satellite would require some 100 tons of annual supply such as fuel and replacement components. The RLV and OTV infrastructure elements would provide the SPS servicing and resupply functions. This operational concept is shown in Figure 10.

SPS	Power output	An SPS would output 1 GW electric power at rectenna site. Diameter of SPS is 2.4 km
	Orbit	Construction and operation orbit is GEO
	Life time	SPS structure is assumed to be semipermanent. Degraded or failed parts are repaired periodically. All components of SPS structure wil be exchanged every 30 years to avoid obsolescence.
	Space laser system	Directly pumped laser using a quantum box-driven semiconductor is proposed.
	Transmission	Laser of about 680 nm wavelength is considered for power transmission. Power density is 1 kW/m2 at rectenna site.
Rectenna	Power generation	Solar cell array is used for laser-electricity convertor. Diameter of rectenna is 1.2 km.

Table 4. Outline of SPS and Rectenna systems

Items	Efficiency (%)	Incident Energy (GW)
Summer Solstice Factor	97	5.6
Seasonal Variation	91	5.5
Laser Conversion	40	5.0
Concentration Efficiency	90	2.0
Atmospheric Losses	80	1.8
Laser-DC Conversion	70	1.4
Total	18	1.0

Table 5. Efficiency

In conceiving the innovative infrastructure, we estimated reductions in development and operational costs. The requisite technologies to achieve such cost reductions, however, were not considered. The following cost estimates assume an architecture composed of one RLV, seven FSs and six OTVs (see Table 1 and Figure 3).

According to the estimates presented in Table 1, the innovative infrastructure would achieve transportation costs of $3,000/kg ($1,360/lb.). This cost structure represents a 75% reduction in the least expensive LEO launch prices, assuming a 50-vehicle annual launch rate (see Figures 4 and 5).

The annual operation cost of existing ELVs is estimated at $6.5 billion; average lift capabilities for expendable vehicles is approximately 300 metric tons. As envisioned, the innovative infrastructure would reduce current launch operations costs by as much as 84% ($1.2 billion) while offering the same lift capacity. When the annual transportation demand rises to more than 1,000 tons, the operation costs could be reduced to one-tenth of the ELV costs, or about $2,500/kg.

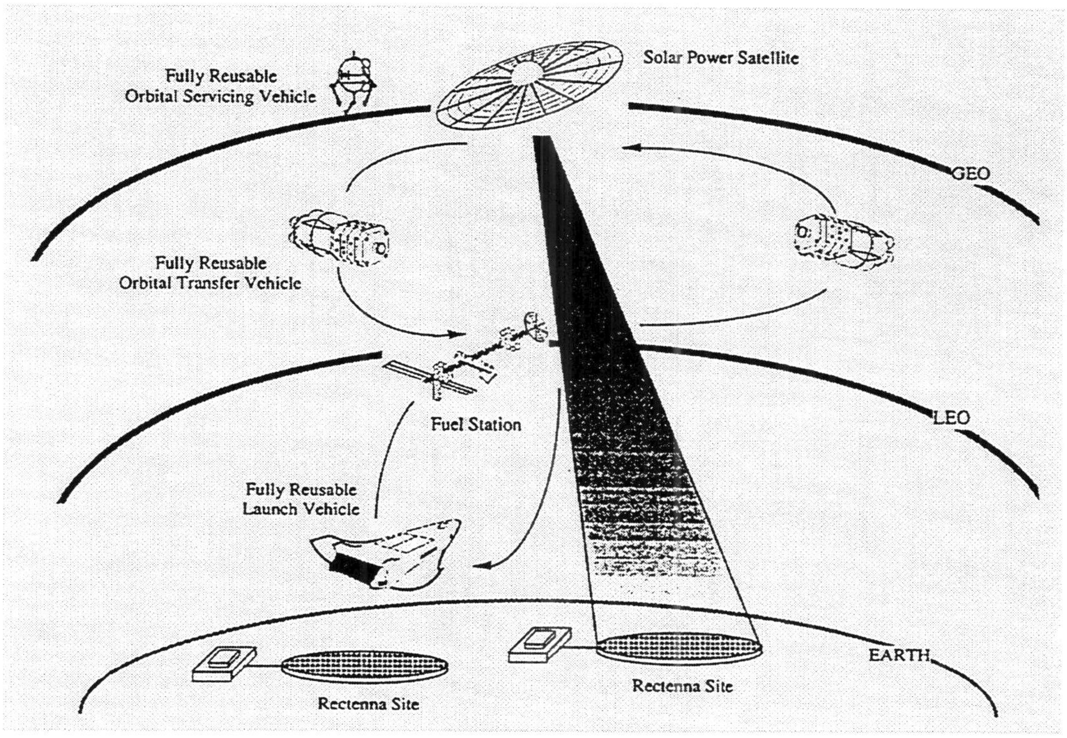

Figure 10. Operational concept

There are four requirements for a viable SPS venture:

- SPS shall be operated for 30 years, including construction.
- Electricity prices shall be set to achieve more than a 20% IRR.
- Total revenue shall exceed the SPS and Rectenna cost.
- Total reliability shall be 99.9%.

Assuming an SPS construction period of 1 year, the SPS would provide electricity at approximately \$0.41/kWh, or about twice the current household electricity charges (about \$0.2/kWh in Japan; see Figure 11). Construction of this system would require five reusable launch vehicles (from Earth to LEO) and approximately five sets of orbital transfer vehicles (from LEO to GEO) annually. Seven reusable fuel stations (seven circular orbits from LEO to GEO) would support these vehicles.

With this pricing regime during the SPS operation, we estimate that the venture would yield annual profits of about \$1,507 M (FY 1991 profits of Tokyo Electric Power Co., Inc. were reported at \$750 M). At this projected revenue base, the SPS investment would break even 6 years after construction (see Figure 12).

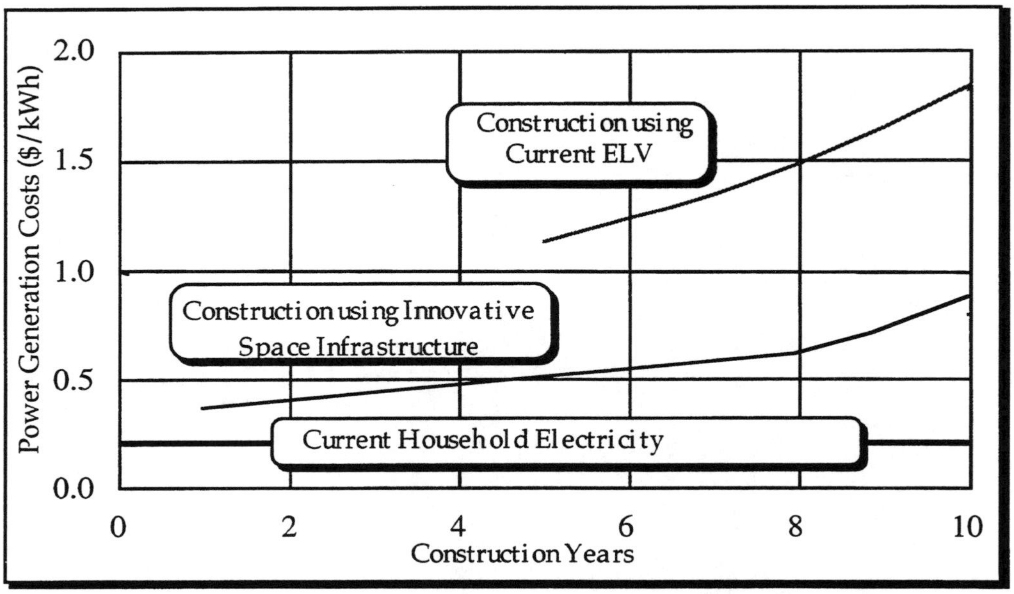

Figure 11. SPS construction years and power generation costs

6. Towards 21st Century Space Activities

About 85% of global space endeavors have been led by government initiatives. While space technologies are often leveraged for dual-uses, most countries have had only sporadic success in applying space technologies to commercial, profit-oriented applications.

Many government and industry officials recognize that the primary task of 21st-century space activities will be to revitalize the market economy. This section will explore whether or not the innovative space infrastructure can create the economic dynamics to cultivate a robust, growing commercial industry.

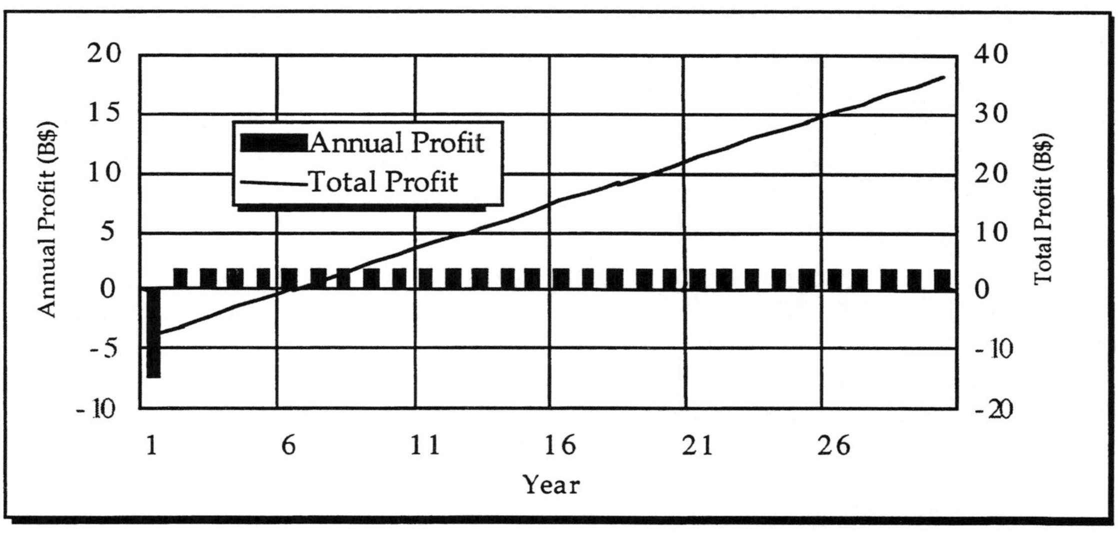

Figure 12. SPS revenue projection
(1 year construction period, $0.41/kWh electricity prices, a 20% IRR)

The terms and conditions are stipulated from a microeconomic approach.

- The capital to develop and integrate the innovative space infrastructure is invested as stipulated in Table 1, and lasts for 10 years.
- The capital to operate that infrastructure is invested as stipulated in Figures 3 and 4.
- The space utilization ventures shall maintain a 20% IRR, and last for 40 years from development to completion of operation. (e.g. actual development of about 15 years, and planned operation of about 30 years, as in the case of the Concorde).
- The venture growth shall be indicated by logistic functions.
- The market growth shall be close to 50% of the targeted market scale after 3 years of the initial investment, and 100% after 10 years.
- The profit rate shall be 5% of sales in the ventures.

Based on these conditions, two cases are evaluated; one is development, construction, operation and servicing under private initiatives, the other is development under a public initiative, with cooperative construction and other elements under private capital.

6.1 Private Initiatives

The venture with a 20% IRR would create a market of $850 B (equivalent to about 20% in Japan's GDP). However, the initial investment (about $50 B) for development of the innovative infrastructure is prohibitively expensive, and the demands are not expected to meet the investment. It is concluded that the creation of that market will be unrealistic.

6.2 Cooperative Initiatives

The preferred cooperation expects the private sector to lead initiatives in other activities than development; however, the construction costs shall be shared with the government for 30 years, in order to reduce uncertainty of demand and the investment burden on operation of the infrastructure. Figure 13 shows the interrelation of the industrial funding ratio and the market sales.

Under the assumption that the government funds 50% of the production costs for 30 years, the annual sales in the potential space-related market would roughly amount to $35 B. This is approximately equivalent to 1% GDP of Japan in FY 1993, and amounts to 1.5 times the sales of the leading Japanese airline companies (in FY 1992 accounts, Japan Airlines reported sales of $13 B, and All

Nippon Airlines $8 B). It is concluded that a high profitability could be achieved with lower risk, and that the payback period would be about 7 years on average (see Figure 14).

In the societies worldwide that increasingly demand improvements in social welfare and prosperity from technology advances, it is important to note that the innovative infrastructure will be realized as a collaborative effort of the public and private sectors. Their contributions are a way of recognizing space endeavors as significant social projects.

7. Conclusion

In order to expand the existing market and explore potential new markets, it is necessary that access to space be affordable enough to provide more opportunities for various missions. From this point, the current architecture of space activities, which is dependent on ELVs, should be reconsidered.

Several factors, such as funding sharing, investment efficiency, elasticity of demand for space utilization and payback period, should be interrelated in order to justify the investment for the innovative space infrastructure commercially.

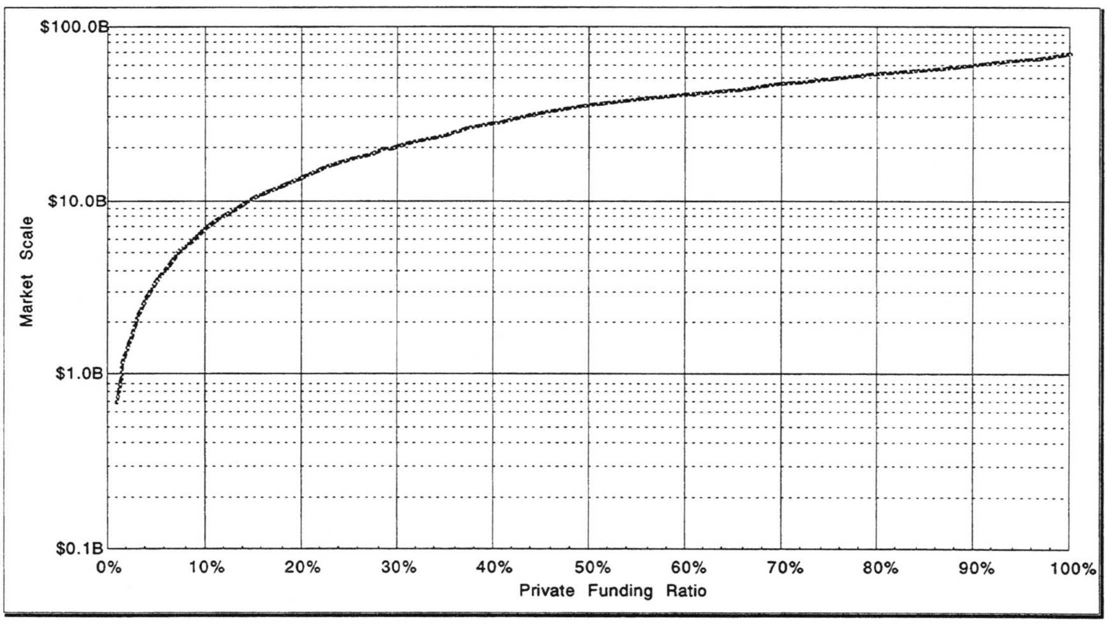

Figure 13. Private funding ratio and market scale

It is obvious that the success of the future space activities depends on less investments for the innovative space infrastructure and more increasing demands for missions. For that purpose, the public and private sectors must cooperate to promote them, even while there remain various issues and options to be addressed in restructuring the space infrastructure.

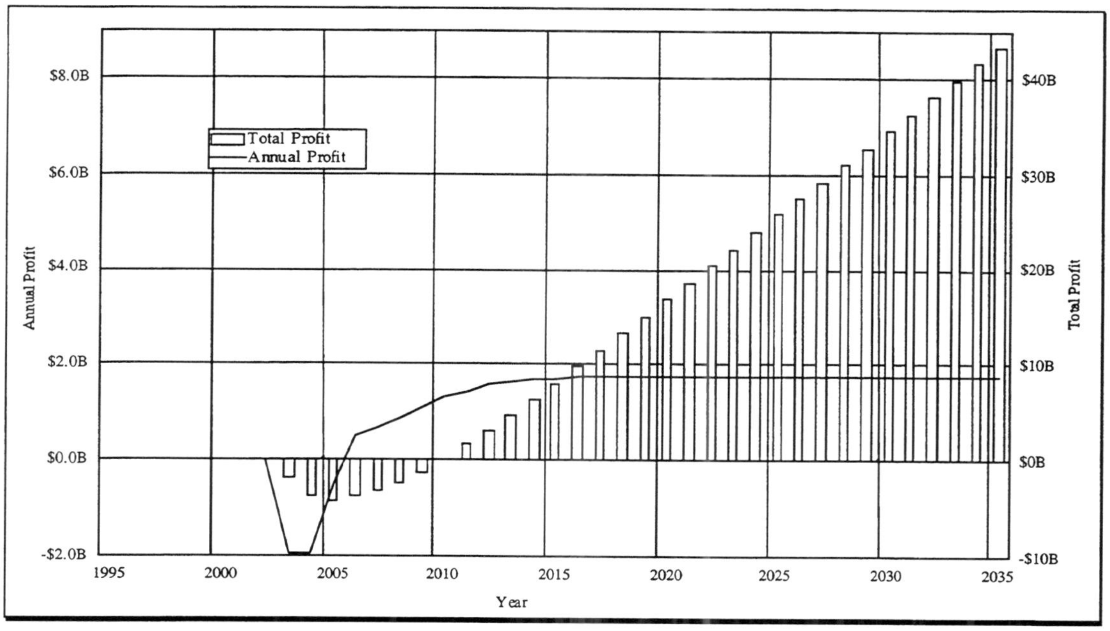

Figure 14. Space-related business perspective
(Private funding 50% of production, a 20% IRR)

In the global trend which consists in shifting priorities from space technology development to economic potential and socioeconomic benefits, the development of an innovative space infrastructure would be fully consistent with this shift of strategic objectives and the creation of new social paradigm.

In other words, Japan must conduct all missions under the policy of 'more benefits return to the public'. Further, as these imperatives are not just Japanese but global, and the investments required are large, Japanese space policy should address the 21st Century space infrastructure in an international context.

To conclude, the innovative space infrastructure is the appropriate starting point for the evolution of Japan's space program in the new era. It provides the opportunity to correct the inadequacies of today's systems, stimulate the expansion of space for human social and economic benefit, and affords an opportunity to encourage international cooperation.

References

1. New Energy and Industrial Technology Development Organization: *Feasibility Study on the Solar Power Satellite System.* NEDO Contract Report, NEDO, Tokyo, Japan, 1991-1993
2. Science and Technology Agency: *Study on Introduction of Space Infrastructure.* STA Contractor Report, STA, Tokyo, Japan, 1992
3. National Space Development Agency of Japan: *Study on Evaluation of Influence Through Introduction of Innovative Space Infrastructure.* NASDA Contractor Report, NASDA, Tokyo, Japan, 1993
4. National Space Development Agency of Japan: *Study on Evaluation of Influence Through Introduction of Innovative Space Infrastructure, Phase II.* NASDA Contractor Report, NASDA, Tokyo, Japan, 1994

5. Tanabe, T., Nakasuka, S. and Sumino, N.: "Towards the realisation of OTV network and its impacts on space activities." *45th Congress of the International Astronautical Federation*, , IAA-94-IAA.1.1.588, Jerusalem, Israel, 1994
6. Nakasuka, S.: "Effect of space fuel station on the transportation system capability and cost - two case studies", *Transactions of the Japan Society for Aeronautical and Space Sciences*, Vol. 34, No. 106, pp. 226-239, 1992
7. Tanabe, T. and Nakasuka, S.: "Fuel station-based transportation network as a future space infrastructure", *Preprints of the World Space Congress*, IAA-92-0168, Washington DC, 1992
8. Tanabe, T. and Nakasuka, S.: "Future Earth-Moon transportation system for manned lunar base project - study of low cost configuration", *Preprints of the 42nd Congress of IAF*, IAF-91-441, Montreal, 1991
9. Koelle, D.: *Transcost*. MBB Report No. URV-185(91), MBB, Munich, Germany, 1991
10. Special Committee on Long-Term Vision Space Activities Commission: *Toward Creation of Space Age in the New Century*. Science and Technology Agency, Tokyo, Japan, 1994
11. Mori, M., Ohue, T. and Nagayama, H.: *Evaluation of Influence Through Introduction of Innovative Space Infrastructure*. Paper presented at the ENECOM-95 Conference, Sendai, 1995
12. Mori, M., Ohue, T. and Nagayama, H.: *Solar Energy Exploitation Using an Innovative Space Infrastructure*. Paper presented at the WPT '95 Conference, Kobe, Japan, 1995

Report on Panel Discussion 4: The Way Forward

G. Haskell, Symposium Convenor & Vice President for Academic Services and Outreach, International Space University

Session Chair: **G. Soffen**, NASA Goddard Space Flight Center, USA

The final panel discussion was guided by **G. Soffen** through four main subjects: population growth, solar power, space exploration and humans in space.

S. Murai re-emphasised the point that population growth is the greatest challenge facing humanity. Growth depends on many factors, but he drew attention to one indicator: forest coverage. Models indicate an "end-point" where less than 10% of the land surface is covered by forest, which could occur in 75 to 100 years. He suggested that space remote sensing techniques could become key elements of a needed global eco-engineering effort.

Participants cited economic insecurity and religious teaching among reasons for pessimism. One speaker was shocked to learn that at least one European country officially encouraged population growth. A single voice expressed some optimism, referring to the amazing adaptability of humans.

G. Khozin took the lead in "passing the baton" to the younger participants, urging that they must not look to the future using paradigms and stereotypes from the past. The future must be shaped, not by narrow individualistic specialists, but by interdisciplinary thinkers concerned for the common good.

The topic of space solar power provided a range of opinions. Several speakers pointed out that many other new energy sources are available. Wind power, gas from agricultural by-products and alcohol as fuel for internal combustion engines were given as examples. Other participants were convinced that the need for space solar power is inescapable. While some speakers were concerned about the cost of space solar power systems, others considered that once the need for this source of energy becomes evident and urgent, the resources will readily be made available. Concern was expressed that solar power from space might be used to increase consumption of resources and lead to an increase in the population.

There was full consensus among the speakers in the panel and from the audience that space exploration - meaning scientific investigation of our near and distant space environment - must and will continue. There was no clear view, however, on how this would be achieved within the shifting balance between public funding and commercially motivated funding for space activities.

In answer to a question from **G. Soffen**, many people in the lecture hall indicated that they would like to fly in space. He asked a second question: why was this aspect of space untouched during the Symposium? Since the discussion panel had already run well beyong the scheduled ending of the Symposium, there was too little time to formulate an answer. Perhaps next year's ISU Symposium on "New Space Markets" will provide the answer.